Minimal processing technologies in the food industry

Related titles from Woodhead's food science, technology and nutrition list:

Food processing technology Principles and practice Third edition
(ISBN: 978-1-84569-216-2)
The first edition of *Food processing technology* was quickly adopted as the standard text by many food science and technology courses. The completely revised and updated third edition consolidates the position of this textbook as the best single-volume introduction to food manufacturing technologies available. The third edition has been updated and extended to include the many developments that have taken place since the second edition was published. In particular, advances in microprocessor control of equipment, 'minimal' processing technologies, functional foods, developments in 'active' or 'intelligent' packaging, and storage and distribution logistics are described. Technologies that relate to cost savings, environmental improvement or enhanced product quality are highlighted. Additionally, sections in each chapter on the impact of processing on foodborne micro-organisms are included for the first time.

Thermal technologies in food processing (ISBN: 978-1-85573-558-3)
Thermal technologies are traditionally a compromise between their benefits in the enhancement of sensory characteristics and preservation, and their shortcomings, for example in reducing nutritional properties. The need to maximise process efficiency and final product quality has led to a number of new developments including refinements in existing technologies and the emergence of new thermal techniques. This collection reviews all these key developments and looks at future trends, providing an invaluable resource for all food processors.

Managing frozen foods (ISBN: 978-1-85573-412-5)
This book examines the key quality factors from raw material selection through processing and storage to retail display. It is a unique overview of this entire industry and provides frozen food manufacturers, distributors and retailers with a practical guide to best practice in maximising quality. *Managing frozen foods* will serve as an invaluable decision-making tool, providing guidance on the selection of raw materials and retail display equipment. The editor concludes the book with an insight into the future of the industry and examines the opportunities offered by recent developments such as anti-freeze proteins and ultrasonic techniques.

'Anyone who properly studies this book and its references would know almost everything there is to know about frozen fish and other foods.'
Professional Fisherman

Details of these books and a complete list of Woodhead's food science, technology and nutrition titles can be obtained by:

- visiting our web site at www.woodheadpublishing.com
- contacting Customer Services (email: sales@woodheadpublishing.com; fax: +44 (0) 1223 893694; tel.: +44 (0) 1223 891358 ext. 130; address: Woodhead Publishing Limited, Abington Hall, Granta Park, Great Abington, Cambridge CB21 6AH, UK)

If you would like to receive information on forthcoming titles in this area, please send your address details to: Francis Dodds (address, tel. and fax as above; e-mail: francis.dodds@woodheadpublishing.com). Please confirm which subject areas you are interested in.

Woodhead Publishing Series in Food Science, Technology and Nutrition:
Number 72

Minimal processing technologies in the food industry

Edited by
Thomas Ohlsson and Nils Bengtsson

CRC Press
Boca Raton Boston New York Washington, DC

WOODHEAD PUBLISHING LIMITED
Oxford Cambridge New Delhi

Published by Woodhead Publishing Limited, Abington Hall, Granta Park, Great Abington, Cambridge CB21 6AH, UK
www.woodheadpublishing.com

Woodhead Publishing India Private Limited, G-2, Vardaan House, 7/28 Ansari Road, Daryaganj, New Delhi – 110002, India
www.woodheadpublishingindia.com

Published in North America by CRC Press LLC, 6000 Broken Sound Parkway, NW, Suite 300, Boca Raton, FL 33487, USA

First published 2002, Woodhead Publishing Limited and CRC Press LLC
Reprinted 2008, 2010

British Library Cataloguing in Publication Data
A catalogue record for this book is available from the British Library.

Library of Congress Cataloging in Publication Data
A catalog record for this book is available from the Library of Congress.

Woodhead Publishing Limited ISBN 978-1-85573-547-7 (print)
Woodhead Publishing Limited ISBN 978-1-85573-679-5 (online)
ISSN 2042-8049 Woodhead Publishing Series in Food Science, Technology and Nutrition (print)
ISSN 2042-8057 Woodhead Publishing Series in Food Science, Technology and Nutrition (online)
CRC Press ISBN 978-0-8493-1454-4
CRC Press order number: WP1454

Printed in the United Kingdom by Lightning Source UK Ltd

Contents

Contributors

Chapters 1, 2 and 3

T. Ohlsson
SIK (Swedish Institute for Food and Biotechnology)
PO Box 5401
SE-402 29
Gothenburg
Sweden

Tel: +46 31 335 56 23
Fax: +46 31 83 37 82
E-mail: to@sik.se

N. Bengtsson
Kyrkvagen 62
435 30 Molnlycke
Sweden

Chapter 4

M. Sivertsvik, J.T. Rosnes and H. Bergslien
Institute of Fish Processing and Preservation Technology (NORCONSERV)
Alex. Kiellands gt. 2
PO Box 327
N-4002 Stavanger
Norway

Tel: +47 51 84 46 00
Fax: +47 51 84 46 15
E-mail: morten.sivertsvik@norconserv.no

Chapter 5

E. Hurme and R. Ahvenainen
VTT Biotechnology
Tietotie 2
PO Box 1500
02044 VTT
Finland

Tel: +358 9 456 5201
Fax: +358 9 455 2103
E-mail: Raija.Ahvenainen@vtt.fi

T. Nielsen
SIK (Swedish Institute for Food and Biotechnology)
Ideon
SE-223 70
Lund
Sweden

Chapter 6

A. Meyer, K. Suhr and P. Nielsen
Food Biotechnology and Engineering Group
BioCentrum-DTU
Technical University of Denmark
DK-2800 Lyngby
Denmark

Tel: +45 4525 2600
Fax: +45 4588 4922
E-mail: anne.meyer@biocentrum.dtu.dk
E-mail: karin.suhr@biocentrum.dtu.dk
E-mail: per.v.nielsen@biocentrum.dtu.dk

F. Holm
FoodGroup Denmark,
Gustav Wieds Vej 10
DK-8000 Aarhus
Denmark

Tel: +45 8620 2000
Fax: +45 8620 1222
E-mail: Fh@foodgroup.dk

Chapter 7

R. Ahvenainen, H. Alakomi, I. Helander, E. Skyttä and T. Sipilainen-Malm
VTT Biotechnology
Tietotie 2
PO Box 1500
02044 VTT
Finland

Tel: +358 9 456 5201
Fax: +358 9 455 2103
E-mail: Raija.Ahvenainen@vtt.fi

Chapter 8

P. Zeuthen
Docent, DTU
Hersegade 7 G
DK-4000,
Lyngby
Denmark

Tel: +46 35 56 65
Fax: +46 35 55 65
E-mail: Peter.Zeuthen@image.dk

Chapter 9

E. Laurila and R. Ahvenainen
VTT Biotechnology
Tietotie 2
PO Box 1500
02044 VTT
Finland

Tel: +358 9 456 5201
Fax: +358 9 455 2103
E-mail: Raija.Ahvenainen@vtt.fi

Chapter 10

M. Gudmundsson and H. Hafsteinsson
Technological Institute of Iceland (MATRA)
Keldnaholt

IS-112 Reykjavik
Iceland

Tel: +354 570 71 00
Fax: +354 570 71 11
E-mail: magnusg@iti.is
E-mail: hannes.hafsteinsson@iti.is

Chapter 11

R. Ahvenainen
VTT Biotechnology
Tietotie 2
PO Box 1500
02044 VTT
Finland

Tel: +358 9 456 5201
Fax: +358 9 455 2103
E-mail: Raija.Ahvenainen@vtt.fi

Foreword

This book emanates from technical reports produced within the network 'Minimal Processing of Foods', running from 1997 to 2000. The network was sponsored by the Nordic Industrial Fund and 38 companies in the five Nordic countries. The editors of this book acted as coordinator (TO) and newsletter editor (NB) in the network.

1

Introduction

T. Ohlsson, SIK (Swedish Institute for Food and Biotechnology), Gothenburg and N. Bengtsson, Consultant, Molnlycke

The term 'minimal processing' has been defined in various ways, for example very broadly as 'the least possible treatment to achieve a purpose' (Manvell, 1996). A more specific definition which addresses the question of purpose describes minimal processes as those which 'minimally influence the quality characteristics of a food whilst, at the same time, giving the food sufficient shelf-life during storage and distribution' (Huis in't Veld, 1996). An even more precise definition, which situates minimal processing methods within the context of more conventional technologies, describes them as techniques that 'preserve foods but also retain to a greater extent their nutritional quality and sensory characteristics by reducing the reliance on heat as the main preservative action' (Fellows, 2000). Minimal processing can, therefore, be seen in the context of the traditional concern of food processing to extend the shelf-life of food. At the same time, whilst they value the convenience that increased shelf-life can bring, consumers have become more critical of the use of synthetic additives to preserve foods or enhance characteristics such as colour and flavour (Bruhn 2000). They have also placed a greater premium on foods which retain their natural nutritional and sensory properties. Minimal processing techniques have emerged to meet this challenge of replacing traditional methods of preservation whilst retaining nutritional and sensory quality.

Although some commentators contrast minimal processing techniques with thermal processing, developments in thermal technologies have been considered 'minimal' where they have minimised quality losses in food compared to conventional thermal techniques (Ohlsson, 1996). This collection therefore opens with a broad-ranging review of the minimal processing of foods with thermal methods. Chapter 2 discusses both high temperature short time (HTST) techniques such as aseptic processing and LTLT (low temperature long time) techniques like sous-vide processing which seek to achieve a better balance between preservation and food quality. The chapter also discusses infrared heating which, like conventional thermal processing, heats a food indirectly. However, arguably the biggest single development in minimal thermal processing has been the application of volume heating methods, which generate heat directly within the product. The second part of the chapter looks at techniques such as ohmic and dielectric (radio frequency and microwave) heating which, potentially, bring new efficiencies to thermal processing.

Chapter 3 discusses a wide range of emerging non-thermal minimal processes, beginning with the two at the most advanced state of development and use in the food industry: irradiation and high pressure processing. In the case of each technique, as with other chapters in the book, the contributors discuss the principles underlying the process, the technology used and applications to particular food products. They also discuss the principal strengths and weaknesses of each technique and likely future developments. They then consider a number of methods based on the pulsed discharge of a high energy capacitor: pulsed white light, ultraviolet light and in particular the use of pulsed electric fields. After coverage of oscillating magnetic fields, the chapter concludes by considering a range of other non-thermal antimicrobial treatments such as ultrasound. In each case, the contributors describe the current state of research and likely prospects for the technology.

The following two chapters discuss developments in packaging. Chapter 4 considers modified atmosphere packaging (MAP), covering principles, technologies and applications in the case of non-respiring and respiring foods such as fruit and vegetables. Complementing this chapter is a review of so-called 'active' and 'intelligent' packaging (of which MAP is sometimes considered an example). Active and smart packaging is distinct in that it responds actively to changes in the food package. As an example, smart packaging can now include materials designed to absorb or emit chemicals during storage, thereby maintaining a preferred environment within the package which maximises product quality and shelf-life. Examples include oxygen, ethylene and water absorbers, and ethanol and carbon dioxide emitters. The chapter describes these kinds of active packaging and others such as packaging materials containing antioxidants and antimicrobials, which help to extend shelf-life, and packaging designed to remove unwanted flavour compounds. One particularly important branch of active packaging is that designed to measure changes during storage and distribution, for example in picking up potential temperature abuse. The chapter therefore discusses current research in such areas as time–temperature indicators, oxygen and carbon dioxide indicators, freshness and doneness indicators.

In response to consumer concerns about synthetic additives, the food industry has developed a range of plant-derived additives with a preservative function. These are reviewed in Chapter 6 which surveys a range of antimicrobial and antioxidant agents. This is followed by two chapters on the critical issue of the safety of minimal processing. In achieving the objective of 'least possible treatment', food manufacturers now combine a number of more gentle processes which collectively preserve a food without any individual process damaging quality. This more complex approach may, however, make it easier for pathogens to survive and pose a risk to consumers. This possibility has led to the development of so-called 'hurdle technology' where an understanding of the complex interaction of variables such as temperature, water activity, pH and preservatives on pathogen survival is used to design a series of hurdles that do not compromise quality but collectively ensure the microbiological safety of food. The principles and application of hurdle technology are discussed together with the range of existing and emerging hurdles. This discussion provides a context for the following chapter which reviews the process of setting safety criteria for minimally-processed foods.

The final three chapters begin with two detailed case studies which show minimal processing in practice at differing points in the supply chain. One chapter discusses minimal processing techniques in the production of fresh fruit and vegetables, whilst a second considers the application of minimal techniques in secondary processing, reviewing the application of high pressure processing and pulsed electric fields to the processing of fish. The final chapter looks at the future of minimal processing. The use of minimal techniques, underpinned by hurdle technology, puts new demands on the supply chain from agricultural production to the point of consumption. These new pressures are discussed together with their implications for the future integration and simplification of the supply chain in achieving safe foods of the quality that consumers expect.

1.1 References

BRUHN, C (2000) 'Food labelling: consumer needs', in Blanchfield, J. Ralph (ed.), *Food labelling*, Woodhead Publishing Limited, Cambridge.

FELLOWS, P (2000) *Food processing technology: principles and practice*, Woodhead Publishing Limited, Cambridge.

HUIS IN'T VELD, J H J (1996) 'Minimal processing of foods: potential, challenges and problems', Paper presented to the EFFoST Conference on the Minimal Processing of Food, Cologne, 6–9 November.

MANVELL, C (1996) 'Opportunities and problems of minimal processing and minimally-processed food', Paper presented to the EFFoST Conference on the Minimal Processing of Food, Cologne, 6–9 November.

OHLSSON, T (1996), 'New thermal processing methods', Paper presented to the EFFoST Conference on the Minimal Processing of Food, Cologne, 6–9 November.

2

Minimal processing of foods with thermal methods

T. Ohlsson, SIK (Swedish Institute for Food and Biotechnology), Gothenburg and N. Bengtsson, Consultant, Molnlycke

2.1 Introduction: thermal methods and minimal processing

Thermal methods are extensively used for the preservation and preparation of foods. Thermal treatment leads to desirable changes such as protein coagulation, starch swelling, textural softening and formation of aroma components. However, undesirable changes also occur such as loss of vitamins and minerals, formation of thermal reaction components of biopolymers and, in minimal processing terms, loss of fresh appearance, flavour and texture.

The classical approach to overcome or at least minimise these undesirable quality changes in thermal processing is the HTST (high temperature short time) concept. It is based on the fact that the inactivation of microorganisms primarily depends on the temperature of the heat treatment, whereas many undesirable quality changes depend primarily on the time duration of the heat treatment (Ohlsson, 1980). High temperatures will give the rapid inactivation of microorganisms and enzymes required in pasteurisation or sterilisation, and short times will give fewer undesired quality changes. However, effective process control is critical if product quality is not to be compromised.

The problem in applying the HTST principle to solid and high viscosity foods is that the parts of the food in contact with the hot surfaces will be overheated during the time needed for the heat to transfer to the interior or coldest spot of the food. This surface overheating will give quality losses that, in severe cases, will counterbalance the advantages of the HTST concept. This problem will be especially pronounced when heat processing packaged or canned products of large dimensions. One way of overcoming this problem in part is by heat processing unpacked foods followed by aseptic packaging.

Electric heating, which directly heats the whole volume of the food, is a method that may overcome these limitations caused by the low heat diffusivity of foods. Direct volume heating methods can be seen as minimal processing methods where they are applied to minimise quality changes in the product (Ohlsson, 1994a). Infrared radiation is used today primarily as a means of rapid cooking of thin unpackaged foods of regular dimensions, but has also demonstrated advantages for pasteurisation of packaged products. One possibility for the future may also be ultrasonic processing, which lies on the borderline between thermal and non-thermal processing (see Chapter 3).

Another quality aspect of thermal processing is that juice loss from meat and fish is strongly dependent on the temperatures reached. Only a few degrees will produce large differences in juice loss, which are important both to the economical yield of the process and consumer perceptions of the juiciness of the product. If effective control is not possible, thermal processing is better done using the LTLT concept, using low temperatures over long times. One option is to vacuum-package food and then apply low temperature heat processing ('sous-vide' processing). This process improves shelf-life and sensory and nutritional quality by controlling causes of negative changes in quality such as exposure to oxygen and extreme temperatures (Bengtsson, 1994).

2.2 Minimal processing by thermal conduction, convection and radiation

This section looks at the following:

- HTST processing using conventional thermal processes
- aseptic processing
- semi-aseptic processes
- sous-vide processing
- infrared heating.

2.3 Heat processing in the package

Traditionally, prolonged shelf-life has been achieved by heat sterilisation or pasteurisation of packaged food products using steam- or water-heated pressure autoclaves (batch or continuous) for sterilisation, and water- and/or steam-heated tunnels or vessels at atmospheric pressure for pasteurisation. Such modern processes are equipped with sophisticated process control of time/temperature/pressure and F- or C-value development, sometimes coupled to on-line process simulation for optimal quality retention and inactivation of bacteria and enzymes. Limitations to achieving truly minimal processing or HTST processing are package size and shape and actual product particle size and the degree of convection inside the package. For small particle size and low viscosity of the surrounding liquid, HTST-like processing can be reached by can rotation. For more solid or high viscosity foods, package depth may have to be reduced to a few centimetres, for example using flexible pouches. In Japan, large-scale processing in pouches applies a sterilisation temperature of up to 128ºC and processing times of only 20 minutes. Correspondingly, package rotation and/or smaller thickness can be applied for milder, HTST-like pasteurisation processes (Ohlsson, 1994b).

There is considerable ongoing research both in improved temperature measurement and in developing a better understanding of thermal processing, particularly using modelling techniques. Both developments are designed to provide improved process control. As an example of the former, time–temperature-integrating (TTI) devices have been developed for the control of sterility in sterilisation and pasteurisation processes. These devices, which consist of beads or small capsules (of diameters down to 3mm) containing a given concentration of a specific enzyme or microorganism, are inserted into food particles before processing. Afterwards they are retrieved for the measurement of remaining activity and the calculation of a corresponding sterilisation or pasteurisation value. The technique can also be applied for products heated in continuous systems, irrespective of what heating method is being used.

2.4 Aseptic processing and semi-aseptic processing

2.4.1 Aseptic processing

HTST processes are possible if the product is sterilised before it is filled into pre-sterilised containers in a sterile atmosphere. This forms the basis of aseptic processing, also known as ultra-high temperature (UHT) processing. Aseptic processing is used to sterilise a wide range of liquid foods, including milk, fruit juices and concentrates, cream, yoghurt, salad dressings, egg and ice cream mix. A number of systems have been introduced for true HTST treatment, using two or more steps for heating, applying steam injection or infusion for the final split-second temperature rise up to some 150ºC, followed by very rapid evaporative and heat exchanger cooling (Kastelein, 1997). Heating equipment for aseptic processing includes the following:

- direct systems:
  - steam injection
  - steam infusion
- indirect systems:
  - plate heat exchangers (including tube-in-tube)
  - tubular heat exchangers (concentric tube or shell-and-tube)
  - scraped surface heat exchangers.

The main advantages and disadvantages of these heating methods are summarised in Table 2.1. Recent developments in these technologies include tube-in-tube heat exchangers, a development of plate heat exchangers, in which the thin-walled plates are formed into a tube and corrugations or twists in the tubes are introduced to promote turbulent flow. One tube is inserted inside another, which may be repeated two or more times and heat transfer takes place across the tube walls. Incoming material is heated by sterilised product to regenerate heat and increase energy efficiency. Steam from hot water is used for the final heating and, after initial cooling by the incoming material, the product is cooled with cold water. This system is relatively low cost and is widely used, although it suffers from some drawbacks (Carlson, 1996):

- the equipment is restricted to relatively low viscosity liquids that do not cause significant fouling
- seal integrity is critical to prevent mixing of incoming material, sterilised product or heating and cooling media
- the seal strength limits the pressure that can be used
- seals may be difficult to clean.

The process has been widely used for fruit juices, milk and dairy products (Lewis and Heppel, 2000).

The problems encountered in straight tube heat exchangers are largely overcome by forming a single tube into a continuous helix or coil, which has a carefully defined ratio between the diameter of the coil and the diameter of the tube. The coil is contained within an insulating material to minimise heat losses. The design of the coil promotes secondary flow of liquid within the tube, and this causes turbulence at relatively low flow rates, and high rates of heat transfer (between two and four times the rates in tube-in-tube or shell-and-tube heat exchangers (Carlson, 1996). This enables processing of heat-sensitive products (e.g. liquid egg) and products that cause fouling of heat exchanger surfaces. The mixing action in the coil gives a uniform distribution of particles, making the equipment suitable for salad dressings, fruit purées and other foods that contain a range of particle sizes, as well as for high viscosity liquids such as cheese sauce. Additionally, the continuous tube has no seals and is easily cleaned in place, and the simple design is virtually maintenance-free.

A major challenge lies in aseptic processing of particulate-containing foods, especially at particle sizes of 5–10 mm or more. The major problem is that of ensuring sterility of the fastest moving particle without heat damage to the slower moving ones. To reduce this problem, processes have been designed where liquid and particles are processed separately and mixed again before the aseptic packaging stage, or where liquid and particles are moved through the heating zone at different rates.

The 'Twintherm' system is a semi-continuous process in which particulate food is heated by direct steam injection in a pressurised, horizontal, cylindrical vessel that rotates slowly. Once the particles have been held for the required time to achieve F_0 values, they are cooled evaporatively, and liquid that has been sterilised in conventional UHT heat exchangers is used to carry the

Table 2.1 Direct and indirect heating systems

Direct systems			
Steam injection		Steam infusion	
Advantages	Limitations	Advantages	Limitations
• One of the fastest methods of heating and the fastest method of cooling and is therefore suitable for more heat-sensitive foods • Volatile removal is an advantage with some foods (for example milk)	• The method is only suitable for low viscosity products • There is relatively poor control over processing conditions • A requirement for potable steam which is more expensive to produce than normal processing steam • Regeneration of energy is less than 50% compared with more than 90% in indirect systems • Flexibility for changing to different types of product is low	• Almost instantaneous heating of the food to the temperature of the steam, and very rapid cooling which results in high retention of sensory characteristics and nutritional properties • Greater control over processing conditions than steam injection • Lower risk of localised overheating of the product • The method is more suitable for higher viscosity foods compared to steam injection	• The method is only suitable for low viscosity products • There is relatively poor control over processing conditions • There is a requirement for potable steam which is more expensive to produce than normal processing steam • Regeneration of energy is less than 50% compared with more than 90% in indirect systems • Flexibility for changing to different types of product is low • Blockage of the spray nozzles • Separation of components in some foods

particles to an aseptic filler. It is claimed to allow more uniform and gentle treatment of particles compared to continuous processes and it is used commercially to produce soups for the European market (Ohlsson, 1992; Alkskog, 1991).

The single flow fraction specific thermal processing (FSTP) system employs a separate holding section which is a cylindrical vessel containing slowly rotating fork blades on a central shaft. These blades form cages which hold the particles as they are rotated around the cylinder from the inlet to discharge pipes. Liquid moves freely through the cages, giving rapid heat transfer. Different

Indirect systems			
Plate heat exchanger		Tube-and-shell heat exchanger	
Advantages	Limitations	Advantages	Limitations
• Relatively inexpensive • Economical in floor space and water consumption • Efficient in energy use (> 90% energy regeneration) • Flexible changes to production rate, by varying the number of plates • Easily inspected by opening the plate stack	• Limited to low viscosity liquids (up to $1.5 Nsm^{-2}$) • Operating pressures limited by the plate gaskets to approximately 700kPa • Liquid velocities at relatively low pressure also low ($1.5–2 ms^{-1}$) • Low flow rates can cause uneven heating and solids deposits on the plates which require more frequent cleaning • Gaskets susceptible to high temperatures and caustic cleaning fluids and are replaced more regularly than in pasteurisation • Careful initial sterilisation of the large mass of metal in the plate stack is necessary for uniform expansion to prevent distortion and damage to plates or seals • Liable to fouling	• Few seals and easier cleaning and maintenance of aseptic conditions • Operation at higher pressures (7,000–10,000 kPa) and higher liquid flow rates ($6 ms^{-1}$) than plate heat exchangers • Turbulent flow at tube walls due to higher flow rates hence more uniform heat transfer and less product deposition	• Limited to relatively low viscosity foods (up to $1.5 Nsm^{-2}$) • Lower flexibility to changes in production capacity • Larger diameter tubes cannot be used because higher pressures needed to maintain the liquid velocity and large diameter pipes have a lower resistance to pressure • Any increase in production rate requires duplication of the equipment • Difficulty in inspecting heat transfer surfaces for food deposits

holding sections can be used for different particle sizes which reduces the risk of over-processing of smaller particles, which is found in more traditional systems. Again the liquid component is sterilised separately in conventional tubular heat exchangers and is then used to carry the sterilised particles to the filler. The system is described in detail by Hermans (1991).

Although it has not yet been possible to take full advantage of the HTST concept for aseptic processing of particulate-containing foods, there remain

Table 2.2 Comparison of conventional canning and aseptic processing and packaging

Criteria	Retorting	Aseptic processing and packaging
Product sterilisation		
Delivery	Unsteady state	Precise, isothermal
Process calculations		
Fluids	Routine – convection	Routine
Particulates	Routine – conduction or broken heating	Complex
Other sterilisation required	None	Complex (process equipment, containers, lids, aseptic tunnel)
Energy efficiency	Lower	30% saving or more
Sensory quality	Unsuited to heat-sensitive foods	Superior – suitable for homogenous heat-sensitive foods
Nutrient loss	High	Minimal
Value added	Lower	Higher
Convenience	Shelf stable	Shelf stable
Suitability for microwave heating	Glass and semi-rigid containers	All non-foil rigid and semi-rigid containers
Production rate	High (600–1,000/min)	Medium (500/min)
Handling/labour costs	High	Low
Downtime	Minimal (mostly seamer and labeller)	Re-sterilisation needed if loss of sterility in filler or steriliser
Flexibility for different container sizes	Need different process delivery and/or retorts	Single filler for different container sizes
Survival of heat-resistant enzymes	Rare	Common in some foods (e.g. milk)
Spoilage troubleshooting	Simple	
Low acid particulate processing	Routine	Not in practice (data from high acid systems being used to design low acid systems)
Post-process additions	Not possible	Possible to add filter sterilised enzymes or probiotics after heat processing

clear advantages in quality compared to canning (see Table 2.2). A breakthrough for the commercial application for particle-containing foods has been made by a US consortium in developing a protocol for validating the successful aseptic processing of particulates. Food and Drug Administration (FDA) approval of this protocol has led to approval of a process for the aseptic processing of a potato soup (Palaniappan and Sizer, 1997). Aseptic processing is now used for foods such as soups, baby foods, fruit and vegetables.

2.4.2 Semi-aseptic processes

There are a number of other similar processes to aseptic processing which may be labelled minimal processing compared to conventional canning, in that heat treatment is done in direct contact between heating medium and food. One

Table 2.3 High heat infusion

Advantages	Disadvantages
• Its capacity to kill heat-resistant spores • Longer operating times, due to reduced fouling • Increased safety, because the flash chamber (and in some cases the homogeniser) is placed prior to UHT treatment • Increased cost efficiency, especially in energy costs and the cost of maintenance, inspection, etc. • The possibility of aseptic flavour addition, giving the technology an advantage in the production of flavoured milk or ice cream mixes, for example • The capacity to manufacture a wider range of products	• Products undergo more severe chemical damage compared to the normal infusion process • In milk production the process also requires downstream homogenisation for the longest possible shelf-life to be obtained

example is the old Flash-18 process, in which people actually worked inside a pressure vessel in which food was heat sterilised in open cookers and packed hot inside the vessel. A modern, automated version of this is the Pressure-Pac system. In the Vatech system, food is packed and autoclaved in only partially sealed cans, so that steam penetrates into the food and drains all liquid from the can, after which the can is sealed under pressure and cooled.

For bulk pasteurisation of vegetables in combination with semi-aseptic modified atmosphere packaging, two interesting methods are claimed to be in commercial use in France (Varoquaux, 1993). One applies vacuum steam followed by vacuum cooling, and the other applies heat below 60°C under CO_2 overpressure. It appears that pasteurisation by heat or other means in combination with packaging under clean room conditions (semi-aseptic) is attracting increasing interest today (Swartzel, 1997). Interesting new process equipment that can be used in such applications are spiral steam heaters and coolers, also using impingement steam jets, and microwave and short-wave infrared tunnels (see below).

In the APV high heat infusion system, the very heat-resistant *Bacillus sporotherm durance* is inactivated by a heating step from 125° to 150°C in 0.2 seconds in a steam infusion chamber, followed by a very precise 2.0 seconds holding time. Upstream, the product is preheated to 70°C followed by vacuum cooling, after which the product is again preheated, now to 125°C, prior to the steam infusion. Compared to other UHT systems this is claimed to avoid any fouling, and energy regeneration is claimed to be up to 75% higher. The advantages and disadvantages of high heat infusion are summarised in Table 2.3. High heat infusion is used mainly for dairy products such as UHT milk, lactose-reduced milk, flavoured milks, various sauces and dressings (Andersen, 2001).

2.5 Sous-vide processing

The actual cooking of raw foods under vacuum in the package – so-called 'sous-vide' cooking – has been developed in France and has gained fairly wide acceptance as a means of producing high quality cooked foods of limited refrigerated shelf-life. As the name implies, the raw material is packed under vacuum, in multi-layer plastic packaging, and cooked in water, by air/steam mixtures or by microwaves at temperatures below 100ºC, removing oxygen. It is cooled rapidly to +3ºC in a blast chiller or in ice water. Loss of nutrients is minimised and an excellent quality obtained. In addition, risks of post-processing contamination are reduced and flexibility and product range in catering increased. A wide range of vegetables, meat and fish products are being processed by this technique, primarily in the institutional market but also increasingly in the consumer markets in France and the Benelux countries (Clausen, 1998; Schellekens, 1996).

The safety and quality of sous-vide products rely on good control and monitoring of hygiene during the entire chain of preparation, packaging, processing and distribution. A Hazard Analysis Critical Control Point (HACCP) approach is strongly recommended. For sous-vide products, the risk of survival and growth of low temperature pathogens such as *Listeria monocytogenes* is a particular risk. Guidelines recommend heating to 70ºC for a minimum of 2 minutes and then storing and distributing a product at temperatures below 3ºC, with a maximum shelf-life of seven days. Because of the potential risk for growth and toxin formation by *Clostridium botulinum*, the following recommendations (FDA) should be met to permit a shelf-life of more than ten days:

- heat treatment to 90ºC for 10 minutes or equivalent
- salt level above 3.5%
- water activity below 0.97
- pH below 5.0
- potential mishandling in distribution and storage monitored by time–temperature integrators.

It has been claimed in France that cooking under vacuum will have beneficial effects also for unpacked produce. Batch vacuum cookers can be combined with semi-aseptic packaging after the cooking operation. The European market for chilled foods as a whole was recently claimed to be worth 9 billion euro and expected to grow to 15 billion euro over the next three to four years. Unfortunately, there appear to be no corresponding statistics available on the size and growth of the market for sous-vide products within the chilled food sector. It is probably still small and limited to the catering and restaurant market rather than retail, but likely to become an important Home Meal Replacement (HMR) category in the future, if its safety record is well kept. So far chilled foods as a whole have maintained an excellent safety record in Europe, and the prevalence of pathogens is considered very low in sous-vide products in particular.

There appears to be a marked difference in attitude towards sous-vide foods in Europe and in North America. The perception of microbial risk seems to dominate in the USA and Canada, leading to more severe heat treatment and shorter recommended chilled storage times or to frozen storage. In Europe, there is more focus on the quality advantages of sous-vide processing, using milder heat treatment and longer storage times (up to 30 days), in the belief that temperature control in the supply chain is less of a problem. Current research is focusing on microorganisms that constitute a safety risk in sous-vide foods, especially *Clostridium botulinum*, and how D- and z-values of such organisms are affected by temperature, pH, composition and additives. For non-proteolytic *Clostridium botulinum,* heat treatments giving a 6-D reduction are recommended. While most efforts have been on microbial safety and shelf-life, more comprehensive data on sensory and nutritional quality and shelf-life are also emerging. They confirm a significant advantage in quality for sous-vide processing compared to alternative techniques, when storage temperatures at or below 3ºC are maintained (anon., 1999).

2.6 Infrared heating

Infrared (IR) waves occupy that part of the electromagnetic spectrum with a frequency beyond that of visible light. When IR waves hit a material, they are either reflected, transmitted or absorbed. Absorbed waves are transformed into heat and the temperature of the material increases.

Long-wave IR heating at wavelengths around 30μm has long been in use for industrial cooking and drying applications, achieving shorter processing times than by convective heating. With the advent of short-wave (1μm) and intermediate-wave (5μm) quartz tubes, IR heating has gained wider acceptance for rapid drying, baking and cooking of foods of even geometry and low thickness. Short-wave IR makes it possible to reach working temperatures in seconds while offering rapid transfer of high amounts of energy and excellent process control. Short-wave IR has a penetration depth of several millimetres in many foods and can therefore be used to about the same effect as microwaves or high frequency for thin materials. In addition, it is not absorbed by transparent plastic packaging and has been successfully applied for the surface pasteurisation of bakery products both in France and in Sweden (anon., 1994). IR is also easy to control and relatively inexpensive with a high degree of energy utilisation. It should be able to compete with other minimal processes for the surface pasteurisation of foods (Driscoll, 1992).

The main commercial applications of IR heating are drying low moisture foods such as breadcrumbs, cocoa, flour, grains, malt and tea. It is also used as an initial heating stage to speed up the initial increase in surface temperature. In this role it is used in baking and roasting ovens and in frying as well as drying. There is some use of the IR technique in drying of fish products (Wei-Renn-Lien and Wen-Rong-Fu 1997). Much research in recent years has been done in

Taiwan, China and Japan (Afzal and Abe 1998). They have studied drying fish, drying rice, rice for parboiling and potato. Particular applications include the following:

- drying vegetables and fish
- drying pasta and rice
- heating flour
- frying meat
- roasting cereals
- roasting coffee
- roasting cocoa
- baking pizza, biscuits and bread.

The technique has also been used for thawing, surface pasteurisation of bread and pasteurisation of packaging materials. The main component in an IR oven is the radiator. Ovens use groups of radiators which can be individually controlled. Radiators may be divided into the following main groups:

- gas-heated radiators (long waves)
- electrically heated radiators:
  - tubular/flat metallic heaters (long waves)
  - ceramic heaters (long waves)
  - quartz tube heaters (medium and short waves)
  - halogen heaters (ultra-short waves).

Experimental IR baking ovens using short-wave IR heating have produced the following results:

- the baking time was 25–50% shorter compared to an ordinary baking oven, depending on the thickness of the product
- energy consumption was comparable to ordinary baking
- weight losses were 10–15% lower
- quality was comparable.

These results show that IR heating for bakery products is very effective compared to other heating techniques. The main advantages of using short-wave IR were found to be as follows:

- high and effective heat transfer
- heat penetration/reduction of baking time
- no heating of air in the oven
- quick regulation and control
- compact and flexible ovens.

2.7 Electric volume heating methods for foods

In the food industry, thermal processing using electric heating is done by applying electromagnetic energy to produce temperature increases in food.

Many of the electric heating methods discussed in this section are not in themselves new, but the knowledge about them as well as their application in the food industry is limited at present (UIE, 1996). Electric heating methods directly transfer energy from the electromagnetic source to the food, without heating up heat transfer surfaces in the heat processing equipment itself. This direct energy transfer is a major advantage as it gives excellent opportunities for high energy utilisation. Looking through the electromagnetic spectrum, we can identify three frequency areas that are employed today in the industry for direct heating of food:

1. The 50/60 Hz area is used for electric resistance heating, sometimes called ohmic heating. In this application, the food itself acts as a conductor between the ground and a charged electrode, normally at 220 or 380 volts.
2. In the high frequency area of 10–60 MHz, foods are placed between electrodes – one of them being grounded – and energy is transferred to the food placed between the electrodes.
3. In the microwave region of 1–3 GHz, energy is transferred to the food through the air by guided waves, controlled by electromagnetic devices called applicators.

The following sections discuss the technologies based on these three frequency areas:

1. Electric resistance/ohmic heating.
2. High frequency/radio frequency heating.
3. Microwave heating.

Some advantages and disadvantages of these technologies are summarised in Table 2.4. In all these electric heating methods, it is important to have an understanding of the interaction between the electromagnetic field at the

Table 2.4 Comparisons between the frequencies of electric volume heating

Ohmic and high frequency		Microwave	
Advantages	Disadvantages	Advantages	Disadvantages
• Better for large, thick foods	• Risk of arching in HF	• Higher heating rate	• Limited penetration
• Lower investment costs	• Larger floor space	• Design freedom	• Higher investment costs
• Easier to understand and control	• Narrow frequency bands	• Less sensitive to food inhomogenity	• More engineering needed
	• Limited R&D support	• Much R&D available	

frequency in question and the material being subjected to the energy. Knowing the electric and dielectric properties of food and other materials in the heating equipment is important in order to understand and control better the application of electric energy for the heating of foods (Ohlsson, 1987). It is also important to understand the impact of these forms of heating on food safety and quality. There have been extensive studies regarding the changes of chemical constituents in the foods as a result of electric or electromagnetic heating. It has been demonstrated that the effect from all practical aspects is the same as for conventional heating to the same temperature. Electric heating equipment for the food industry has also to be designed and operated according to international and national safety standards. The levels of allowable leakage vary over the frequency range according to these standards. Equipment for measuring and monitoring electromagnetic energy leakage from electric equipment is readily available (IEC, 1982).

2.8 Electric resistance/ohmic heating

2.8.1 Fundamentals

In electric resistance heating, the food itself acts as a conductor of electricity taken from the mains supply: that is 50 Hz in Europe and 60 Hz in the USA. The food may also be immersed in a conducting liquid, normally a weak salt solution of similar conductivity to the food. Heating is accomplished according to Ohm's law, where the conductivity, or the inverse, the resistivity, of the food will determine the current that will go between the ground and the electrode. Normally, voltages up to 5000 V are applied. The conductivity of foods increases considerably with increasing temperature. To reach high temperatures it is therefore necessary to increase the current or the voltage or to use longer distances between the electrodes and ground.

In ohmic heating processes, foods are made part of an electric circuit through which alternating current flows, causing heat to be generated within the foods due to the electrical resistance of the foods. Therefore, in a liquid–particulate food mixture, if the electrical conductivity of the two phases is comparable, heat could be generated at the same or comparable rate in both phases in ohmic heating. In other conditions, heat can be generated faster in the particulate than in the liquid. Ohmic methods thus offer a way of processing food at the rate of HTST processes, but without the limitation of conventional HTST on heat transfer to particulates.

2.8.2 Equipment

The best-known electric resistance heating system is the APV ohmic heating column, where electrodes are in direct contact with the food, which is transported in a vertical concentric tube. Electrodes, normally four, are connected to earth and line voltage. The inside of the tube is lined with high temperature and electrically inert plastic material. The electrodes themselves are constructed of noble metal-coated titanium to avoid adverse electrochemical reactions and dissolution of metal ions into the food. If stainless steel electrodes are used instead, such as in the Reznik process for liquid egg pasteurisation, a higher frequency has to be used to avoid electrode reactions.

2.8.3 Applications

The APV ohmic system has been installed for pasteurisation and sterilisation of a number of food products with excellent resulting quality, particularly when plug flow is being achieved. The majority of these installations are found in Japan for the production of fruit products (Tempest, 1996). There are also other installations, e.g. for prepared food in the UK. The ohmic heating system shows excellent retention of particle integrity for particles up to 2 cm in diameter, due to the absence of mechanical agitation, typical for traditional heat exchanger-based heating systems. A special reciprocal piston pump is also used to maintain high particle integrity. A long traditional tubular heat exchanger is used for cooling. Extensive microbiological evaluation of the system has shown that the method can produce sterile products reliably (Ruan *et al.*, 2001).

Other industrial cooking operations for electric resistance heating involve rapid cooking of potatoes and vegetables for blanching and in preparing foods for the catering sector. One of the major problems with these applications is that of ensuring that the electrode materials are inert and do not release metal ions into the conducting solutions and eventually to the foods. Ensuring that the electrode material is isolated from the food is also one of the problems of electric resistive heating. Another problem is the need to control effectively the electric conductivity of all constituents of the food product, as this determines the rate of heating for the different constituents. This often requires well-controlled pre-treatments to eliminate air in foods and control salt levels in foods and sauces, etc. (Zoltai and Swearingen, 1996).

Currently, at least 18 ohmic heating operations have been supplied to customers in Europe, Japan and the USA. The most successful of these systems have been for the processing of whole strawberries and other fruits for yoghurt in Japan, and low-acid ready-to-eat meals in the USA (anon., 1996). Currently there are two commercial manufacturers of ohmic heating equipment: APV Baker Ltd, Crawley, UK, and Raztek Corp., Sunnyvale, CA, USA. In the USA, a consortium of 25 partners from industry (food processors, equipment manufacturers, and ingredient suppliers), academia (food science, engineering, microbiology and economics) and government was formed in 1992 to develop products and evaluate the capabilities of the ohmic heating system. A 5kW pilot-scale continuous-flow ohmic system manufactured by APV Baker Ltd was evaluated by the consortium at Land-O'Lakes, Arden Hills, Minnesota, from 1992 to 1994. A wide variety of shelf-stable low- and high-acid products, as well as refrigerated extended shelf-life products, were developed. They were found to have texture, colour, flavour and nutrient retention that matched or exceeded those of traditional processing methods such as freezing, retorting and aseptic processing. The consortium concluded that the technology was viable. In addition to the technical evaluation, an economic study was also initiated. Ohmic operational costs were found to be comparable to those for freezing and retorting of low-acid products (Zoltai and Swearingen, 1996).

Although the technological approaches associated with aseptic food processing (i.e. pumps, fillers, heater electrode, etc.) have developed significantly, the identification, control and validation of all the critical control points required to demonstrate that an ohmically processed multi-phase food product has been uniformly rendered commercially sterile have yet to be generally accepted. Consequently, the FDA has not yet approved processes involving continuous ohmically processed multi-phase food products. The ohmic heating process has the promise to provide food processors with the opportunity to produce new, high-value-added, shelf-stable products with a quality previously unrealised with current sterilisation techniques. Applications that have been developed include aseptic processing of high-value-added ready-prepared meals for storage and distribution at ambient temperature; preheating of food products prior to in-can sterilisation; and the hygienic production of high-value-added ready-prepared foods for storage and distribution at chilled temperatures. Ohmic heating can also be used for heating of high-acid food products such as tomato-based sauces prior to hot-filling, with considerable benefits in product quality. Other potential applications include rapid heating of liquid food products, which are difficult to heat by conventional technologies, blanching, evaporation, dehydration, fermentation and extraction (Parrott, 1992).

The advantages of ohmic heating technology claimed in the previous research are summarised as follows (Skudder, 1988, and Kim *et al.*, 1996):

- heating food material by internal heat generation without the limitation of conventional heat transfer and some of the non-uniformity commonly associated with microwave heating due to limited dielectric penetration – heating takes place volumetrically and the product does not experience a large temperature gradient within itself as it heats
- achieving higher temperature in particulates over liquids in liquid–particulate mixtures, which is impossible for conventional heating
- reducing risks of fouling on heat transfer surface and burning of the food products, resulting in minimal mechanical damage and better nutrient and vitamin retention
- high energy efficiency because 90% of the electrical energy is converted into heat
- optimisation of capital investment and product safety as a result of high solids loading capacity
- ease of process control with instant switch-on and shut-down
- reducing maintenance cost (no moving parts)
- ambient-temperature storage and distribution when combined with an aseptic filling system

- a quiet, environmentally friendly system.

The disadvantages of ohmic heating are associated with its unique electrical heating mechanisms. The heat generation rate is affected by the electrical heterogeneity of the particle, its heat channelling properties, the complex relationship between temperature and electrical field distributions, and particle shape and orientation. All these make the process complex and contribute to non-uniformity in temperature, which may be difficult to monitor and control. The three major challenges limiting the commercialisation of ohmic heating are as follows:

1. Lack of a complete model that takes into account differences in electrical conductivity between the liquid and solid phases and the responses of the two phases to temperature changes, which affect relative heating rates and distribution.
2. Lack of data concerning critical factors affecting heating, including residence time, orientations, loading levels, etc.
3. Lack of applicable temperature validating techniques for locating cold/hot spots.

The development of magnetic resonance imaging (MRI) now allows more accurate and dynamic mapping of temperature distribution of ohmically heated food systems, and the development of accurate models (Ruan *et al.*, 2001). Further developments here will meet these remaining challenges and allow the potential of ohmic heating to be more fully exploited.

2.9 High frequency or radio frequency heating

2.9.1 Fundamentals

High frequency (HF) heating is done in the MHz part of the electromagnetic spectrum. The frequencies of 13,56 and 27,12 MHz are set aside for industrial heating applications. Foods are heated by transmitting electromagnetic energy through the food placed between an electrode and the ground. The HF energy used will allow for transfer of energy over air gaps and through non-conducting packaging materials. To achieve sufficiently rapid heating in foods, high electric field strengths are needed. HF heating is accomplished by a combination of dipole heating, when the water dipole tries to align itself with the alternating electric field, and electric resistance heating from the movement of the dissolved ions of the foods. In the lower temperature range, including temperatures below the freezing point of foods, dielectric heating is important, whereas for higher temperatures electric conductivity heating dominates. The conductivity losses or the dielectric loss factor increases with increasing temperature which may lead to problems of runaway heating when already hot parts of the food will absorb a majority of the supplied energy. Dielectric property data of foods are reasonably abundant in the low temperature range, but few data are available for temperatures above normal room temperature.

2.9.2 Equipment

An important part of HF heating equipment is the design of the electrodes. A number of different configurations are currently used, depending on factors such as field strength needed and the configuration of the sample. For high moisture applications, the traditional electrode configuration is mostly used. For low moisture applications, such as dried foods and biscuits, electrodes in the form of rods, giving stray fields for foods placed on a conveyor belt, are often used. Electrodes can be designed to create uniform electric field patterns and thus uniform heating patterns for different types of food geometry. These dedicated designs are supported by computer simulation techniques such as finite element method (FEM) software packages (Metaxas, 1996).

The HF power is generated in an oscillating circuit, of which the food is a part. It consists of a coil, the condensator plates with the food in between, an amplifier in the form of a triode and an energy source. It is important to monitor and maintain a given frequency, as this may otherwise vary as the food is heated, bearing in mind that the food itself is part of the oscillating circuit. The control function is today improved by the introduction of so-called 50 ohms technology, which allows a separate control for tuning the load circuit.

2.9.3 Applications

Radio frequency (RF) heating has been used in the food processing industry for many decades. In particular, RF post-baking of biscuits and cereals and RF drying of foods are well-established applications. More recently, RF thawing equipment has demonstrated substantial benefits over conventional techniques and over comparable microwave tempering systems. Furthermore, as public concern over food safety issues continues to grow, and as the demand for convenience foods increases, RF pasteurisation and sterilisation processes are becoming more important.

The post-baking of biscuits (Holland, 1974) is one of the most accepted and widely used applications of RF heating in the food processing industry. The addition of a relatively small RF unit to the end of a conventional baking line results in substantial increase in product throughput, together with improvements in product quality. The same process has also been applied to cereal, pastry and bread products. More recently, the RF system has been incorporated directly into the hot air oven, allowing RF-assisted baking of a wide range of products to be carried out in a very compact unit. The principal role of RF heating in baking is the removal of moisture, particularly at the end of the process when conventional heating is inefficient. RF drying is intrinsically self-levelling (Jones and Rowley, 1997), with more energy being dissipated in wetter regions than in drier ones. This RF levelling leads to improvements in product quality and more consistent final products. As well as baking applications, RF drying applications in the food industry include the drying of food ingredients (e.g. herbs, spices, vegetables), potato products (e.g. French fries) and a number of pasta products.

A rediscovered, and rapidly growing, application of RF in the food processing industry is its use for the bulk defrosting of meats and fish. Conventionally, large blocks of meat are thawed slowly, often over a period of days. The volumetric nature of RF heating allows the thawing process to be accelerated, while still maintaining control of the temperature distribution within the food product. Typically, RF defrosting times of 1–2 hours are possible. However, with very regularly shaped material, such as blocks of fish fillets, heating to -2°C can be achieved in less than 10 minutes (Bernhard, 2001). While microwaves are suited for heating only, RF can be sucessfully used for complete thawing. Recently, HF cooking equipment for pumpable foods has also been developed. These devices involve pumping a food through a plastic tube placed between two electrodes, shaped to give a uniform heating. Excellent temperature uniformity has been demonstrated in these applications, e.g. for continuous cooking of ham and sausage emulsions (Tempest, 1996).

In common with microwave and ohmic heating, the volumetric nature of RF heating gives rise to a number of significant advantages over more traditional, surface heating techniques. The most important of these to the food industry are as follows:

- *Improved food quality.* The main reason for using RF heating in food processing (rather than any other thermal technology) is improved food quality. First, the volumetric process leads to more uniform heating, removing the risk of overheating food surfaces while trying to heat the centre of products. Second, the selective nature of RF heating, with energy being dissipated according to the local loss factor, can produce very uniform products, even when there are relatively large variations in the unprocessed food.
- *Increased throughput.* Conventional surface heating often has to heat foods relatively slowly to avoid the risk of overheating the surface. Moreover, once the surfaces of foods have dried out, they often form a good thermal barrier layer, making it even more difficult to heat the centres. By contrast, volumetric RF heating avoids these effects, allowing production lines to operate much faster.
- *Shorter process lines.* As an alternative to increased throughput, food processing lines that include RF systems can be significantly shorter for a given throughput.
- *Improved energy efficiency.* Since the RF energy is dissipated directly within the product being heated, processing lines using this technology can be very efficient, particularly when the increased throughput is also taken into account.
- *Improved control.* Since the power dissipated within the food is due to the presence of an electric field, if this field is changed, or switched off, the

heating of the food products responds almost instantaneously. In this way, RF heating can be controlled very precisely, again leading to improvements in food quality.

In food processing, RF heating also has a number of advantages over the alternative volumetric technologies, namely microwave and ohmic heating. The main ones are as follows:

- *Contactless heating.* Although the heating mechanism is essentially the same as with ohmic heating, RF does not require the electrodes to be in contact with the food. This removes the constraint that the food product has to be pumpable, and allows RF heating to be applied to solid as well as liquid heating.
- *Increased power penetration.* The longer wavelength at radio frequencies compared with microwave frequencies, and the dielectric properties of foods, mean that RF power will penetrate further into most products than microwave power. For example, the penetration depth (the distance for the power to fall to $1/e$ of its initial value) in unfrozen meat products is typically only a few centimetres at microwave frequencies, but in the order of a metre or so at RF frequencies.
- *Simpler construction.* Large RF applicator systems are generally simpler to construct than microwave ones. In particular, the longer wavelength at radio frequencies allows relatively large entry and exit ports to be designed – 2 m wide ports are not untypical. Moreover, the geometries of RF applicator systems naturally lend themselves to industrial food processing applications.
- *Improved moisture levelling.* In food products, the variation of the dielectric loss factor with moisture content is generally greater at radio than at microwave frequencies. Consequently, the use of RF heating for baking and drying applications leads to improved moisture levelling and correspondingly higher quality final products.

When compared with conventional heating techniques and, to some extent, with ohmic heating, the main disadvantages of RF heating relate to equipment and operating costs:

- *Equipment and operating cost.* For an equivalent power output, RF heating equipment is more expensive than conventional convection, radiation or steam heating systems. It is also more expensive than an equivalent ohmic heating system. However, in some applications, improvements in product quality and throughput often more than justify the initial capital investment. As an electroheat technology, the unit energy costs of an RF system will be higher than an equivalent conventional heating system. Nevertheless, when factors such as increased energy efficiency and increased throughput are taken into account, the total energy cost may be comparable to (or even less than) a conventional system.

- *Reduced power density.* Given that the electric field is limited to avoid the occurrence of an electrical breakdown, then the power density can be much higher at microwave than radio frequencies. The main consequences of this are that RF systems are usually significantly larger than microwave heating systems of the same power rating, and that faster heating rates can often be achieved with a microwave system.

The current and increasing demand for high quality food products will mean that RF post-baking and RF-assisted baking will continue to be important stages in the processing of biscuit, cereal and pastry products. Similarly, public concern over food hygiene issues will continue to require rapid and safe food thawing techniques, such as RF or microwave meat and fish defrosting systems. Increasing public awareness of general food safety issues and the rising demand for convenience, pre-packaged foods will also lead to a growth in the demand for microwave pasteurisation and sterilisation techniques (Rowley, 2001).

2.10 Microwave heating

2.10.1 Fundamentals

Microwaves used in the food industry for heating are the Industrial, Scientific and Medical (ISM) frequencies 2 450 MHz or 915 (896) MHz, corresponding to 12cm or 34cm in wavelength. The majority of foods contain a substantial proportion of water. The molecular structure of water consists of a negatively charged oxygen atom, separated from positively charged hydrogen atoms, and this forms an electric dipole. When a microwave or RF electric field is applied to a food, dipoles in the water and in some ionic components such as salt attempt to orient themselves to the field (in a similar way to a compass in a magnetic field). Since the rapidly oscillating electric field changes from positive to negative and back again several million times per second, the dipoles attempt to follow and these rapid reversals create frictional heat. The increase in temperature of water molecules heats surrounding components of the food by conduction and/or convection. Because of their widespread domestic use, some popular notions have arisen that microwaves 'heat from the inside out'. What in fact occurs is that outer parts receive the same energy as inner parts, but the surface loses its heat faster to the surroundings by evaporative cooling. It is the distribution of water and salt within a food that has the major effect on the amount of heating, although differences also occur in the rate of heating as a result of the shape of the food.

The depth of penetration of both microwaves and RF energy is determined by the dielectric constant and the loss factor of the food. These properties have been recorded for some foods (Kent, 1987). They vary with the moisture content and temperature of the food and the frequency of the electric field. In general, the lower the loss factor (i.e. greater transparency to microwaves) and the lower the frequency, the greater the penetration depth. The penetration ability of microwaves in foods is limited. For normal 'wet' foods the penetration depth is approximately 1–2 cm at 2450 MHz. At higher temperatures, the electric resistance heating from the dissolved ions will also play a role in the heating mechanisms, normally further reducing the penetration depth of the microwave energy. The limited penetration depth of microwaves implies that the distribution of energy and heat within the food can vary. The control of the heating uniformity of microwave heating is difficult. These difficulties in controlling heating uniformity must be seen as the major limitation for industrial application of microwave heating (Ohlsson, 1983).

2.10.2 Equipment

The transfer of microwave energy to food is done by contactless wave transmission. The microwave energy feed system is designed to control the uniformity during the heating operation. Microwave equipment consists of a microwave generator (termed a *magnetron*), aluminium tubes named *wave guides*, and a metal chamber for batch operation, or a tunnel fitted with a conveyor belt for continuous operation. Because microwaves heat all biological tissues, there is a risk of leaking radiation causing injury to operators, particularly to the eyes. Chambers and tunnels are therefore 'sealed' by absorbers or microwave traps to prevent the escape of microwaves. Detailed descriptions of component parts and operation of microwave heaters are given by Copson (1975) and Buffler (1993).

The magnetron is a cylindrical diode ('di' meaning two and 'electrode'), which consists of a sealed copper tube with a vacuum inside. The tube contains copper plates pointing towards the centre like spokes on a wheel. This assembly is termed the 'anode' and has a spiral wire filament (the cathode) at the centre. When a high voltage (e.g. 4000 V) is applied, the cathode produces free electrons, which give up their energy to produce rapidly oscillating microwaves, which are then directed to the wave guide by electromagnets. The wave guide reflects the electric field internally and thus transfers it to the heating chamber. It is important that the electric field is evenly distributed inside the heating chamber to enable uniform heating of the food. In batch equipment a rotating antenna or fan is used to distribute the energy, or the food may be rotated on a turntable. Both methods reduce shadowing (areas of food that are not exposed to the microwaves). In continuous tunnels a different design of antennae is used to direct a beam of energy over the food as it passes on a conveyor. It is important that the power output from the magnetron is matched to the size of the heating chamber to prevent flash-over. Power outputs of continuous industrial equipment range from 30kW to over 600kW.

Many different designs are used in industrial applications, starting from the traditional multi-mode cavity oven, via direct radiation wave guide applicators to sophisticated periodic structures (Metaxas, 1996). Design of applicators needs to be done with proper care taken with regard to the interaction between parameters important to heating uniformity such as:

- food composition and geometry
- packaging geometry and composition
- microwave energy feed system (applicator design).

As pointed out by Ryynänen and Ohlsson (1996), the influence of food geometry and the actual layout of the components on a plate for reaching good heating uniformity is often poorly understood. The microwave energy feed system controls the electric field polarisation. This in its turn affects the tendencies for overheating of food edges, which is one of the most severe problems of uneven microwave heating of foods (Sundberg *et al.*, 1996). The very high frequency used in microwave heating allows for very rapid energy transfer and thus high rates of heating. This is a major advantage, but can also lead to problems of non-uniform heating when too high energy transfer rates are used.

2.10.3 Applications

There are various industrial applications of microwave heating:

- baking and cooking
- tempering
- drying
- pasteurisation and sterilisation.

Rosenberg and Bögl (1987a) provide a good summary of the use of microwaves in baking bread, cakes and pastry. The major contribution of microwave technology is to accelerate baking, leading to an enhanced throughput with negligible additional space required for microwave power generators. Such accelerated baking can be achieved without loss of product quality, allowing appropriate degrees of crust formation and surface browning. The fast combined process also allows the use of flour with high α-amylase and low protein content (for example from European soft wheat). In contrast to conventional baking the microwave heating inactivates enzymes fast enough (due to a fast and uniform temperature rise in the whole product) to prevent the starch from extensive breakdown, and develops sufficient CO_2 and steam to produce a uniform porous texture (Decareau, 1986). Today the main use of microwaves in the baking industry is microwave finishing. While the conventional oven technique is used at the beginning with high moisture dough, microwaves improve the end baking, where the low heat conductivity would lead to considerably higher baking times.

Microwaves can also be applied as part of a parallel process. One example is the frying of doughnuts with microwave assistance (Schiffmann, 1986), resulting in a shorter frying time and a lower fat uptake. Pre-cooking processes can also be accelerated with the help of microwaves, as has been established for pre-cooking of poultry (Decareau, 1986), meat patty and bacon. In bacon precooking, convective air flow removes the surface water using microwaves as the main energy source, thus rendering the fat and coagulating the proteins by an increased temperature. This process yields a valuable byproduct, rendered fat of high quality, which is used as food flavorant (Schiffmann, 1986).

Another widely used industrial microwave application is the tempering of foods (Metaxas, 1996, mentions a figure of 250 units all over the world). Tempering is defined as the thermal treatment of frozen foods to raise the temperature from below -18°C to temperatures just below the melting point of ice (approximately -2°C). At these temperatures the mechanical product properties are better suited for further machining operations (e.g. cutting or milling). The time for conventional tempering strongly depends on the low thermal conductivity of the thawed surface layer and the low permissable surface temperature and can be in the order of days for larger food pieces such as blocks of butter, fish, fruits or meat. As a result this conventional process needs large storage rooms, there is a significant drip loss and the danger of microbial growth. By using microwaves (mostly with 915 MHz due to their larger penetration depth) the tempering time can be reduced to minutes or hours (Edgar, 1986) and the required space is diminished to one-sixth of the conventional system (Metaxas, 1996). Another advantage is the ability to use the microwaves at low air temperatures, thus reducing or even stopping microbial growth. It is important to control heating uniformity and the end temperature to avoid localised melting which would be coupled to a thermal runaway effect. The best homogeneity in this application is reached in a multi-source multi-mode cavity, equipped with mode stirrers (Metaxas, 1996).

Microwaves can help in the acceleration of conventional drying where this is limited by low thermal conductivities, especially in products of low moisture content. The use of microwaves can prevent damage to nutrients and sensory quality caused by long drying times or high surface temperatures, including limiting hardening in food. Microwave drying can be subdivided into two types:

1. Drying at atmospheric pressure.
2. Drying with applied vacuum conditions.

Combined microwave air-dryers are widely used, and can be classified into a serial or a parallel combination of both methods. In the serial process, microwaves are mainly used to finish partly dried food or food of low moisture content, where an intrinsic levelling effect is advantageous. Areas of high water content transform more of the microwave energy into heat and are selectively dried. Well-studied examples for serial hot air and microwave dehydration are pasta drying (Decareau, 1985) and the production of dried onions (Metaxas and Meredith, 1983). The finish drying of potato chips has also been tried with varying success (O'Meara, 1977). The combination of microwave and vacuum drying also has potential. Although the microwave-assisted freeze drying is well studied, as can be read in detail in Sunderland (1980), until now practically no commercial industrial application can be found, due to high costs and a small market for freeze-dried food products (Knutson *et al.*, 1987).

It seems that microwave vacuum drying with pressures above the triple point of water has more commercial potential. The benefit of using microwave energy is overcoming the disadvantage of very high heat transfer and conduction resistances, leading to higher drying rates. These high drying rates also allow retention of water-insoluble aromas and lead to less shrinkage (Erle, 2000). The use of vacuum pressures helps retain product quality, since the reduced pressure limits the product temperatures to lower values, as long as a certain amount of free water is present. This enables the retention of temperature-sensitive substances like vitamins and colours. Commercial applications of microwave vacuum dehydration include the drying of grains in short times without germination (Decareau, 1985). A relatively new and successful combination of microwave vacuum drying (called puffing) and air drying is used mainly to produce dried fruits and vegetables, with improved rehydration properties (Räuber, 1998). Conventional pre-drying stabilises the product through case hardening, and the microwave vacuum process opens the cell structures (puffing) due to the fast vaporisation, generating an open pore structure. The consecutive post-drying reduces the water content to the required moisture.

Since microwave energy can heat many foods (containing water or salts) effectively and fast, its use for pasteurisation and sterilisation has also been intensively studied. The advantages of using microwaves in microorganism deactivation include high and homogeneous heating rates (in some foods) and the corresponding short process times, which can yield a very high quality. Many references can be found in the review by Rosenberg and Bögl (1987b). Microwave pasteurisation or sterilisation applications include pre-packed food like yoghurt or pouch-packed meals as well as the continuous pasteurisation of fluids like milk (Decareau, 1985; Rosenberg and Bögl, 1987b). Conveyor belt systems have been planned for packed food systems (Harlfinger, 1992). For continuous fluid pasteurisation or sterilisation, tubes intersecting wave guides or small resonators have been developed (Sale, 1976). Whereas the pasteurisation process can take place at atmospheric pressures, in the case of sterilisation only temperatures of more than 100°C may be used in order to achieve satisfactory short sterilisation times and to maintain high product quality. For products that contain free water, like many food products, the reachable temperature at atmospheric pressure is limited to the boiling point at around 100°C. Therefore the pressure during the sterilisation process has to be increased. The consequence is the need for special compression and decompressing systems, such as sliding gates, which have to be connected to the microwave heater.

Pasteurisation with microwave heating is also possible for pumpable foods (Püschner, 1964). Microwaves are directed to the tube where the food is transported and heated directly across the tube cross-section. Again, uniformity of heating must be ensured, which requires selection of the correct dimensions of the tube diameter and the proper design of the applicators (Ohlsson, 1993). Systems where the food is transported through the heating zone by a screw are also available (Berteaud, 1995).

Sterilisation using microwaves has been investigated for many years, but commercial use has only come in recent years, in Europe and Japan. Microwave pasteurisation and sterilisation promise to give very rapid heat processing, which should lead to only small quality changes due to the thermal treatment, according to the HTST principle. However, it has turned out that very high requirements on heating uniformity must be met in order to fulfil these quality advantages (Ohlsson, 1991). It is critical to know and control the lowest temperatures within the product, where the microorganism destruction has the slowest rate. To minimise temperature variation and also for process economy reasons, microwaves should, preferably, be used in combination with conventional heating, using rapid volumetric heating for the final burst of 10–30ºC to achieve HTST-like processing. However, current commercial equipment does not have any marked advantage in quality over optimised conventional processing, but there is a clear possibility of future improvements.

In a European project, with the participation of SIK, Campden & Chorleywood RA, the Frauenhofer Institute and industrial partners, industrial microwave sterilisation was investigated. Computer simulation programs, developed for field distribution calculations, permitted the optimisation of microwave applicator design as well as of product shape and layout. Well-functioning process validation methods were developed, based on spore suspensions enclosed in small alginate beads. The project demonstrated the benefits of microwave sterilisation in terms of sensory and nutritional quality compared to conventional retorting and demonstrated that microwave sterilisation of foods should be a viable industrial process, as demonstrated by heating experiments in batch as well as continuous pilot equipment (EC, 2000).

The major advantages of microwave heating are speed of heating, time savings and volume instead of surface heating. Disadvantages of current industrial microwave applications range from high energy costs, which have to be counterbalanced by higher product quality, to the lack of complete microwave heating models to predict heating effects on food. The latter disadvantage has been partly overcome by increases in computing power which makes it possible to compute more and more realistic models by numerical methods (Regier and Schubert, 2001). The best test for such models are experiments to establish real temperature distributions within the product, which are particularly important in pasteurisation and sterilisation applications. While more conventional temperature probe systems like fibreoptic probes, liquid crystal foils or infrared photographs only give incomplete information about the temperature distribution within the whole sample, MRI has the potential to give very useful information about heating patterns (Nott *et al.*, 1999). Together with better modelling techniques, improved instrumentation will give microwave techniques an additional boost in industrial food production.

2.11 Inductive electrical heating

A process for inductive heating of pumpable liquids has been developed at the Technical University of Munich. In this process, the food is pumped through coils of plastic tubing wound around the magnetic core of a strong electromagnet. The alternating magnetic field induces high frequency alternating electric fields and currents in the food, to raise its temperature very quickly to sterilisation temperatures (for example from 100°C to 140°C in 30 seconds). The heating is claimed to be very even throughout and very accurate, with a minimum of adverse effects on sensory and nutritional properties (Anon., 2000).

2.12 Future trends

Progress in pasteurisation and sterilisation in the future will involve continued emphasis on process control and product safety. On-line sensors and integrated production information systems that allow for more flexibility and increased productivity will be seen. Today there are contradictory trends for increased safety and for less visible preservation methods and reductions in salt, sugar and acid content in food products. This points towards the need for more research and development on the engineering aspects of food pasteurisation and sterilisation and on the combination of methods, both thermal and non-thermal, for optimal results.

2.13 References and further reading

AFZAL, T.M. and ABE, T. 1998. Diffusion in potatoes during far infrared radiation drying. *Journal of Food Engineering* **37** 4, 353–65.

ALKSKOG, L. 1991. Twintherm – a new aseptic particle processing system. Paper presented at the 'News in Aseptic Processing and Packaging' seminar, Helsinki, January 1991.

ANDERSEN, J. 2001. Instant and high heat infusion. In P. Richardson (ed.) *Thermal Technologies in Food Processing*. Woodhead Publishing, Cambridge, pp. 229–40.

ANON. 1994. Les infrarouges, une chaleur facile. *Cahiers des industries alimentaires* **33**, Sept. 1994, 3–10.

ANON. 1996. Ohmic heating garners 1996 industrial achievement award. *Food Technology* **20**, 114–15.

ANON. 1999. Third European Symposium on Sous-Vide, 25–26 March, Leuven, Belgium.

ANON. 2000. Unkonventionelle Haltbarmachungsverfahren unter der Lupe. Best before ... *Food Design* 1/2000.

BENGTSSON, N. 1994. Other new developments in thermal and non-thermal methods of pasteurisation and sterilisation and their packaging implications. Proceedings of IoPP Packaging Technology Conference, Chicago, 11–12 November.

BERNHARD, J.P. 2001. Saircm. Personal Communication.

BERTEAUD, A-J. 1995. Thermo-star. Bulletin from MES, 15 Rue des Solets, RUNGIS, France.

BUFFLER, C.R. 1993. *Microwave Cooking and Processing: Engineering*

Fundamentals for the Food Scientist. AVI/Van Nostrand Reinhold, New York.

CARLSON, B. 1996. Food processing equipment: historical and modern designs. In J.R.D. David, R.H. Graves and V.R. Carlson (eds) *Aseptic Processing and Packaging of Food.* CRC Press, Boca Raton, pp. 51–94.

CLAUSEN, I. 1998. Sous-Vide: Er det fremtidens tillberedningsmetode? *Plus Process* **2**, 12–14.

COPSON, D.A. 1975. *Microwave Heating.* AVI, Westport, CN, pp. 262–85.

DECAREAU, R.V. 1985. *Microwaves in the Food Processing Industry.* Academic Press Inc., Orlando.

DECAREAU, R.V. 1986. Microwave food processing equipment throughout the world. *Food Technology* **40**, 99–105.

DRISCOLL, J. 1992. Infrared heating and food processing. *Nutrition & Food Science* **23** 1, 19–20.

EC 2000. Food Project Results. AIR Agro-Industrial Research. EUR 19070. Contract Air2-CT93-1054. In A. Luchetti (ed.) *Microwave Sterilisation: Process validation nutritional wholesomeness, product quality and the consumer*, pp. 40–3.

EDGAR, R. 1986. The economics of microwave processing in the food industry. *Food Technology* **40**, 106–12.

ERLE, U. 2000. Untersuchungen zur Mikrowellen-Vakuumtrocknung von Lebensmitteln. PhD thesis, Universität Karlsruhe.

FRYER, P. and LI, Z. 1993. Electrical resistance heating of foods. *Trends in Food Science & Technology* **4** 11, 364–9.

HARLFINGER, L. 1992. Microwave sterilization. *Food Technology* **46**, 57–61.

HERMANS, W. 1991. Single flow fraction specific thermal processing of liquid foods containing particulates. Paper presented at the 'News in Aseptic Processing and Packaging' seminar, Helsinki, January 1991.

HOLLAND, J.M. 1974. Dielectric post-baking in biscuit making. *Baking Industries Journal* **8**, 6, 29–30.

IEC 1982. *Safety in Electroheat Installations.* Publication 519. IEC, Geneva, Switzerland.

JONES, P.L. and ROWLEY, A.T. 1997. Dielectric dryers. In C.J. Baker (ed.) *Industrial Drying of Foods.* Chapman & Hall, London.

KASTELEIN, J. 1997. Aseptic production of food products. *The European Food & Drink Review*, Spring, 27–30.

KENT, M. 1987. *Electrical and Dielectric Properties of Food Materials.* Science and Technology Publishers, Hornchurch, UK.

KIM, H.J., CHOI, Y.M., YANG, T.C.S., TAUB, I.A., TEMPEST, P., SKUDDER, P.J., TUCKER, G. and PARROTT, D.L. 1996. Validation of ohmic heating for quality enhancement of food products. *Food Technology* **50**, 253–61.

KNUTSON, K.M., MARTH, E.H. and WAGNER, M.K. 1987. Microwave heating of food. *Lebensmittel-Wissenschaft und Technologie* **20**, 101–10.

LEWIS, M.J. and HEPPEL, N.J. 2000. *Continuous Flow Thermal Processing of Foods.* Aspen Publications, Gaithesbury, MD.

LILLARD, H.S. 1994. Decontamination of poultry skin by sonication. *Food Technology* **48**, 72–3.

METAXAS, A.C. 1996. *Foundation of Electroheat: A Unified Approach.* Wiley, Chichester.

METAXAS, A.C. and MEREDITH, R.J. 1983. *Industrial Microwave Heating.* Polu Pelegrinus, London.

NOTT, K.P., HALL, L.D., BOWS, J.R., HALE, M. and PATRICK, M.L. 1999. Three-dimensional MRI mapping of microwave induced heating patterns. *International Journal of Food Science and Technology* **34**, 305–15.

OHLSSON, T. 1980. Temperature dependence of sensory quality changes during thermal processing. *Journal of Food Science* **45** 4, 836–9.

OHLSSON, T. 1983. Fundamentals of microwave cooking. *Microwave World* **4** 2, 4–9.

OHLSSON, T. 1987. Dielectric properties – industrial use. In *Physical Properties of Foods* – 2. Elsevier Applied Science, London, pp. 199–212.

OHLSSON, T. 1991. Microwave processing in the food industry. *European Food and Drink Review* **7**, 9–11.

OHLSSON, T. 1992. R & D in aseptic particulate processing technology. In A. Turner (ed.) *Food Technology International Europe.* Sterling, London, pp. 49–53.

OHLSSON, T. 1993. In-flow microwave heating of pumpable foods. Paper presented at International Congress on Food and Engineering, Chiba, Japan, 23–27 May.

OHLSSON, T. 1994a. Minimal processing – preservation methods of the future: an overview. *Trends in Food Science and Technology* **5** 11, 341–4.

OHLSSON, T. 1994b. Progress in pasteurisation and sterilisation. Developments in Food Engineering. In T. Yano, R. Matsume and K. Nakamura (eds) *Proceedings of the 6th International Congress on Engineering and Food.* Blackie Academic & Professional, London.

O'MEARA, J.P. 1977. Why did they fail? A backward look at microwave applications in the food industry. *Journal of Microwave Power & Electromagnetic Energy* **8** 2, 167–72.

PALANIAPPAN, P. and SIZER, C.E. 1997. Aseptic process validated for foods containing particulates. *Food Technology* **8**, 60–8.

PARROTT, D.L. 1992. Use of ohmic heating for aseptic processing of food particulates. *Food Technology* **46**, 68–72.

POVEY, M.J.W. and MASON, T.J. (eds) 1998. *Ultrasound in Food Processing.* Blackie Academic & Professional, London.

PÜSCHNER, H. 1964. *Wärme durch Mikrowellen.* Philips Techn. Bibliotek, Eindhoven, Holland.

RASO, J., CONDON, S. and SALA TREPAT, F.J. 1994. Mano-thermo-sonication: a new method of food preservation? In *Food Preservation by Combined Processes.* Flair. Conc. Action No. 7, Subgroup B. EUR 15776EN.

RÄUBER, H. 1998. Instant-Gemüse aus dem östlichen Dreiländereck. *Gemüse,* 10'98.

REGIER, M. and SCHUBERT, H. 2001. Microwave processing. In P. Richardson (ed.) *Thermal Technologies in Food Processing.* Woodhead Publishing, Cambridge, pp. 178–207.

ROSENBERG, U. and BÖGL, W. 1987a. Microwave thawing, drying and baking in the food industry. *Food Technology* **41**, 85–91.

ROSENBERG, U. and BÖGL, W. 1987b. Microwave pasteurization, sterilization, blanching and pest control in the food industry. *Food Technology* **41**, 92–121.

ROWLEY, A. 2001. Radio frequency heating. In P. Richardson (ed.) *Thermal Technologies in Food Processing.* Woodhead Publishing, Cambridge, pp. 163–77.

RUAN, R., YE, X., CHEN, P., DOONA, C. and TAUB, I. 2001. Ohmic heating. In P. Richardson (ed.) *Thermal Technologies in Food Processing.* Woodhead Publishing, Cambridge, pp. 241–65.

RYYNÄNEN, S. and OHLSSON, T. 1996. Microwave heating uniformity of ready meals as affected by placement, composition, and geometry. *Journal of Food Science* **61** 3, 620–4. SIK-Publikation nr 735.

SALE, A.J.H. 1976. A review of microwave for food processing. *Journal of Food Technology* **11**, 319–29.

SCHELLEKENS, M. 1996. Sous vide cooking: state of the art. In Proceedings of Second European Symposium on Sous Vide, Leuven, Belgium, 10–12 April.

SCHIFFMANN, R.F. 1982. Method of Baking Firm Bread. US Patent 4,318,931.

SCHIFFMANN, R.F. 1986. Food product development for microwave processing. *Food Technology* **40**.

SCHIFFMANN, R.F., MIRMAN, A.H. and GRILLO, R.J. 1981. Microwave Proofing and Baking Bread Utilizing Metal Pans. US Patent 4,271,203.

SKUDDER, P.J. 1988. Ohmic heating: new alternative for aseptic processing of viscous foods. *Food Engineering* **60**, 99–101.

SUNDBERG, M., RISMAN, P.O., KILDAL, P-S. and OHLSSON, T. 1996. Analysis and design of industrial microwave ovens using the finite difference time domain method. *Journal of Microwave Power and Electromagnetic Energy* **31** 3, 142–57.

SUNDERLAND, J.E. 1980. Microwave freeze drying. *Journal of Food Process Engineering* **4**, 195–212.

SWARTZEL, K.R. 1997. Past and present challenges for aseptics. Presentation at the International Symposium on Advances in Aseptic Processing and Packaging Technologies, Hsinchu, Taiwan, 4–5 November.

TEMPEST, P. 1996. Electroheat technologies for food processing. *Bulletin of APV Processed Food Sector.*

UIE 1996. Electricity in the food and drinks industry. UIE, B.P. 10.

VAROQUAUX, P. 1993. Recent developments in the processing of fruit and vegetables in France. *The European Food and Drink Review*, Autumn, 33–4, 37.

WEI-RENN-LIEN and WEN-RONG-FU. 1997. Small fish dehydration by far infrared heating. *Food Science Taiwan* **24** 3, 348–56 [in Chinese].

ZHAO, Y. 2000. Using capacitive (radio frequency) dielectric heating in food processing and preservation: a review. *Journal of Food Process Engineering* **23**, 25–55.

ZOLTAI, P. and SWEARINGEN, P. 1996. Product development considerations for ohmic heating. *Food Technology* **50** 5, 263–6.

3

Minimal processing of foods with non-thermal methods

T. Ohlsson, SIK (Swedish Institute for Food and Biotechnology), Gothenburg and N. Bengtsson, Consultant, Molnlycke

3.1 Introduction

Prolonged shelf-life in foods has traditionally been associated with thermal processing, alone or in combination with chemical or biochemical preservation methods. Heat processing, however, tends to reduce product quality and freshness, to some extent, as the price for extended shelf-life. The ideal processing method would therefore be one that inactivated microorganisms and halted deteriorative reactions by other means than heating or freezing – in other words, an essentially non-thermal method.

This chapter deals with the following non-thermal methods:

- irradiation
- high pressure
- pulsed electric fields
- pulsed white light
- ultrasound
- ultraviolet radiation.

The chapter concentrates on physical alternatives to thermal processing. Other chemical and biochemical preservation methods such as modified atmosphere packaging (MAP), the application of antimicrobial substances and protective cultures, are dealt with in other chapters.

Hardly any non-thermal physical method of minimal processing is sufficiently effective on its own in inactivating microorganisms and enzymes at intensities that will not reduce sensory and nutritional quality. Combination with other 'hurdles' will be required, be it refrigeration, heat, MAP, chemical substances or other preservation techniques, to result in sufficient shelf-life and product safety at

Table 3.1 Non-thermal methods in fairly advanced development

Method	Application	Development stage
Pulsed white light	Surface pasteurisation of packaged foods, pasteurisation of transparent liquids, sterilisation of packaging	Imminent commercial applications
Pulsed electric field (PEF)	Pasteurisation of liquid foods	Pilot plant stage
UV excimer lasers in combination with peroxide	Sterilisation of packaging	Commercial or pilot plant
UV-C lamps	Sterilisation of air particle filters and of equipment	Commercial
Pulse power system	Submerged arc discharge	Claimed commercial

minimal loss of fresh food quality. The concept of hurdle technology is dealt with in Chapter 7.

Of the different non-thermal processes being studied, only two as yet have been developed commercially on any scale: ionising radiation and high pressure. While the former has been in use for some 30 years on a limited scale, high pressure processing was first introduced in Japan in the early 1990s. Ultraviolet cold laser irradiation is also a commercial process, but only for the sterilisation of packaging in combination with hydrogen peroxide, and not for food treatment. Table 3.1 lists non-thermal treatments or processes in a fairly advanced stage of development, of which pulsed white light is on the borderline of commercial application. Table 3.2 lists methods in early stages of development, some of which may never reach commercial application because of problems of cost, reliability or safety. In the following sections, the main emphasis will be on the processes already in commercial application or likely to become so in the foreseeable future.

3.2 Ionising radiation

3.2.1 Mechanisms of action and dose levels

Under irradiation, molecules absorb energy and form ions or free radicals which are highly reactive, breaking a small percentage of chemical bonds. The cellular destruction caused by disrupting genetic material in the living cell is the major effect of radiation on microorganisms in foods. However, at permitted dose levels, the effect is too low to induce any radioactivity in the food.

In ionising radiation, it is important to distinguish between radicidation, radurisation and radappertisation, depending on the objective and dose of the treatment:

Table 3.2 Non-thermal methods in early development

Method	Application	Food types	Development stage
Ultrasound	Inactivation of microorganisms and enzymes	Liquids	Laboratory and pilot plant stage
Oscillating magnetic fields (OMF)	Partial inactivation of vegetative flora	Liquids and packaged solids	Laboratory scale
Supercritical CO_2 micro-bubble method	Inactivation of micro-organisms and enzymes	Liquids	Early laboratory stage
OH-radicals by pulsed discharge plasma	Inactivation of bacterial spores	Surfaces	Early laboratory stage
Hydrodyne method – explosion shock waves in liquid media	Inactivation of microorganisms and tenderisation of meats	Packaged solids and liquids	Pilot scale
Photosensitisers and release of active oxygen	Inactivation of microorganisms in smart packaging	Packaged food Surface and head-space	Early laboratory stage
Air ion bombardment	Inactivation of microorganisms	Surfaces	Early laboratory stage
Contact glow discharge electrolysis	Inactivation of vegetative organisms	Liquids	Early laboratory stage

- Radicidation is carried out to reduce viable non-spore-forming pathogenic bacteria, using a dose between 0.1 kGy and 8 kGy (1 Gy = one joule/kg).
- Radurisation has the objective of reducing viable spoilage organisms, using 0.4 kGy to 10 kGy.
- Radappertisation is applied to kill both vegetative bacteria and spores, using dose levels from 10 kGy to 50 kGy.

For commercial applications, radicidation and radurisation dominate for pasteurising foods, often in combination with some thermal treatment. Radappertisation is used on a limited scale for wet foods, since it affects sensory properties negatively, unless applied after freezing of the food.

3.2.2 Applications

Ionising radiation is widely used for the sterilisation of medical equipment and also packaging material for aseptic processing of foods. Development in food applications has been slow, mainly because of resistance from consumers and consumer pressure groups. Otherwise, the effectiveness and safety of irradiation has been very well documented, as evidenced by the recommendations of the

Table 3.3 Applications of food irradiation

Application	Dose range (kGy)	Examples of foods	Countries with commercial processing
Sterilisation	7–10	Herbs, spices	Belgium, Canada, Croatia, Czech Republic, Denmark, Finland, Israel, Korea (Rep.), Mexico, South Africa, USA, Vietnam
	Up to 50	Long-term ambient storage of meat (outside permitted dose)	None
Sterilisation of packaging materials	10–25	Wine corks	Hungary
Destruction of pathogens	2.5–10	Spices, frozen poultry, meat, shrimps	Belgium, Canada, Croatia, Czech Republic, Denmark, Finland, France, Iran, Netherlands, South Africa, Thailand, Vietnam
Control of moulds	2–5	Extended storage of fresh fruit	China, South Africa, USA
Extension of chill life from 5 days to 1 month	2–5	Soft fruit, fresh fish and meat at 0–4°C	China, France, Netherlands, South Africa, USA
Inactivation/control of parasites	0.1–6	Pork	–
Disinfestation	0.1–2	Fruit, grain, flour, cocoa beans, dry foods	Argentina, Brazil, Chile, China
Inhibition of sprouting	0.1–0.2	Potatoes, garlic, onions	Algeria, Bangladesh, China, Cuba

World Health Organisation (WHO), and the fact that permission to use irradiation has been granted by public health authorities in some 40 countries (WHO, 1981, 1998; anon., 1984; anon., 1988). Irradiation is regarded by many experts as the best-researched food processing method of all.

Irradiated foods are available in some 30 countries, the most common application being for the sterilisation of spices in preference to ethylene oxide. Other applications include fruits and vegetables, rice, potatoes, onions, sausages and dried fish. A summary of applications of food irradiation around the world is provided in Table 3.3. Applications accepted by the Food and Drug Administration (FDA) in the USA and permitted dosages are listed in Table 3.4. The most recent addition is that of uncooked raw meat. The potential of irradiation has gained wide publicity in the USA as a result of the programme launched by the US government in 1997 to improve food safety and reduce the high levels of food poisoning incidents in the USA. However, it must be

Table 3.4 Applications of ionising radiation accepted in the USA by the Food and Drug Administration

Product	Dose (kGy)	Purpose	Date
Wheat, wheat flour	0.2–0.5	Insect disinfestation	1963
White potatoes	0.05–0.15	Sprout inhibition	1964
Pork	0.3–1	*Trichinella spiralis* control	22.07.85
Enzymes (dehydrated)	10 max.	Microbial control	18.04.86
Fruit	1 max.	Disinfestation, ripening delay	18.04.86
Vegetables, fresh	1 max.	Disinfestation	18.04.86
Herbs	30 max.	Microbial control	18.04.86
Spices	30 max.	Microbial control	18.04.86
Vegetable seasonings	30 max.	Microbial control	18.04.86
Poultry, fresh or frozen	3 max.	Microbial control	02.05.90
Animal feed and pet food	2–25	*Salmonella* control	28.09.95
Meat, uncooked, chilled	4.5 max.	Microbial control	02.12.97
Meat, uncooked, frozen	7.0 max	Microbial control	02.12.97

Source: *Food Technology*, January 1998, p. 56

remembered that irradiated foods were not produced commercially in the USA until 1992. Commercial use of irradiation in the USA and elsewhere is still limited, despite its advantages with regard to food safety, product shelf-life and sensory quality. In addition, irradiation equipment available has, as a rule, not been easily adapted to on-line volume processing of food materials, and production costs have been high. It has been estimated that half a million tonnes of food products and ingredients are irradiated worldwide annually (Farkas, 1999). Most of these foods are destined for further food processing. Only in a few countries are commercial quantities sold to the consumer on a regular basis.

The main advantages of irradiation are as follows:

- there is little or no heating of the food and therefore negligible change to sensory characteristics
- packaged and frozen foods may be treated
- fresh foods may be preserved in a single operation, and without the use of chemical preservatives
- energy requirements are very low
- changes in nutritional value of foods are comparable with other methods of food preservation
- processing is automatically controlled and has low operating costs.

A major disadvantage is the high capital cost of irradiation plant, but concern over the use of food irradiation has also been expressed by some (for example, Webb and Lang, 1990, and Webb and Henderson, 1986). Other concerns, for example over operator safety, are discussed by Webb and Lang (1987) and by Welt (1985). They describe the main problems as follows:

- the process could be used to eliminate high bacterial loads to make otherwise unacceptable foods saleable
- if spoilage microorganisms are destroyed but pathogenic bacteria are not, consumers will have no indication of the unwholesomeness of a food
- there will be a health hazard if toxin-producing bacteria are destroyed after they have contaminated the food with toxins
- the possible development of resistance to radiation in microorganisms
- loss of nutritional value
- until recently, inadequate analytical procedures for detecting whether foods have been irradiated
- public resistance due to fears of induced radioactivity or other reasons connected to concerns over the nuclear industry.

Although irradiation is a powerful decontamination technique, the dose that can be applied to a particular food is sometimes limited by the impact of high doses on sensory quality. Changes in flavour after irradiation have been noted, for example, in chicken, beef, pork and seafood (Sudarmadji and Urbain, 1972). A more effective way of balancing microbiological safety and sensory quality is to combine irradiation with other preservation techniques such as heating, hydrostatic pressure, MAP or the use of chemical preservatives (Farkas, 1990; Wills *et al.*, 1973; Grant and Patterson, 1991). In the case of fruits and vegetables, it is almost impossible to eliminate foodborne pathogens completely by washing or by chemical means. Ionising irradiation is very effective in decontaminating fruit and vegetables, especially in combination with a chlorine wash.

3.2.3 Processing equipment

An irradiation plant consists of an irradiation chamber containing the radiation source into which products are moved by conveyor belt for treatment, with separate loading and unloading areas. The irradiation chamber needs shielding to prevent radiation from penetrating beyond the chamber. Irradiation sources are γ-rays from Cobalt 60 or Caesium 137 and machine-generated X-rays and electron beams. Looking first at the equipment side, the majority of installed plants for food processing have been heavy, solidly shielded irradiators, using Cobalt 60 as the radiation source. Cobalt 60 has a half-life of only five years, requiring refuelling fairly frequently. Also, nuclear fission occurs continuously, irrespective of whether any material is being processed or not, requiring therefore a high degree of utilisation for economy. These units are not very well suited for on-line processing in food factories, and are commonly located centrally for serving the needs of many customers, irradiating medical supplies and packaging as well as food.

More recently, electron accelerators have come into use for irradiating foods. Major advantages over Cobalt 60 sources are smaller equipment, requiring much less shielding, and greater compatibility with on-line processing, such as in a French plant for killing Salmonella in de-boned chicken meat at 3 tons/hr, using an SPI Circe II accelerator at a radiation dose of 5kGy (Rice, 1993). Electron accelerators also do not contain any nuclear material, and the radiation is cut off when the electricity is switched off. One disadvantage is that penetration depth is limited to slightly more than 3cm, making it unsuitable for irradiating sack loads or pallet loads of produce.

The Gray Star Company in the USA has presented a new and interesting development in irradiation sources. Their new source is based on Caesium 137, having a half-life of 30 years and emitting photons at lower energy levels, requiring less shielding and lighter construction. Weight and size can thus be reduced considerably, permitting more practical transport to the irradiation site. Such a unit can be installed directly in the food factory for irradiating one pallet of product at a time. This development tends to make stationary nuclear type units more competitive with electron accelerators in practical handling and cost, while maintaining the advantage of processing material of large dimensions (anon., 1997). A radiation flux distributor has also been developed to maintain even distribution of the dose over the entire load, as checked by automatic dosiometry to ensure that all food has received the intended dose of irradiation. A complete commercial system for e-beam irradiation has also been developed in Sweden, requiring only 10% of the space of a traditional system (anon., 2001b). There are now over 150 gamma irradiation plants and hundreds of electron accelerator facilities, although many of these are used for non-food products.

3.2.4 Cost

The cost of investment and processing using irradiation is considered high. For commercial facilities with high throughput, total investment and operating costs appear to be similar for electron beam and Cobalt 60 sources. However, since a Cobalt 60 source cannot usually be integrated into a processing line, electron accelerators will have a cost advantage. This may, however, be offset by the development of the Caesium 137 source, mentioned above, which is easier to transport, and requires much less shielding and reloading of radioactive material than a Cobalt 60 source.

3.2.5 Outlook

Irradiated foods marketed in several countries are judged superior by consumers and have sold well. In the USA four retail chains now sell irradiated produce as high quality products, in spite of obligatory radiation labelling. Leading food technologists in the USA believe that public opinion is on the point of turning in favour of irradiated foods, when consumers are well informed of its specific advantages. A strong contributing factor in the USA is the increased level of concern about food poisoning with estimates of as many as 9,000 deaths annually and several millions of food poisoning cases, causing damage in the order of $30 billion a year. Increased concern about food safety has made consumers much more willing to consider preservation technologies such as irradiation.

The great virtue of irradiation, as pointed out by WHO, FDA and USDA (United States Department of Agriculture), is that it is effective in killing disease-causing bacteria and delaying food spoilage without negative effects on food quality. It is in fact the only low temperature method to inactivate some pathogenic microorganisms effectively. Where consumers have had a chance to test irradiated foods and form their own opinion, these foods have been well received. When informed of the benefits, consumers have even been willing to pay a premium price for irradiated foods on the grounds of greater safety (Olson, 1998). As a result of this shift in attitudes, a network of pasteurisation centres for meat is being set up in the USA (by Ion Beam Applications of Belgium), using x-ray or electron beam (e-beam) technology. In Iowa, an e-beam irradiation plant has been constructed (by Titan Corporation) for pasteurising up to 400 million pounds of produce a year, establishing an alliance with a number of food companies for testing new products and processing commercial quantities of produce. All this clearly indicates that a final breakthrough for irradiation seems to be on its way in the USA.

In expanding the use of irradiation, suitable dose levels have to be established experimentally for each new product. Not all foods are suitable for irradiation, such as shell eggs (where irradiation produces viscosity changes) or milk (where it causes off flavour). More work on alternative packaging materials needs to be done since some materials, such as PVC, are unsuitable. According to an Institute of Food Technologists (IFT) expert panel on food safety and nutrition (Olson, 1998):

> Irradiation of food can effectively reduce or eliminate pathogens and spoilage micro-organisms while maintaining wholesomeness and sensory quality. Selection of appropriate treatment conditions can minimise or prevent objectionable changes in food quality. Methods to detect foods that have been irradiated are becoming internationally accepted.

3.3 High pressure processing

3.3.1 Mechanism

By subjecting foods to high pressures in the range 3,000–8,000 bars, microorganisms and enzymes can be inactivated without the degradation in

Table 3.5 Current applications of high pressure processing

Product	Manufacturer	Process conditions
Jams, fruit dressing, fruit sauce topping, yoghurt, fruit jelly	Meidi-ya Company, Japan	400 Mpa, 10–30 min, 20°C
Grapefruit juice	Pokka Corp., Japan	120–400 Mpa, 2–20 min, 20°C + additional heat treatment
Mandarin juice	Wakayama Food Ind., Japan	300–400 Mpa, 2–3 min, 20°C
Non-frozen tropical fruits	Nishin Oil Mills, Japan	50–200 Mpa ('freeze' at −18°C)
Tenderised beef	Fuji Ciku Mutterham, Japan	100–50 Mpa, 30–40 min, 20°C
Avocado	Avomex, USA	700 Mpa, 600–800 L/h
Orange juice	UltiFruit, France	500 Mpa, 5 or 10 min cycles, includes a 1 min hold

Source: Campden New Technology bulletin no. 14

flavour and nutrients associated with traditional thermal processing. The technology was first commercialised in Japan in the early 1990s for the pasteurisation of acid foods for chilled storage. In spite of massive research efforts, particularly in Europe and the USA, commercial development outside Japan has been slow so far, mainly because of the very high investment and processing costs of high pressure (HP) processing as well as regulatory problems in regions such as Europe. Some current applications of HP processing are shown in Table 3.5.

The HP process is non-thermal in principle, even if the pressure increase in itself causes a small rise in temperature. HP affects all reactions and structural changes where a change in volume is involved, as in the gelation of proteins or starch. The mechanism behind the killing of microorganisms is a combination of such reactions, the breakdown of non-covalent bonds and the puncturing or permeabilisation of the cell membrane. Vegetative cells are inactivated at about 3,000 bars at ambient temperature, while spore inactivation requires much higher pressures (6,000 bars or more) in combination with a temperature rise to 60–70°C. Certain enzymes are inactivated at 3,000 bars, while others are very difficult to inactivate at all within the pressure range that is practically available today. Moisture level is extremely important in this context, little effect being noticeable below 40% moisture content.

Pulsed or oscillating pressurisation will be more effective in spore inactivation than continuous pressure. Rapid decompression increases the impact force on the spore coat much more than the preceding compression, and makes possible sterilisation at lower pressures. At low pressure, 500–3,000 bars, considerable germination of spores can occur, strongly influenced by temperature and pH, which then allows organisms to be killed by moderate pressures. In the case of a large variety of moist products, pressurisation to above 100 MPa in less than 30 seconds, mainly at a temperature around 90°C, with a holding time of only a few minutes, resulted in complete inactivation of even thermo-resistant spores (Hoogland *et al.*, 2001). The combination of Nisin, high pressure and lowered temperature may allow for considerable reduction in processing time and/or pressure in HP treatment. Microbial kill is completed without the frequently encountered survival of some pathogens.

3.3.2 Equipment

The main components of an HP system are as follows:

- a pressure vessel and its closure
- a pressure generation system
- a temperature control device
- a materials handling system (Mertens, 1995).

Most pressure vessels are made from a high tensile steel alloy 'monoblocs' (forged from a single piece of material), which can withstand pressures of 400–600 MPa. For higher pressures, pre-stressed multi-layer or wire-wound vessels are used (Mertens, 1995). Vessels are sealed by a threaded steel closure; a closure with an interrupted thread, which can be removed more quickly; or by a sealed frame which is positioned over the vessel. In operation, after all air has been removed, a pressure-transmitting medium (either water or oil) is pumped from a reservoir into the pressure vessel using a pressure intensifier until the desired pressure is reached. This is termed 'indirect compression' and requires static pressure seals. Another method, termed 'direct compression', uses a piston to compress the vessel, but this requires dynamic pressure seals between the piston and internal vessel surface, which are subject to wear and are not used in commercial applications.

Temperature control in commercial operations can be achieved by pumping a heating/cooling medium through a jacket that surrounds the pressure vessel. This is satisfactory in most applications as a constant temperature is required, but if it is necessary to change the temperature regularly, the large thermal inertia of the vessel and relatively small heat transfer area make this type of temperature control very slow to respond to changes. In such situations, an internal heat exchanger is fitted.

There are two methods of processing foods in high pressure vessels: in-container processing and bulk processing. Because foods reduce in volume at the very high pressures used in processing (for example, water reduces in volume by approximately 15% at 600 MPa), there is considerable stress and distortion to the package and the seal when in-container processing is used. It is likely that conventional plastic and foil pouches will prove suitable and research is continuing on the optimum design of the package, seal integrity and other suitable packaging materials. Materials handling for in-container processing is achieved using automatic equipment, similar to that used to load/unload batch

Table 3.6 Advantages and limitations of in-container and bulk high pressure processing

In-container processing		Bulk processing	
Advantages	Limitations	Advantages	Limitations
Applicable to all solid and liquid foods	Complex materials handling	Simple materials handling	Only suitable for pumpable foods
Minimal risk of post-processing contamination	Little flexibility in choice of container	Greater flexibility in choice of container	Aseptic filling of containers required – potential post-processing contamination
No major developments needed for high pressure processing	Greater dead-time in use of pressure vessel	Maximum efficiency (> 90%) in use of high pressure vessel volume	All pressure components in contact with food must have aseptic design and be suitable for cleaning-in-place (CIP) and sterilising-in-place
Easier cleaning		Minimum vessel dead-time (no opening/closing of vessel needed, faster loading/unloading)	

retorts. Bulk handling is simpler, requiring only pumps, pipes and valves. A comparison of the advantages and limitations of in-container and bulk processing is shown in Table 3.6.

Semi-continuous processing of fruit juices at 4,000–6,000 lh^{-1} using pressures of 400–500 MPa for 1–5 min at ambient temperature is used by one company in Japan, whereas another uses a similar process operating at 120–400 MPa followed by a short heat treatment before the juice is packaged. The process is highly energy efficient although at present the capital costs of equipment remain high. It is possible that such liquid foods could also be used as the pressurising fluid by direct pumping with HP pumps. Such systems would reduce the capital cost of a pressure vessel and simplify materials handling. If liquids were also rapidly decompressed through a small orifice, the high velocity and turbulent flow would increase the shearing forces on microorganisms and thus increase their rate of destruction (Earnshaw, 1992). Developments in HP processing, reported by Knorr (1995a), include combined freeze concentration, pressure freezing and HP blanching. Initial results suggest that pressure-blanched fruits are dried more rapidly than those treated by conventional hot water blanching.

HP equipment has long been in use in commercial production of quartz crystals and ceramics. This equipment is also suitable for food processing with some modification. Among the many equipment manufacturers, the following may be mentioned: Mitsubishi Heavy Industries and Kobe Steel Ltd in Japan;

ACB, Flow Pressure Systems, GEC Alstom, National Forge and Stanstead Fluid Power in Europe; and Engineered Pressure Systems, NKK Corporation and Flow International in the USA. Pressure chambers for food processing are available up to 500 litres volume and for pressures up to 8,000 bars. For cost reasons, there is a practical limitation to 6,000 bars, which will be sufficient for most applications.

For technical reasons, all available units are *batch systems*, even if development work is being undertaken to develop truly *continuous systems*. By combining a number of units in a staggered fashion, semi-continuous production can be achieved. The pressurising medium is usually water, and foods are packed in flexible packaging with little or no headspace to withstand and evenly distribute the pressure. Most systems are *vertical*, some with an external HP intensifier to minimise the number of sensitive HP components in the hydraulic system. Recently, the ACB company has developed a semi-continuous *horizontal* pressure vessel with a double set of pistons for loading and unloading in a straight line. Commercial lines are designed to be automated to streamline production and minimise time for loading, pressurisation, holding, depressurising, unloading and drying.

Examples of semi-continuous systems have been developed by, for example, the companies Alstom and Flow Pressure Systems. In the Flow Pressure semi-continuous system, the liquid to be processed is pumped into one or several so-called isolators (pressure vessels in which a separator partitions the food liquid from the ultra-high pressure (UHP) water source). After pressure treatment, the liquid is pumped into a hold tank and aseptic filling station. In the Alstom system, the pressure chamber is filled with the liquid to be treated and compressed directly by a mobile piston (pushed by pressurised water) up to a maximum pressure of 5,000 bars. After a predetermined holding time, pressure is released and the liquid pumped by the piston to a holding vessel. Several pressure chambers can be served in parallel by the same main pressure generator, so that a continuous downstream flow can be maintained. Since the pressure chamber is completely filled with product, the capacity per cycle is considerably increased compared to the processing of already packaged products in a conventional batch system, and cycle time is reduced by about 30% (Träff, 1998). Many efforts are being made to substitute batch processing with a truly continuous HP process. Unilever, for example, has patented a continuous system in which the material to be treated is passed down an open, narrow tube in a steady flow. A pressure differential of 100 MPa or more is maintained between entrance and exit ends of this tube (Agterof *et al.*, 2000).

Hoogland *et al.* (2001) have described the development, within a Dutch consortium, of a more cost-efficient new generation of HP processing equipment. By using composite materials instead of steel, the cost of the pressure vessel is reduced. The use of internal pressure intensifiers, pressurised by external pumps, further reduces cost. With the new system, now at pilot plant stage, cycle times are being reduced to 2–5 minutes. Another advantage of using composite materials for the pressure chamber is that the chamber wall, dissipating some of the adiabatic heat generated when pressurising the food load, will not cool the product surface. Since pressure and product temperature have a synergistic inactivation effect, cooling at the chamber wall could compromise the inactivation process.

3.3.3 Costs

Overall estimates by several equipment manufacturers point towards investment costs for a commercial system in the range of 0.5–2 million euros and production costs at 4,000 bars of 0.1–0.2 euro/kg of processed goods. An HP plant for fruit juice pasteurisation is about 20 times the cost of an equivalent heat exchanger system (Manvell, 1996). Actual costs will depend on chamber capacity, fill density, time–pressure–temperature combinations in processing and the degree of utilisation of the line. Investment cost will be about 75% of total production costs. The economic benefits of scale and effective utilisation can be seen in calculations that suggest that filling the pressure chamber to 50% of its capacity is twice as expensive as using 85% of chamber capacity. Using a 400 litre press will give a running cost only 35% of that when using a 150 litre press (Manvell, 1996). Costs can also be saved by using a common hydraulic system for two chambers. A new steel has also been developed which has three times the strength of that being used at present in pressure chambers and equipment.

3.3.4 Applications

A range of current commercial applications is shown in Table 3.5. In Japan, fairly small capacity presses are being used, and the commercial product range is limited mainly to marmalade, fruit jellies, citrus juices and fruit yoghurts. In Europe, there are currently two well-established commercial applications, for fruit juice in France and for ham in Spain, both pasteurisation processes. In Mexico, a US company processes avocado purée for the US market at as high a pressure as 7,000 bars, inactivating bacteria and enzymes for a resulting refrigerated shelf-life above 30 days. In the USA HP processing is also in operation for the shucking and pasteurisation of oysters and, more recently, for the pasteurisation of ham products. Farkas (1999) reported microbial stability for up to two years for HP processed dishes of prepared foods in the absence of oxygen and at chilled temperature.

In Japan, there is now more interest in the potential of HP to produce entirely novel food products and textures than as a substitute for other preservation processes. Nearly a dozen commercial rice-based foods have been introduced, foremost HP pre-cooked rice for microwave preparation in the home, as well as salmon, meats and hams of novel texture. Other interesting HP application areas, in addition to food preservation, are the following:

- tempering of chocolate
- gelatinisation of starches and proteins
- blanching of vegetables
- tenderisation of meats
- coagulation and texturisation of fish and meat minces
- freezing and thawing (very rapid and without any temperature gradient)
- increased water absorption rate and reduced cooking time for beans.

Processing at 103 MPa and 40–60°C for 2.5 min improves the eating quality of meat and reduces cooking losses. The extent of tenderisation depends on all three factors involved: pressure, temperature and holding time. Commercially produced products include pressure-processed salted raw squid and fish sausages (Hayashi, 1995). Other possible applications are improved microbiological safety and elimination of cooked flavours from sterilised meats and pâté. These effects are reviewed by Johnston (1995).

Starch molecules are similarly opened and partially degraded, to produce increased sweetness and susceptibility to amylase activity. Other research has found that the appearance, odour, texture and taste of soybeans and rice did not change after processing, whereas root vegetables, including potato and sweet potato, became softer, more pliable, sweeter and more transparent (Galazka and Ledward, 1995). Fruit products are reported to retain the flavour, texture and colour of the fresh fruit. Other changes are described by Knorr (1995b). Other applications include tempering chocolate, where the high pressures transform cocoa butter into the stable crystal form, and preservation of honey and other viscous liquids, seafoods, dairy products such as unpasteurised milk and mould-ripened cheese.

There is active research into other applications, some of which are summarised below (Hayashi, 1995):

- the texture of surimi gels could be significantly influenced by the selection of pressure, time and temperature
- pre-cooked meats, processed in 8 oz plastic bowls, retained high acceptability for at least one year at 80°F
- a high quality tomato juice with improved physical characteristics could be produced by HP
- at 4,000–9,000 bars and 15–30°C, HP processing of luncheon meats (franks, Fleischwurst, sliced ham) demonstrated markedly prolonged shelf-life with an eating quality relative to heat pasteurised or irradiated samples.

3.4 Methods based on pulsed discharge of a high energy capacitor

A number of methods fall into this category:

- pulsed white light
- ultraviolet (UV) light

- laser light
- pulsed electric field (PEF)
- pulsed or oscillating magnetic fields

In the case of pulsed white and ultraviolet light, the DC power loaded capacitor is discharged over a flashlight. In the case of PEF it is discharged over a pair of electrodes and, in the case of pulsed magnetic fields, over a magnetic coil. Of these methods, pulsed white light has already been commercialised for the sterilisation of surfaces and transparent liquids. PEF is the subject of considerable research interest in Europe and the USA and is expected to reach the commercial stage in a few years' time. In contrast, there is much less research on oscillating magnetic fields, indicating a waning interest.

3.5 Pulsed white light

3.5.1 Equipment

The high current discharge through gas-filled flashlights results in millisecond flashes of broad spectrum white light, about 20,000 times more intense than sunlight. Conversion efficiency of electricity to light is about 50%. The spectral distribution is 25% UV, 45% visible light and 30% infrared. The rate of flashes is 1–20 flashes/sec, a few flashes being generally sufficient for the pasteurising or sterilising treatment. This means that treatment time is very short and throughput high. Depending on the application, wavelengths that would adversely affect food flavour or quality are filtered off.

The flashlights are arranged in arrays, adapted to the particular application, be it the continuous sterilisation of packaging film in aseptic processing, the sterilisation on-line of transparent liquids or the surface pasteurisation of solid foods in plastic packaging. Most plastic packaging materials transmit broadband light well, exceptions being PET, polycarbonate, polystyrene and PVC. For complex surfaces, such as those of solid foods like meat and fish, it will be difficult to illuminate or reach all parts of the surface to obtain a sterilising effect.

3.5.2 Mechanism

The UV content in combination with the extremely high intensity (10^{13} watts during some milliseconds) inactivate microorganisms by a combination of photochemical and photothermal effects. The antimicrobial effects of light at UV wavelengths are due to absorption of the energy by highly conjugated double carbon bonds in proteins and nucleic acids, which disrupts cellular metabolism. Pulsed white light is not strictly a non-thermal process, but the thermal action, due to its very short duration, is limited to the extreme outer surface. Momentary surface temperatures of 50–100°C are probably reached, but without any overall resulting temperature rise.

3.5.3 Cost

For the treatment of clear liquids, processing costs of 0.4 US cents/litre are claimed, about equal to the cost of conventional thermal processing. For treating solid surfaces at 4 J/cm^2, costs have been estimated at about 1 cent/m^2 which again are comparable to some conventional thermal processes.

3.5.4 Applications

This processing method is effective particularly on dry surfaces like packaging or dehydrated food surfaces. As has been noted, surfaces need ideally to be smooth where there are no fissures which might protect microorganisms from the light. A few flashes will give 9 log reduction of CFU/cm^2 for vegetative bacteria and 7–8 log reduction for bacterial and mould spores, the latter being slightly more resistant. Inoculated spores of *Bacillus cereus* on packaging material at 10–1,000 CFU/cm^2 were, for example, completely inactivated at a light intensity above 2 J/cm^2. Likewise, *Salmonella* contamination on the surface of shell eggs was completely removed (Dunn *et al.*, 1997).

On complex surfaces such as of meats, fish, chicken, etc., only 2–3 log cycles reduction were reached within the maximum of 12 J/cm^2 permitted by the FDA, limiting shelf-life to around two weeks (Chaillou, 1999; Mimouni, 1999). For sliced white bread and for cakes in plastic packaging, shelf-life was prolonged from a few days to more than two weeks, inactivating mould spores. For tomatoes a 30-day shelf-life was reached. Good results were also obtained pasteurising bottled beer. The French company TVTE has further developed Russian patents on pulsed light, using impulse Xenon lamps and flashes above 100 kW/m^2. According to Chaillou (1999) and Mimouni (1999), the major present application for UV-intensive pulsed light processing is the decontamination of pharmaceuticals, water, air and food packaging as well as baked goods. The shelf-life of bread, cakes, pizza and bagels, packaged in clear film, was extended to 11 days at room temperature after treatment by pulsed light. Shrimps had an extension of shelf-life to seven days under refrigeration. In Europe the process is now near commercial application for the pasteurisation of drinking water on-line and for the sterilisation of packaging material.

3.6 Ultraviolet light

Antimicrobial treatment with UV can be regarded as a special case of pulsed light treatment (which contains 25% UV), but without pulsing, with a minimum of visible light and at a much lower power intensity. The UV spectrum is commonly divided into three intervals:

1. UV-A (λ 320–400 nm).
2. UV-B (λ 280–320 nm).
3. UV-C (λ 200–80 nm).

The main germicidal effect lies in the UV-C interval. At UV-C irradiation at intensities in the order of 1,000 J/m^2 or more, bacteria, yeast and viruses suffer as much as a 4 log reduction. The mechanism causing cell death is related to the absorption of UV by DNA/RNA. The necessary treatment time is much longer than for the pulsed light treatment.

UV-C is being used commercially today for disinfecting air particle filters and for the surface decontamination of entire processing areas (after cleaning). Because of the very low penetration depth of UV, it is a surface treatment only. UV may be permitted in some countries for the surface treatment also of certain foods, but can easily cause off flavours and colour changes. A UV excimer laser, operating at 248 nm, is being used commercially, and in pilot scale studies, for the sterilisation of packaging materials in combination with peroxide, to achieve a 5 log reduction in about one second, due to the higher intensities possible when using a laser (anon., 1998).

3.7 Laser light

Laser systems have demonstrated a capacity for achieving rapid microbial inactivation on surfaces or in clear liquids. Laser sterilisation can be combined with photo sensitisers. Laser light is monochromatic and requires scanning, while flashlight is broadband and covers a large area. Laser systems cost about £100,000 for 1 kW while flashlight systems cost about £30,000. Both systems are viable for industrial decontamination of surfaces (Watson, 1999).

3.8 Pulsed electric field (PEF) or high electric field pulses (HEFP)

The concept of applying very intensive electric field pulses to inactivate microorganisms in foods was patented by Maxwell laboratories in the mid-1980s and developed to pilot plant scale in cooperation with Tetra Pak, the US Army Natick Research Centre and others. However, experience with strong field pulses for treatment of biological material for other purposes than food preservation stretches back some 30 years. At field strengths in the order of 15–30 kV/cm, microbial cells are killed by electroporation of their membranes, without any appreciable temperature rise or chemical or physical changes in the food material. Recently the interest in this potential food preservation method has increased significantly and is now the subject of much international research cooperation.

3.8.1 Mechanisms

When an electric field is applied to a food in a short pulse (1–100μs), there is a pronounced lethal effect on microorganisms. The precise mechanisms by which microorganisms are destroyed by electric fields are not well understood, but are likely to include the following:

- formation of pores in cell membranes when the applied electric field causes the electrical potential of the membrane to exceed the natural level of 1 V (Zimmermann *et al.*, 1974) – the pores then cause swelling and rupturing of the cells
- electrolysis products or highly reactive free radicals produced from components of the food by the electric arc, depending on the type of electrode material used and the chemical composition of the food
- induced oxidation and reduction reactions within the cell structure that disrupt metabolic processes (Gilliland and Speck, 1967)
- heat produced by transformation of induced electric energy.

The degree of inactivation of microorganisms by PEF is greater at higher electric field intensities and/or with an increase in the number and duration of the pulses. Other factors that influence the degree of inactivation include the temperature of the food, its pH, ionic strength and electrical conductivity (Vega-Mercado *et al.*, 1999). Studies in which foods were inoculated with target microorganisms and processed by PEF have resulted in cell reductions of up to six log cycles. PEF treatment has no effect on bacterial spores and only limited effect on enzyme activity. It is not in itself a food preservation process, which gives sufficient commercial shelf-life, but has to be combined with other 'hurdles', such as low storage temperature. Significant shelf-life extensions have then been obtained with a minimum of quality loss.

PEF is not entirely a non-thermal process, as some heat will also be generated from electric currents, depending on e-fields and the electric conductivity of the food. However, since treatment time is usually only a matter of a few seconds, heat generation will have limited influence on the killing effects of the treatment. There appears to be a clear synergistic effect between PEF and Nisin (just as between high pressure processing (HPP) and Nisin) (Wouters and Smelt, 1997).

3.8.2 Equipment

A simple electrical system is used, consisting of a high voltage DC source (pulse generator), a capacitor bank and a switch to discharge energy to electrodes around a treatment chamber. The switching mechanism is devised to control voltage pulse frequency and duration as well as waveform, all important to the antimicrobial effect of treatments. A cooling system to offset the moderate temperature rise may also be included. The risk of dielectric breakdown, arcing and resulting electrolytic side effects limit applications to liquid foods containing small particles only and no air bubbles. Another limitation is that liquids with high electric conductivity are unsuitable for treatment. At the present, much remains to be done on design of the treatment chamber, switching devices and electrodes and on scaling up towards commercial size continuous processing. So far, only pilot equipment up to 300 L/h capacity is available.

In PEF processing, the identification of critical process and design parameters is crucial, as well as determining the mechanisms and kinetics of microbial and enzymatic inactivation. Computer simulation techniques have been developed for determining e-field distribution as a function of treatment chamber design, to achieve pasteurisation at minimum risk for dielectric breakdown, while keeping energy requirement and temperature rise at a minimum. By such modelling the maximum and minimum e-fields have successfully been reduced and increased, respectively, for a continuous flow treatment chamber. Scale-up calculations have also been carried out for setting equipment specifications for large-scale processing. The world's first commercial scale, transportable PEF processing system was recently installed at the Ohio State University Department of Food Science and Technology by DTI (Diversified Technologies Inc.) in association with the OSU Pulsed Electric Field Consortium. This unit has a capacity of 1,000–6,000 litres/h and is combined with an aseptic packaging unit (Ho and Mittal, 2000).

3.8.3 Applications

A considerable volume of work has been reported on the effect of processing variables on the inactivation of different strains of microorganisms in PEF treatment. Nevertheless, according to Wouters and Smelt (1997), there is still 'a limited availability of systematic, quantified and kinetic data regarding food safety and quality after PEF treatment'. From the standpoint of regulatory acceptance, knowledge is also lacking concerning possible electrochemical reactions, and on the influence of scale-up on variations in field distribution and microbial reduction.

At Ohio State University, a pioneer in PEF equipment design, apple juice, cider and orange juice have been successfully processed in pilot scale equipment, combining extended shelf-life with fresh-like quality. Experience with actual food products has been very positive for pumpable, free-flowing liquids without particulates, extending shelf-life by a week or more. Some of these applications are summarised in Table 3.7 of Vega-Mercado *et al.* (1999). However, until recently only small-scale pilot or laboratory equipment has been available for testing. Some examples include the following:

- PEF pasteurisation of skim milk with shelf-life extended to 10–14 days
- for liquid whole egg, more than four weeks of shelf-life was reached at a superior quality compared to thermal pasteurisation
- for fresh-squeezed orange juice, more than six weeks' refrigerated shelf-life was obtained, with less loss of flavour compounds and vitamin C than with heat processing – scale-up studies gave equivalent results in a 15 L/h laboratory system and in a 150 L/h pilot plant
- by minimising salt content (and electrical conductivity) and reducing particle size in commercial cheese sauce and salsa, these products were made suitable for PEF treatment

- for apple juice, PEF treatment extended shelf-life at +4°C to 3–4 weeks (Vega-Mercado *et al.*, 1999).

3.8.4 Cost estimates and outlook for PEF

Based on many years of experimental work, the capital cost for future commercial PEF systems has been estimated to be about twice that of corresponding thermal systems. One estimate has suggested initial investment costs ranging from $450,000 to as much as $2 million (Vega-Mercado *et al.*, 1999). These potential costs have been considered economic considering the expected quality advantages from non-thermal (or low thermal) PEF treatment. The expected increase in product cost to the consumer was estimated at about 3%. PEF products marketed ten years from now are likely to be liquids such as fruit juices, purées, salsas, milk and dairy products and liquid eggs.

3.9 Oscillating magnetic fields

It has long been known that magnetic fields can affect biological material, including microorganisms. Such fields can, under some conditions, damage microorganisms, but in others they show a stimulating effect or no effect at all on cell growth. The actual mechanisms involved are not really known. Real evidence of the potential effectiveness of oscillating magnetic fields (OMF) for non-thermal pasteurisation of foods has, so far, only been given in a US patent from 1985. For effective growth inhibition it appears that magnetic field strengths in the order of 5–50 Tesla (about 1,000 times stronger than the Earth's magnetic field) and oscillating frequencies between 5 kHz and 500 kHz have to be used. It is claimed that 1–10 pulses of less than a millisecond duration will be sufficient for treatment with a temperature rise below 5°C (Hofman, 1985). The reduction in bacterial counts obtained was only in the order of 2–3 log cycles, which is on the low side even for a pasteurisation process. Continued research will be needed to demonstrate if, and how, the efficiency of the process can be improved, and to clarify the mechanisms involved. For the present, it is uncertain whether the process will really have commercial potential, either in terms of performance or cost, in spite of its interesting feature as a non-thermal process, which permits treatment of liquid and solid foods in sealed packages.

3.10 Other non-thermal antimicrobial treatments

As seen from Tables 3.1 and 3.2, about a dozen more non-thermal treatments have been proposed for minimal processing to extend the shelf-life of refrigerated foods or to sterilise packaging material for such processing. Of these, Pulse Power submerged arc discharge is claimed to be in limited commercial use. The other methods are all in early stages of development, many of which are unlikely ever to get beyond laboratory or pilot plant testing. They demonstrate, however, the active search today for non-thermal methods with minimal negative effects on food quality and freshness. Among these methods, ultrasound, pulsed power and air ion bombardment show the greatest potential.

3.11 Ultrasound

Considering the mechanisms at work in power ultrasound treatment, it must be regarded as both thermal and non-thermal. Ultrasound consists of vibrations like sound waves, but at frequencies above the range of human hearing (18 kHz–500 MHz). In liquids and wet biological material these vibrations produce cycles of compression and expansion and the phenomenon of cavitation. When cavitation bubbles implode they generate spots of extremely high pressure and temperature which can disrupt cell structure and inactivate microorganisms.

High intensity (10–1,000 W/cm^2) ultrasound is in practical use today for cleaning surfaces and for changing the properties of foods, such as emulsification and meat tenderisation. The effects of ultrasound on meat proteins produce tenderisation in meat tissues after prolonged exposure, and the release of myofibrillar proteins which, in meat products, result in improved water binding capacity, tenderness and cohesiveness (McClements, 1995). Good results have been reported on the decontamination of poultry processing lines by a combination of ultrasound and chemical sanitisers.

Attempts are also being made to pasteurise or sterilise foods by ultrasound alone or in combination with other preservation techniques. The shear forces and rapidly changing pressures created by ultrasound waves are effective in destroying microbial cells, especially when combined with other treatments, including heating, pH modification and chlorination (Lillard, 1994). The mechanism of cell destruction and effects on different microorganisms have been reviewed by Rahman (1999). The application of ultrasound and heat has been termed thermosonication. Pressurising the food liquid during thermosonication makes it possible to maintain cavitation at temperatures above the boiling point at atmospheric temperature, effectively inactivating spore formers like *Bacillus cereus*. The lethality increase is claimed to be ten-fold compared to non-pressure thermosonication. A combined heat and ultrasound treatment under pressure, termed mano-thermo-sonication (MTS) is described by Sala *et al.* (1995). The initial studies indicated that the lethality of MTS treatments was 6–30 times greater than that of a corresponding heat treatment at the same temperature and was greater for yeasts than for bacterial spores. MTS effectiveness depended on the intensity, amplitude and time of ultrasonication and the applied pressure. It is likely that ultrasound reduces the heat resistance of microorganisms by physical damage to cell structures, caused by extreme pressure changes, and disruption of cellular protein molecules. This makes them more sensitive to denaturation by heat. Similar changes to protein structures in enzymes may partly explain the synergistic effect of ultrasound and heat on enzyme inactivation (Sala *et al.*, 1995). There may thus be future applications for ultrasound to reduce the intensity of conventional heat treatments (e.g. thermosonication as a minimal pasteurisation process) and thus improve the sensory characteristics and nutritional properties of foods produced by traditional heat processes.

According to research at the Campden and Chorleywood Food Research Association (CCFRA), heat transfer is markedly enhanced when combining heat treatment with low frequency–high intensity ultrasound (20–100 kHz), intensifying the inactivation of both spoilage and pathogenic organisms. Inactivation was increased up to twenty-fold over heating alone at temperatures of 55°C or 60°C (Leadly and Williams, 2001).

Rahman (1999) has also reviewed research into the use of ultrasound to assist drying and diffusion (acoustic drying). In some foods (for example, gelatin, yeast and orange powder), the rates of drying are increased by two to three times. This is thought to be due both to the creation of microscopic channels in solid foods by the oscillating compression waves, and by changing the pressure gradient at the air/liquid interface, which increases the rate of evaporation. Acoustic drying has the potential to be an important operation because heat-sensitive foods can be dried more rapidly and at a lower temperature than in conventional hot air driers. Additionally, unlike high velocity air drying, the food is not blown or damaged by acoustic drying.

3.12 Pulse power system

Scientific Utilization in the USA has developed a non-thermal commercial system in operation for the pasteurisation of orange juice. The system applies a submerged arc discharge, which generates a small amount of ozone and UV and a significant shock wave which, in combination, inactivate microorganisms (Mermelstein, 1998c). However, more research is required before the process can gain FDA approval.

3.13 Air ion bombardment

Researchers in the UK have demonstrated that a direct stream of negative air ions will kill foodborne pathogens plated on an agar substrate (Sutherland and Copas, 1997). Determining factors were exposure time, distance from emitter, ion charge and rate of ion flow. The main components are a high voltage generator, a sharply pointed emitter of negative ions and an earthed conducting surface on which the food to be treated is positioned. The main microbiocidal effect is believed to be caused by the superoxide radical anion which damages the microbial cell membrane. Preliminary work has demonstrated that a few minutes' treatment effectively killed yeast and both Gram- and Gram+ bacteria, including spore formers. It appears to be a simple and fairly inexpensive treatment with, potentially, similar applications as for the intensive light pulses. Research work with actual foods is now under way.

3.14 Plasma sterilisation at atmospheric pressure

Mizuno *et al.* (1997) have presented a novel method of non-thermal sterilisation, using OH-radicals produced by pulsed discharge plasma at atmospheric pressure. Argon is used as carrier gas for nebulised hydrogen peroxide, which is subjected to pulsed discharge plasma in a concentric cylindrical electrode (15–22 kV, 230–240 Hz). The authors believe that the method can successfully be applied for high-speed continuous sterilisation at room temperature of dry food materials and surfaces. There is a growing field of research into the application of plasma for sterilisation and decontamination, demonstrating a remarkable ability of relatively cold ionised gases to kill bacterial cells rapidly.

3.15 Conclusion

Of the non-thermal methods developed to commercial application, *irradiation* is by far the best from the viewpoint of effectiveness, product quality and safety. Unfortunately, there have been psychological barriers among consumers, consumer groups and legislators to its application on foods, especially in the Nordic countries. Judging from recent experience in the USA, these barriers seem to be diminishing, and irradiation looks likely to develop significantly. There is already widespread acceptance, for example, of irradiation of spices as a safer alternative to ethylene oxide. Consumers seem increasingly prepared to accept irradiation as a safeguard against food poisoning from fresh-like minimally processed foods too.

High pressure has an important role to play in the pasteurisation and textural modification of foods, maintaining freshness in taste and appearance. An important commercial drawback is the cost of the technology and the degree to which consumers will regard the superior quality sufficient to merit a premium price. The fate of the first entries of HP foods into markets in the USA, France and Spain will be very important from this viewpoint.

Of the other techniques, *pulsed light* is likely to find a number of commercial applications in the surface pasteurisation of packaged food, packaging and water in the near future. *Pulsed electric fields* seems to be only halfway to commercial use, requiring much continued research and development before regulatory approval and industrial-scale application is a reality, possibly in three to seven years' time. Future developments in *ultrasound* and *air ion bombardment* are well worth watching.

3.16 References and further reading

AGTEROF *et al.* 2001. US Patent.

ANON. 1984. Codex General Standard for Irradiated Foods and Recommended International Code of Practice for the Operation of Radiation Facilities Used for the Treatment of Food. Food and Agricultural Organisation of the United Nations, Rome.

ANON. 1988. *Food Irradiation: A Technique for Preserving and Improving the Safety of Food.* WHO, Geneva.

ANON. 1997. Transportable irradiator to be evaluated by USDA. *Food Techn.*, vol. 51, no. 6, June, p. 96.

ANON. 1998. The inactivation and recovery of subtilis spores on packaging surfaces by UV excimer lamps. (UK) *Food Link News*, no. 24, July.

ANON. 2001a. It's good to work under pressure. *Food Manufacture*, February, pp. 37–8.

ANON. 2001b. Newsflash. *Food Technology*, vol. 55, no. 9, p. 102.

AUTIO, K. (ed.) 1998. Proceedings from the VTT Symposium on Fresh Novel Foods by High Pressure, Helsinki, Finland, 21–22 September.

BARBOSA-CÁNOVAS, G.V., POTHAKAMURY, U.R. and SWANSON, B.G. 1998. *Nonthermal Preservation of Foods.* Marcel Dekker, New York.

CHAILLOU, M. 1999. Decontamination and sterilisation by pulsed light, Paper 6 in proceedings from conference 'La Conservation de Demain', 2nd edition. Bordeaux Pessac, France, 13–14 October.

DUNN, J., BUSHNELL, A., OTT, T. and CLARK, W. 1997. Pulsed white light food processing. *Cereal Foods World*, vol. 52, no. 1, January, pp. 56–61.

EARNSHAW, R.G. 1992. High pressure technology and its potential use. In: A. Turner (ed.) *Food Technology International Europe.* Sterling Publications International, London, pp. 85–8.

EUROPEAN CONFERENCE ON EMERGING FOOD SCIENCE AND TECHNOLOGY, Tampere, Finland, 22–24 November 1999. Programme – abstracts of papers and posters.

FARKAS, J. 1990. Combination of irradiation with mild heat treatment. *Food Control*, vol. 1, pp. 223–9.

FARKAS, J. 1999. Radiation processing: an efficient means to enhance the bacteriological safety of foods. *New Food*, vol. 2, no. 2, November, pp. 31–3.

GALAZKA, V.B. and LEDWARD, D.A. 1995. Developments in high pressure food processing. In: A. Turner (ed.) *Food Technology International Europe.* Sterling Publications International, London, pp. 123–5.

GILLILAND, S.E. and SPECK, M.L. 1967. Mechanism of the bactericidal action produced by electrohydraulic shock. *Appl. Microbiol.*, vol. 15, 1038–44.

GRANT, I.R. and PATTERSON, M.F. 1991. Effect of irradiation and modified atmosphere packaging on the microbiological safety of minced pork stored under temperature abuse conditions. *Int. J. Food Sci. Technol.*, vol. 26, pp. 521–33.

HAYASHI, R. 1995. Advances in high pressure processing in Japan. In: A.G. Gaonkar (ed.) *Food Processing: Recent Developments*. Elsevier, London, p. 85.

HO, S. and MITTAL, G.S. 2000. High voltage pulsed electric field for liquid food pasteurisation. *Food Rev. Int.*, vol. 16, no. 4, pp. 395–434.

HOFMAN, G.A. 1985. Deactivation of Micro-organisms by an Oscillating Magnetic Field. US Pat. 4524079.

HOOGLAND, H., DE HEIJ, W. and VAN SCHEPDATE, L. 2001. High pressure sterilisation: novel technology, new products, new opportunities. *New Food*, vol. 1, no. 4, pp. 21–6.

JOHNSTON, D.E. 1995. High pressure effects on milk and meat. In: D.A. Ledward, D.E. Johnson, R.G. Earnshaw and A.P.M. Hastings (eds) *High Pressure Processing of Foods*. Nottingham University Press, pp. 99–122.

KNORR, D. 1995a. Hydrostatic pressure treatment of food: microbiology. In: G.W. Gould (ed.) *New Methods of Food Preservation*. Blackie Academic and Professional, London, pp. 159–75.

KNORR, D. 1995b. High pressure effects on plant derived foods. In: D.A. Ledward, D.E. Johnson, R.G. Earnshaw and A.P.M. Hastings (eds) *High Pressure Processing of Foods*. Nottingham University Press, pp. 123–36.

LEADLY, C. and WILLIAMS, A. 2001. Current and potential applications for power ultrasound in the food industry. *New Food*, vol. 4, no. 3, pp. 23–6.

LEISTNER, L. and GORRIS, L.G.M. 1994. Food preservation by combined processes. Final Report FLAIR Concerted Action No. 7, Subgroup B. European Commission. EUR15776EN.

LILLARD, H.S. 1994. Decontamination of poultry skin by sonication. *Food Techn.*, vol. 48, May, pp. 72–3.

LOAHARANU, P. 1994a. Status and prospects of food irradiation. *Food Techn.*, vol. 48, May, pp. 124–31.

LOAHARANU, P. 1994b. Cost/benefit aspects of food irradiation. *Food Techn.*, vol. 48, January, pp. 104–8.

MCCLEMENTS, D.J. 1995. Advances in the application of ultrasound in food analysis and processing. *Trends in Food Science and Technology*, vol. 6, pp. 293–9.

MANVELL, C. 1996. Opportunities and problems of minimal processing and minimally processed foods. Paper presented at EFFoST Conference on Minimal Processing of Foods, Cologne (November).

MELLBIN, P. 1998. High electric field pulses – application and equipment. Presentation at the Nordic Network Seminar on Non-thermal Processing Technologies, SIK, Göteborg, Sweden, 9 March.

MERMELSTEIN, N.H. 1997. High-pressure processing reaches the US market. *Food Techn.*, vol. 51, no. 6, June, pp. 95–6.

MERMELSTEIN, N.H. 1998a. High pressure processing begins. *Food Techn.*, vol. 52, no. 6, June, pp. 104–6.

MERMELSTEIN, N.H. 1998b. Processing papers cover wide range of topics: process related papers at this year's IFT Annual Meeting & Food Expo, *Food*

Techn., vol. 52, July, pp. 50–4.

MERMELSTEIN, N.H. 1998c. Interest in pulsed electric field processing increases. *Food Techn.*, vol. 52, January, pp. 81–2.

MERTENS, B. 1995. Hydrostatic pressure treatment of food: equipment and processing. In: G.W Gould (ed.) *New Methods of Food Preservation.* Blackie Academic and Professional, London, pp. 135–58.

MIMOUNI, A. 1999. Applications of the pulsed light process in the food industry, Paper 7 in proceedings from conference 'La Conservation de Demain', 2nd edition. Bordeaux Pessac, France, 13–14 October.

MIZUNO, A., KURAHASHI, M., IMANO, S., ISHIDA, T. and NAGATA, M. 1997. Sterilization using OH radicals produced by pulsed discharge plasma in atmospheric pressure. IEEE Industry Appl. Soc. Annual Meeting, New Orleans, Louisiana, USA, 5–9 October.

NAKAMURA, I. and TOSHIKO, O. 1997. Sterilisation of the microbial cells in an aqueous solution by contact glow discharge electrolysis. *Nippon Shokuhin Kagaku Kaishi*, vol. 44, no. 8, pp. 594–6.

OLSON, D.G. 1998. Irradiation of food. *Food Techn.*, vol. 52, no. 1, January, pp. 56–61.

POVEY, M.J.W. and MASON, T.J. 1998. *Ultrasound in food processing.* Blackie Academic & Professional, London.

PROCEEDINGS OF THE BORDEAUX CONFERENCE ON FOOD PRESERVATION FOR TOMORROW, October 1999.

PROCEEDINGS FROM THE EFFOST CONFERENCE ON MINIMAL PROCESSING OF FOODS, Cologne, 6–9 November 1996.

PROCEEDINGS OF PEF WORKSHOP III. Ohio State University, 30–31 March 1998.

RAHMAN, M.S. 1999. Light and sound in food preservation. In: M.S. Rahman (ed.) *Handbook of Food Preservation.* Marcel Dekker, New York, pp. 669–86.

RICE, J: 1993. E-B ionization zaps Salmonella. *Food Processing*, July, pp. 12–16.

SALA, F.J., BURGOS, J., CONDON, S., LOPEZ, P. and RASO, J. 1995. Effect of heat and ultrasound on micro-organisms and enzymes. In: G.W. Gould (ed.) *New Methods of Food Preservation.* Blackie Academic and Professional, London, pp. 176–204.

SHIMUDA, M. and OSAJIMA, Y. 1998. Non-heating inactivation of microorganisms and enzymes: application of supercritical carbon dioxide micro bubble method to food industry. *Nippon Shokuhin Kagaku Kaishi*, vol. 45, no. 5, pp. 334–9.

SMELT, J.P.P.M. 1998. Recent advances in the microbiology of high pressure processing. *Trends in Food Sci. & Techn.*, vol. 9, pp. 152–8.

SUDARMADJI, S. and URBAIN, W.M. 1972. Flavour sensitivity of selected animal protein foods to gamma radiation. *J. Food Sci.*, vol. 37, pp. 371–2.

SUTHERLAND, J.P. and COPAS, J. 1997. Use of positive and negative air ion bombardment to control microbial growth. *The European Food & Drink Review*, Summer, pp. 69–71.

TAYER, D.W. and RAJKOWSKI, K.I. 1999. Developments in irradiation of fresh fruits and vegetables. *Food Techn.*, vol. 53, no. 11, November, pp. 62–4.

TRÄFF, A. 1998. A 600 MPa press for food processing. Presentation at the Nordic Network Seminar on Non-thermal Processing Technologies, SIK, Göteborg, Sweden, 9 March.

VEGA-MERCADO, H., GONGORA-NIETO, M.M., BARBOSA-CÁNOVAS, G.V. and SWANSON, B.G. 1999. Non-thermal preservation of liquid foods using pulsed electric fields. In: M.S. Rahman (ed.) *Handbook of Food Preservation*. Marcel Dekker, New York, pp. 487–520.

WATSON, J. 1999. Laser technology. In: Proceedings from the European Conference on Emerging Food Science and Technology, Tampere, Finland, 22–24 November, p. 62.

WEBB, T. and HENDERSON, A. 1986. *Food Irradiation: Who Wants It?* London Food Commission, London.

WEBB, T. and LANG, T. 1987. *Food Irradiation: The Facts*. Thorsons, Wellingborough.

WEBB, T. and LANG, T. 1990. *Food Irradiation: The Myth and the Reality*. Thorsons, Wellingborough.

WELT, M.A. 1985. Barriers to widespread approval of food irradiation. *J. Ind. Irradiat. Technol.*, vol. 3, no. 1, pp. 75–86.

WHO. 1981. Wholesomeness of irradiated food. Report of a Joint FAO/IAEA/WHO Expert Committee. WHO Technical Report Series 659. WHO, Geneva.

WHO. 1998. Food Safety: Joint FAO/IAEA/WHO Study Group on High-Dose Irradiation. *Weekly Epidemiological Record*, no. 3, 16 January, WHO, Geneva.

WILLS, P.A., CLOUSTON, J.G. and GERRATY, N.L. 1973. Microbiological and entomological aspects of the food irradiation program in Australia. In Radiation Preservation of Food, Proceedings of a Symposium, Bombay, 1972. IAEA, Vienna, pp. 231–59.

WOUTERS, P.C. and SMELT, J.P.P.M. 1997. Inactivation of microorganisms with pulsed electric fields: potential for food preservation. *Food Biotechnology*, vol. 11, no. 3, pp. 193–229.

ZIMMERMANN, U., PILWAT, G. and RIEMANN, F. 1974. Dielectric breakdown on cell membranes. *Biophys. J.* vol. 14, pp. 881–9.

4

Modified atmosphere packaging

M. Sivertsvik, J. T. Rosnes and H. Bergslien, Institute of Fish Processing and Preservation Technology (NORCONSERV), Stavangar

4.1 Introduction

During the last few decades there has been a trend towards an increased demand for chilled food products with prolonged shelf-lives. Consumers prefer pre-processed products that are fresh or fresh-like, convenient and easy to prepare and without additives. Pre-packed products offer today's busy consumers a considerable saving in time and money. This has promoted the development of alternative technologies for foodstuff packaging, distribution and storage, such as modified atmosphere packaging (MAP), resulting in products with an increased shelf-life and higher quality. MAP has become a commercial and economic reality in markets that have a well-established *and* controlled cold chain and which can sustain a high-priced quality product.

The use of carbon dioxide (CO_2) to inhibit bacterial growth is not a new technology. In 1877 Pasteur and Joubert observed that *Bacillus anthracis* could be killed by using CO_2 (Valley, 1928) and five years later the first article on the preservative effect of carbon dioxide on food was published by Kolbe (1882), showing an extended storage life for beef placed inside a cylinder filled with a CO_2 atmosphere. In the 1920s work at the Low Temperature Research Station in Cambridge, UK, showed that storing apples in atmospheres containing lowered levels of oxygen and increased levels of carbon dioxide could increase their shelf-life. In the 1930s beef carcasses were transported in atmospheres containing CO_2, which approximately doubled the storage life previously obtained (Davies, 1995), but it is only in the last two decades that MAP has become a more widely commercially used technology for the storage and distribution of foods.

4.2 MAP principles

The principle of MAP is the replacement of air in the package with a fixed gas mixture. Once the gas mixture is introduced, no further control of the gas composition is exercised, and the composition will inevitably change. MAP is also known as 'gas packaging' or 'gas exchange packaging', but these are not recommended on packaging labels since consumers often perceive the word *gas* negatively. Lately MAP has also been referred to as *protective atmosphere packaging* or, when used in labelling, 'Packaged in a protective atmosphere'.

MAP should not be confused with *controlled atmosphere packaging/storage* (CAP/S) where the atmosphere composition is maintained and controlled throughout the storage time. CAP is mostly used in the transport and storage of selected post-harvest products, and this technology is not covered within this chapter. Vacuum packaging is also occasionally looked upon as a type of MAP, but this is not generally regarded as modified atmosphere packaging since the atmosphere is not altered but only removed from the package.

The most apparent advantage of MAP is the achieved increased shelf-life, but MAP also has several other advantages and also disadvantages (see Table 4.1). The effectiveness of MAP on extending shelf-lives, however, is dependent on several factors: type of food, initial quality of the raw material, gas mixture, storage temperature, hygiene during handling and packaging, gas/product volume ratio and the barrier properties of the packaging material.

Table 4.1 Advantages and disadvantages with modified atmosphere packaging (adapted from Farber, 1991; Davies, 1995; Sivertsvik, 1995)

Advantages	Disadvantages
• Shelf-life increase by possibly 50–400% • Reduced economic losses due to longer shelf-life • Decreased distribution costs, longer distribution distances and fewer deliveries required • Provides a high quality product • Easier separation of sliced products • Centralised packaging and portion control • Improved presentation – clear view of product and all-around visibility • Little or no need for chemical preservatives • Sealed packages, barriers against product re-contamination and drip from package • Odourless and convenient packages	• Added costs • Temperature control necessary • Different gas formulations for each product type • Special equipment and training required • Product safety to be established • Increased pack volume – adversely affects transport costs and retail display space • Loss of benefits once the pack is opened or leaks • Dissolved CO_2 into the food could lead to pack collapse and increased drip

4.3 MAP gases

The three major gases used in the MAP of foods are oxygen (O_2), nitrogen (N_2) and carbon dioxide (CO_2). For most food products different combinations of two or three of these gases are used, chosen to meet the needs of the specific product. Usually for non-respiring products, where microbial growth is the main spoilage parameter, a 30–60% CO_2 split is used, the remainder being either pure N_2 (for O_2 sensitive foods) or combinations of N_2 and O_2. For respiring products levels around 5% CO_2 and O_2 are usually used with the remainder being N_2 in order to minimise the respiration rate.

Several other gases such as carbon monoxide (used to maintain the red colour of red meats), ozone, ethylene oxide, nitrous oxide, helium, neon, argon (increases shelf-lives for some fruits and vegetables), propylene oxide, ethanol vapour (used on some bakery products), hydrogen, sulphur dioxide and chlorine have been used experimentally or on a restricted commercial basis to extend the shelf-life of a number of food products (Day, 1993). However, regulatory constraints, safety concerns and negative effects on sensory quality and/or economic factors hamper the use of these gases. More information about the most promising uses for these gases is given later for the relevant food products.

4.3.1 Carbon dioxide (CO_2)

CO_2 is the most important gas in the MAP of foods, due to its bacteriostatic and fungistatic properties. It inhibits the growth of the many spoilage bacteria and the inhibition rate is increased with increased CO_2 concentrations in the given atmospheres. CO_2 is highly soluble in water and fat, and the solubility increases greatly with decreased temperature. The solubility in water at 0°C and 1 atm is 3.38g CO_2/kg H_2O; however, at 20°C the solubility is reduced to 1.73g CO_2/kg H_2O (Knoche, 1980). Therefore the effectiveness of the gas is always conditioned by the storage temperature, resulting in increased inhibition of bacterial growth as the temperature is decreased (Ogrydziak and Brown, 1982; Gill and Tan, 1980; Haines, 1933). The solubility of CO_2 leads to dissolved CO_2 in the food product (Knoche, 1980), according to:

$$CO_2(g) + H_2O(1) \leftrightarrow HCO_3^- + H^+ \leftrightarrow CO_3^{2-} + 2H^+ \tag{4.1}$$

or for pH values less than 8, typical for most foods, the concentration of carbonate ions may be neglected (Dixon and Kell, 1989):

$$CO_2 + H_2O \leftrightarrow H_2CO_3 \leftrightarrow HCO_3^- + H^+ \tag{4.2}$$

The concentration of CO_2 in the food is dependent on the product's water and fat content, and on the partial pressure of CO_2 in the atmosphere, according to Henry's law (Ho *et al.*, 1987). Devlieghere *et al.* (1998a, 1998b) have demonstrated that the growth inhibition of microorganisms in a modified atmosphere is determined by the concentration of dissolved CO_2 in the product. After the packaging has been opened, the CO_2 is slowly released from the product and continues to exert a useful preservative effect for a certain period of time, referred to as CO_2's residual effect (Stammen *et al.*, 1990).

The action of CO_2 on the preservation of foods was originally thought to be caused by the displacement of some or all of the O_2 available for bacterial metabolism, thus slowing growth (Daniels *et al.*, 1985). However, experiments with storage of bacon and pork showed a considerable increase in shelf-life under pure CO_2 atmospheres, compared to storage in normal air atmospheres (Callow, 1932). The preservative effect was not due to the exclusion of O_2, since storage in 100% N_2 offered no advantage over normal air storage. The same results were also seen on pure cultures of microorganisms isolated from spoiled pork.

A drop in surface pH is observed in modified atmosphere (MA) products due to the acidic effect of dissolved CO_2, but this could not entirely explain all of CO_2's bacteriostatic effects (Coyne, 1933). It was shown that CO_2 was more effective at lower temperatures and that the change in pH caused by the CO_2 did not account for the retardation of growth. In a study on several pure cultures of bacteria isolated from fish products, CO_2 atmospheres were found to inhibit the growth of the bacteria markedly, whereas normal growth patterns were observed under air or N_2 atmospheres (Coyne, 1932). It was also observed that bacterial growth was inhibited even after the cultures were removed from the CO_2 atmosphere and transferred to an air environment, interpreted as a residual effect of CO_2 treatment. Bacterial growth was distinctly inhibited when atmospheres with 25% CO_2 were used and almost no growth was observed under higher CO_2 concentrations for four days at 15°C. The obtained results could neither be explained by the lack of O_2 nor the pH effect. Coyne suggested the possibility that an intracellular accumulation of CO_2 would upset the normal physiological equilibrium in other ways, i.e. by slowing down enzymatic processes that normally result in the production of CO_2. Thus the effect of CO_2 on bacterial growth is complex and four activity mechanisms of CO_2 on microorganisms have been identified (Farber, 1991; Dixon and Kell, 1989; Daniels *et al.*, 1985; Parkin and Brown, 1982):

1. Alteration of cell membrane function including effects on nutrient uptake and absorption.
2. Direct inhibition of enzymes or decreases in the rate of enzyme reactions.
3. Penetration of bacterial membranes, leading to intracellular pH changes.
4. Direct changes in the physico-chemical properties of proteins.

A probable combination of all these activities accounts for the bacteriostatic effect. A certain amount (depending on the foodstuff) of CO_2 must dissolve into the product to inhibit bacterial growth (Gill and Penney, 1988). The ratio between the volume of gas and the volume of the food product (G/P ratio) should usually be between 2:1 or 3:1 (volume of gas two or three times the volume of food). This high G/P ratio is also necessary to prevent package collapse because of the CO_2 solubility in wet foods. Dissolved CO_2 fills much less volume compared to CO_2 gas, and after packaging a product in a CO_2

atmosphere, under-pressure is developed within the package and package collapse may occur. The CO_2 solubility could also alter the food water-holding capacity and thus increase drip (Davis, 1998). Exudation pads should be used to absorb drip loss from products.

4.3.2 Nitrogen (N_2)

N_2 is an inert and tasteless gas, and is mostly used in MAP as a filler gas because of its low solubility. N_2 is almost insoluble in water and fat and will not absorb into the food product, and therefore counteracts package collapse as caused by dissolved CO_2. N_2 is used to displace O_2 from air in packages with O_2-sensitive products, to delay oxidative rancidity, and as an alternative to vacuum packaging, to inhibit the growth of aerobic microorganisms.

4.3.3 Oxygen (O_2)

The use of O_2 in MAP is normally set as low as possible to inhibit the growth of aerobic spoilage bacteria. Its presence may cause problems with oxidative rancidity (e.g. fatty fish like salmon and mackerel). However, high levels of O_2 are used in red meat products to maintain the red colour of the meat; O_2 (around 30%) in the atmosphere for lean fish species has been used to reduce drip loss and colour changes; for respiring products O_2 is included in the atmosphere to prevent anaerobic respiration; and recently high levels (80–90%) of O_2 have shown promising results for extending the shelf-lives of selected fruits and vegetables.

Originally, O_2 was introduced into the packaging atmosphere of selected products in order to reduce the risk of anaerobic pathogenic bacterial growth, but this process has now been generally discredited (ACMSF, 1992). It is now recognised that the growth of *Clostridium botulinum* in foods does not depend upon the total exclusion of oxygen, nor does the inclusion of O_2 as a packaging gas ensure that the growth of *C. botulinum* is prevented.

4.4 Gas mixtures

Typical gas mixtures recommended for packaging of different foods could be divided into groups, depending on which spoilage mechanisms they are meant to inhibit. If the spoilage is mainly microbial (usually the most important spoilage parameter for high water activity (a_W) foods), the CO_2 level in the mixture should be as high as possible, limited only by the negative effects of CO_2 on the specific food: package collapse, product exudate and off-taste. Usually a mixture of around 30–60% CO_2 and 40–70% N_2 is adequate. The CO_2 level could be set even higher for bulk packages where packaging collapse is no problem. For food products where the main spoilage parameter is oxidative rancidity the gas mixture should be O_2 free. Depending on the product, gas mixtures of 100% N_2

(or vacuum packaging) or CO_2/N_2 mixtures (where microbial spoilage is an important parameter) are used for oxygen-sensitive products. For respiring foods, too high a CO_2 level or too low O_2 concentrations are not preferred. Gas mixtures with approx. 5% of CO_2 and O_2, combined with high-permeable packaging materials are used. Sometimes MA for respiring foods could be established inside the permeable package without the use of an external gas mixture simply by modifying the gas (air) inside the package with natural respiration and permeation.

Maintenance of the correct gas mixture within MA packages is essential to ensure product quality, appearance and shelf-life extension. Analyses of the gases within MA packages can indicate faults in heat seals, packaging materials, MAP machinery or gas mixing prior to flushing. Corrective action must be taken if the gas analyses of MA packages show that the gas compositions lie outside the limit of tolerances. The use of continuous gas analysers is recommended, but often not applicable, so head-space gas analysis of the finished packages must be carried out at regular intervals.

Seal integrity of MA packages is a critical point since it determines whether an MA pack is susceptible to external microbial contamination and air dilution of the contained gas mixture. Great care should be taken to ensure that the seal area is not contaminated by the product, product drip or moisture. The seal integrity of MA packages should be inspected at regular intervals. Many different seal integrity tests exist, both destructive and non-destructive (Day, 1992).

4.5 Packaging and packages

MAP is used in at least three major packaging areas: retail packages, bulk and master bags. Retail packages are usually made of barrier packaging materials on a deep drawing (form–fill–seal) or flow-pack machine. Retail packages are also made from pre-formed trays with barrier lidding film. Bulk packages are larger units meant for transportation and storage, usually as large barrier bags. Master bags are also large MAP units with several non-barrier retail packages inside. The purpose of master bags is to provide for the transportation and storage of products, and they are usually opened prior to retail sale. The non-barrier retail packages are then removed from the master bag and sold under non-MAP conditions.

The packaging gases permeate through the plastic films at different rates according to the type of polymer and the temperature. Packaging materials with sufficient barrier properties need to be used according to the shelf-life and storage temperature characteristics of the specific products. Most packaging materials used for MA products consist of several layers in order to achieve low water vapour transmissions, high gas barriers and mechanical strength to withstand handling during the packaging and distribution process. Typical MA materials consist of laminations of polyester and polyethylene, nylon and polyethylene, polyvinyl dichloride, orientated polypropylene and many more. The cost of packaging materials is usually directly related to the barrier properties, which force the producer to evaluate the necessary quality of the material versus its cost. Obviously a baguette requiring a shelf-life of several months in a 100% CO_2 environment at an ambient temperature will need a better barrier than, for example, a raw fish product stored at 1°C for only 10 days.

For respiring products, the packaging material permeability has to accommodate the gaseous exchange of respiration. Thus the packaging material permeability has to be chosen according to the product's respiration rate. The transparency of the packaging materials should also be considered since some products are sensitive to light. In these cases, packaging materials with light barriers (e.g. metallised film or UV-barrier layer) should be used.

4.6 MAP of non-respiring foods

4.6.1 Fish and other seafood

Fish and shellfish are highly perishable, due to their high a_W, neutral pH and the presence of autolytic enzymes which cause the rapid development of undesirable odours and flavours. The quality deterioration of chilled fish is usually dominated by microbial activity, but oxidative activities also play an important role for some fish species. The deterioration is highly temperature dependent (Rosnes *et al.*, 1994) and can be inhibited with the use of low storage temperatures (e.g. fish stored on ice). A further inhibition and an increased shelf-life could be obtained through the use of low storage temperatures *combined* with high CO_2 contents in the atmosphere surrounding the product (i.e. MAP). It is evident that MAP can extend the shelf-life of fish and shellfish products. For raw fish an increase of 50–100% in storage life is usually observed, and for cooked shellfish a shelf-life extension of 100–200% can be obtained under ideal storage conditions (Sivertsvik *et al.*, 2002; Stammen *et al.*, 1990).

Fish normally have a particularly heavy microbial load owing to the method of capture and transport to shore, slaughtering methods, evisceration and the retention of skin in retail portions. Microbial activity causes the breakdown of fish proteins and trimethylamin oxide (TMAO), with a resulting release of undesirable fishy odours such as trimethylamine (TMA). The main spoilage bacteria in cod packaged in high levels of CO_2 has been assumed to be *Photobacterium phosphoreum* (Dalgaard *et al.*, 1993), which is different from the main spoilage bacteria of traditionally ice-stored fish: *Shewanella putrefacies* (Gram *et al.*, 1989). Thus a different bacterial flora will develop in MAP conditions.

A gas mixture containing 40–60% CO_2, 40–60% N_2 and no O_2 is recommended for fatty fish products, since the oxidative rancidity of unsaturated fat in fatty fish also results in other additional offensive odours and flavours, apart from microbial spoilage. Vacuum packaging could also be an alternative to MAP for fatty fish such as salmon, providing similar sensory shelf-lives when the primary sensory spoilage parameter is oxidative rancidity (Rosnes *et al.*, 1997; Randell *et al.*, 1999). But the microbial quality is still better under MAP conditions compared to vacuum packaging.

For white fish, crustaceans and molluscs, a gas mixture containing either 40% CO_2/30% O_2/30% N_2 or 40% CO_2/60% N_2 is recommended. The level of CO_2 and gas/product volume ratio is the decisive factor for determining the observed shelf-life extension. The use of 30% O_2 in the package reduces the drip. The drip from MAP lean fish could also be reduced significantly by dipping fillets in 20% NaCl solution for about 20 seconds prior to packaging (Bjerkeng *et al.*, 1995), hence oxygen-free gas mixtures could be used for all types of raw fish. The pre-treatment with salt solution does not negatively influence sensory parameters, the dipping resulting in a salt content of around 1% in the fish fillet. A gas/product ratio of 3:1 is usually recommended for MAP of raw fish.

The inclusion of CO_2 is necessary for inhibiting common aerobic spoilage bacteria, such as the *Pseudomonas* species and *Acinetobacter*/*Moraxella* species. However, for retail packages of fish and other seafood products, too high a proportion of CO_2 in the gas mixture can induce packaging collapse and excessive drip. In fishery products eaten without prior heating, such as crab and cooked fish, an acidic, sherbet-like flavour can be noted when high partial pressures of CO_2 are used.

MAP can be combined with superchilling processes to further extend the shelf-life and safety of fresh fish. In this technique, also known as partial freezing, the temperature of the fish is reduced to 1–2°C below the initial freezing point and some ice is formed inside the product (Sikorski and Pan, 1994). A shelf-life extension of about seven days is obtained for superchilled fish compared to traditionally ice-stored fish of the same type (LeBlanc and LeBlanc, 1992). The superchilling process will store refrigeration capacity inside the product to help keep the core temperature low during chilled storage. Superchilling combined with MA packaging is a mild preservation system that can maintain a high microbiological and sensory quality of whole salmon (Sivertsvik *et al.*, 2000) and salmon fillets (Fig. 4.1) for more than three weeks. When combining MAP with superchilling, a synergistic hurdle effect of the very low storage temperature, CO_2, and the increased CO_2 solubility into the food is achieved, which MAP or superchilling used on their own cannot achieve.

CO_2/N_2 mixtures with up to 100% CO_2 are recommended for the bulk transportation of fresh fish. Freshness of chilled fish is often evaluated on the red colour of the gills, turning grey or brown during storage. Absence of O_2 will cause discoloration of the gills, but can be avoided by using small amounts of carbon monoxide (CO) in the atmosphere (Rosnes *et al.*, 1998; see also section 4.6.2 about red meat and discoloration). Bulk storage of salmon fillets (*Oncorhynchus kisutch* and *O. keta*) in air-tight bulk containers has been reported to have an acceptable sensory shelf-life of 21 days in 90% CO_2 atmospheres at 0°C (Barnett *et al.*, 1982). In bulk storage of whole cod (*Gadus morhua*) in CO_2, a shelf-life increase of at least four days was observed when compared to cod stored in air (Einarsson and Valdimarsson, 1990).

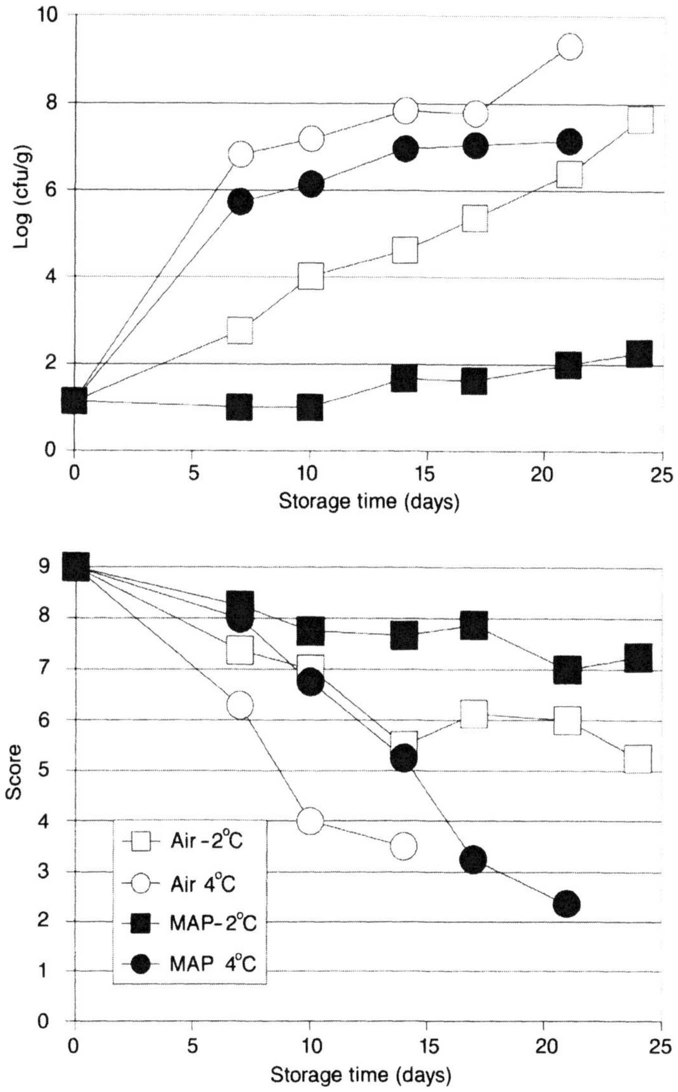

Fig. 4.1 Modified atmosphere packaging (60% CO_2: 40% N_2) versus over-wrap air packaging on salmon fillets stored at chill temperature (4°C) and under superchilling conditions (−2°C). Effect on (a) psychrotrophic plate count, (b) cooked flavour score, sensory evaluation (Sivertsvik *et al.*, unpublished results).

A further use of MAP is for the packaging and distribution of live blue mussels. The mussels are packaged in plastic barrier bags with a 50 : 50 mixture of O_2 and CO_2 together with some seawater. The O_2 keeps the mussels alive while the CO_2 has the dual effect of inhibiting microbial growth and creating under-pressure inside the package (because of the solubility), and thus keeps the mussels closed until the packages are opened (European Patent Application, 2001).

Only the highest quality fish and seafood products should be used to benefit from the extended shelf-life advantages of MAP. The achievable shelf-life will depend on the species, fat content, initial microbial load, gas mixture and the temperature of storage. The maintenance of recommended chilled temperatures and good hygiene and handling practices throughout the entire capture-to-consumption chain is essential for ensuring the safety and extended shelf-life of fish and seafood products.

4.6.2 Meat

The two principal spoilage mechanisms affecting the shelf-life of raw red meat are microbial growth and colour changes (oxidation of the red oxymyoglobin pigment) (Gill, 1996). When red meat is kept under proper chilled conditions, the controlling factor influencing the shelf-life of the product is the rate of oxidation of the red oxymyoglobin pigment into its brown oxidated form, metmyoglobin (Stiles, 1991). For this reason, high concentrations of O_2 are necessary for the MAP of red meats in order to maintain the desirable bright red colour for a longer period. Highly pigmented red meats, such as venison and wild boar, require higher concentrations of O_2.

Aerobic spoilage bacteria, such as *Pseudomonas* species, which are normally predominant on red meats, are inhibited by CO_2 (Gill and Molin, 1991). Consequently, to create the dual effect of red colour stability and microbial inhibition, gas mixtures containing 20–30% CO_2 and 70–80% O_2 and a gas/product ratio of 2:1 are recommended for extending the chilled shelf-life of red meats from two to four days up to five to eight days. In Norway an alternative gas mixture is used for red meat: 60–70% CO_2, 30–40% N_2 and $< 0.5\%$ carbon monoxide (CO). This mixture provides a unique combination of long microbiological shelf-life and a stable, cherry-red colour of the meat (Sørheim *et al.*, 1997). The shelf-life of meat packaged in the CO mixture is longer than that of meat packaged in the commonly used atmospheres with high O_2, because of higher CO_2 levels and the lack of O_2 in the package. The consumption of meat packaged in atmospheres containing $< 0.5\%$ CO only results in negligible levels of carboxyhaemoglobin entering into the bloodstream, and it is highly improbable that the CO will present a toxic threat to consumers (Sørheim *et al.*, 1997).

For non-red meats, microbiological growth and oxidative rancidity limit the shelf-life, and gas mixtures of CO_2 and N_2 should be used. Meat with higher levels of unsaturated fat (like pork) are more prone to oxidative rancidity, hence the complete removal of O_2 is more important. The microbial spoilage is usually dependent on the pH in the meat; a lower pH level provides for a longer shelf-life. Beef, for example, can therefore be vacuum packaged because of its low muscle pH, whereas lamb, which has higher muscle pH, must be packaged in a CO_2 enriched atmosphere in order to achieve a comparable shelf-life (Church and Parsons, 1995).

The maintenance of the recommended chilled temperatures and the 'good hygiene and handling' procedures throughout the slaughtering, processing, packaging, distribution and retail chain are also of vital importance in ensuring the safety and extended shelf-life of all meat products.

4.6.3 Poultry

In contrast to red meats, poultry does not undergo irreversible discoloration of the meat surface in the presence of O_2 (Stiles, 1991). The spoilage of raw poultry is mainly caused by microbial growth, particularly growth of *Pseudomonas* species and *Achromobacter* species. These aerobic spoilage bacteria are effectively inhibited by the use of CO_2 in MAP. Levels of CO_2 in excess of 20% are required to extend the shelf-life of poultry significantly. Package collapse and excessive drip can be a problem for raw poultry, so if higher levels of CO_2 are being used the gas/product ratio also should be increased. In cases where package collapse is not a problem (e.g. bulk or master bags), 100% CO_2 is recommended. In both retail and bulk MA packages, N_2 is used as an inert filler gas.

The achievable shelf-life of MA packed raw poultry and game bird products will depend on the species, fat content, initial microbial load, packaging type, gas mixture and temperature of storage (Stiles, 1991).

4.6.4 Cooked, cured and processed foods and ready meals

For cooked, cured and processed foods the main spoilages are microbial growth, colour changes and oxidative rancidity. For cooked products the heating process should be sufficient to kill vegetative bacterial cells, to inactivate degradative enzymes and to stabilise colour. Consequently, spoilage of cooked meat products is primarily due to post-cooking contamination by microorganisms. MAP with mixtures of CO_2/N_2 should be used without the use of O_2, combined with good hygiene and handling procedures. Cooked shrimp and crab can benefit greatly from such a gas mixture (Sivertsvik, 1995). Whole cooked shrimp stored in 60% CO_2/40% N_2 atmospheres had the same sensory and microbiological quality as freshly cooked shrimp even after 14 days of chilled storage, (Sivertsvik *et al.*, 1997). The gas/product ratio of cooked products would depend on the level of CO_2 and the food's a_w, but a ratio of 2:1 is usual.

For processed food products with relatively high levels of salt and/or other preservatives (e.g. smoked salmon), which effectively inhibit a wide range of spoilage microorganisms, oxygen-free MAP or vacuum packaging could be used to inhibit oxidative rancidity. Cooked, cured and processed meat products containing high levels of unsaturated fat are liable to be spoiled by oxidative rancidity, but MAP with CO_2/N_2 mixtures inhibits this undesirable reaction, and delays the development of oxidative warmed-over flavour (Church and Parsons, 1995). Sliced, cooked, cured and processed meat products are often vacuum packed for retail sale, but the same shelf-life could be achieved using MA and with the additional benefit of an easier separation of meat slices.

The principal spoilage mechanism for ready meals and other cook-chill products is microbial growth, which is primarily due to post-cooking contamination and/or poor temperature control. The pasteurisation process should kill vegetative bacterial cells, inactivate degradative enzymes, and restore the colour. However, heat-resistant spores, such as those from *Clostridium* species and *Bacillus* species, will survive the cooking process and may germinate if the recommended chilled temperature is not maintained. Other possible food poisoning hazards can arise from post-cooking contamination as a result of poor hygiene and handling practices and/or faulty seal integrity. Poor temperature control will exacerbate the problem of microbial growth. Therefore it is recommended that strict control of temperature, hygiene and handling should be maintained throughout the shelf-life. The use of additional barriers to microbial growth (such as acidification, use of preservatives and/or reduction in a_W), as and when appropriate, is strongly recommended.

Ahvenainen *et al.* (1990) studied the influences of MAP on selected ready-to-eat foods. MAP with >20% CO_2 retarded the growth of mould and discoloration of ham pizzas. MAP also created benefits for the packaging of vegetable salads with herring, increasing the shelf-life by a few days. Mayonnaise-based potato salad could not be improved by gas packaging, as atmospheres with >20% CO_2 caused a strong, objectionable odour and taste in the salad.

4.6.5 Dairy products

The principal spoilage mechanisms affecting dairy products are microbial growth and oxidative rancidity. The type of spoilage affecting dairy products will depend on the intrinsic properties of the different products. For example, low a_W products such as hard cheeses are generally spoiled by mould, whereas higher a_W products such as creams and soft cheeses are susceptible to yeast and bacterial spoilage, oxidative rancidity and physical separation.

MAP can extend the shelf-lives of dairy products, but often similar shelf-lives to those with vacuum packaging are achieved. Hard cheeses are generally packed in CO_2/N_2 gas mixtures, which are very effective at inhibiting mould growth. CO_2 levels should be at least 30%. Similarly, soft cheeses are packed in CO_2/N_2 gas mixtures, which can also inhibit bacterial spoilage and oxidative rancidity. MAP is particularly effective for crumbly cheeses such as Cheshire and grated cheeses, where vacuum packaging would cause undesirable compression (Fierheller, 1991). MAP is not suitable for mould-ripened cheeses since the gas mixtures would inhibit the desirable mould growth. Some dairies flush CO_2 into the head-space of yoghurt and sour cream packages to increase the shelf-life (Fierheller, 1991).

4.6.6 Bakery products

The principal spoilage mechanisms for non-dairy bakery products are mould growth, staling and moisture migration. Water activity (a_W) and storage temperature are the two most important parameters influencing the shelf-life (Ooraikul, 1991). Yeasts may cause a problem in certain filled (cakes) or chilled (pizza crust) products. Bacterial growth in bakery products is restricted and rarely a problem, because of the low a_W. However, it is possible that *Staphylococcus aureus* and *Bacillus* species may be able to grow in certain products and hence pose a potential food poisoning hazard. Consequently, good hygiene and handling practices must be observed throughout processing, packaging and storage.

The use of MAP can significantly extend the shelf-lives of non-dairy bakery products. Since moulds are aerobic microorganisms, CO_2/N_2 gas mixtures very effectively inhibit them. A gas/product ratio of 2:1 is recommended. Materials with high water vapour barriers are used to prevent moisture migration.

MAP appears to have little effect on the rate of staling. Staling rates increase at chilled temperatures and hence most cold-eating bakery products are normally stored at ambient temperatures. For hot-eating bakery products, such as pizza bases, the staling process, which is caused by starch retro gradation, is reversed during the reheating cycle.

For bakery products containing fruit as an ingredient (e.g. strawberry layer cake, apple pie, cherry cream cheese cake) yeast is a major cause of spoilage. Yeasts being facultative microorganisms will grow under anaerobic conditions and oxygen-free MAP has no effect on their growth. A possible method to control yeast growth in these products is to modify the head-space atmosphere using ethanol vapour. The ethanol could be sprayed on the product prior to packaging or be released into the atmosphere after packaging using ethanol emitters (Ethicap, Freund Industrial Co., Tokyo, Japan). Ethanol emitters can also delay the staling of bakery products (Hurme *et al.*, 1998).

4.6.7 Dry foods

Because of low a_W, bacterial growth is not an important parameter for dry foods (Fierheller, 1991). The principal spoilage mechanism affecting dried foods containing a high proportion of unsaturated fatty acids is oxidative rancidity. MAP using N_2 very effectively inhibits this deleterious reaction. Due to the very long achievable shelf-lives in MAP for dried foods, MAP materials must have very high moisture and gas barriers. Some products also benefit from using packaging materials with light barriers such as colour printed, pigmented or metallised films. Certain dried foods, such as dried baby milk, are particularly susceptible to oxidative rancidity and residual O_2 levels should be below $< 0.2\%$. In order to achieve very low residual O_2 levels, O_2 scavengers may be incorporated into MA packages. These O_2 scavengers may also be used for other low a_W foods such as bakery products or other very O_2 sensitive foods (more information about active packaging technologies is found in Chapter 5).

Fresh pasta is often packaged in MAP. There are numerous products of this type with ground meat (lasagne, ravioli) or without (pasta, gnocchi) (Coulon and Louis, 1989). For pasta without meat, an atmosphere of $> 80\%$ CO_2 is recommended in order to achieve a shelf-life of two weeks at 4°C. Less CO_2 in the atmosphere has to be used for pasta with meat, because of the problems associated with package collapse.

4.7 MAP of respiring foods

4.7.1 Fresh prepared produce

Unlike other chilled perishable foods that are MA packed, fresh fruit and vegetables continue to respire after harvesting, and consequently any subsequent packaging must take into account this respiratory activity. The products of aerobic respiration are CO_2 and water vapour, whereas fermentation products such as ethanol, acetaldehyde and organic acids are produced during anaerobic respiration. The achievable shelf-life of MA packed produce is inversely proportional to the respiration rate.

The important parameters influencing the shelf-life of fresh prepared produce are intrinsic (the respiration rate) and extrinsic (storage temperature, relative humidity, initial microbial load, packaging film and equipment, packaging filling weight, volume and area, and light). The respiration rate is affected by product type and size, degree of preparation, product variety and growing conditions, maturity and tissue type, atmospheric composition and temperature (Kader *et al.*, 1989).

The principal spoilage mechanisms affecting whole and prepared fresh produce are microbial growth, enzymatic browning and moisture loss. MAP resulting in depleted O_2 and/or enriched CO_2 levels is very effective at inhibiting these spoilage mechanisms, as well as reducing respiration, delaying ripening, decreasing ethylene production and sensitivity, retarding textural softening, reducing chlorophyll degradation and alleviating physiological disorders (Day, 1998).

Hence, the depletion of O_2 and the enrichment of CO_2 are natural consequences of the process of respiration when fresh fruit or vegetables are stored in a hermetically sealed package. Such modification to the atmospheric composition results in a decrease in the respiration rate of the plant material. If the produce is sealed in an impermeable film, in-pack O_2 levels will fall to very low concentrations, whereby anaerobic respiration will be initiated. Anaerobic conditions will usually result in undesirable fermentation reactions (undesirable odours and flavours) and a marked deterioration in product quality (softening). In addition, there is a risk of growth of anaerobic pathogens such as *Clostridium botulinum*.

A minimum level of 2–3% O_2 is recommended to ensure aerobic conditions. Packaging films with the correct intermediary permeability should be used. A desirable equilibrium modified atmosphere (EMA) is established when the rate of O_2 and CO_2 transmission through the package equals the product's respiration rate. EMA can passively evolve within a hermetically sealed package that has sufficient permeability without the introduction of a gas mixture. Since both the permeability of the gases through the packaging material and the respiration rate are dependent on the temperature, small changes in the storage temperature could alter the EMA leading to either anaerobic conditions or higher O_2 levels and consequently faster spoilage.

By gas flushing packages of fresh produce with 5% CO_2/5% O_2/90% N_2, it is possible to establish a beneficial EMA more quickly than by a passively generated EMA. A gas/product ratio of 2:1 is recommended. Such a procedure may be necessary to delay enzymatic browning reactions, which could result in spoilage before a passively generated EMA has been established.

The disadvantage of EMA of fresh produce is the complexity of the processes. Anaerobic conditions could easily be developed and microperforated films with high permeability are rather expensive when compared to other packaging materials (Day, 1998). The microperforated films could also lead to the ingress of microbes during wet handling of the finished packages, tainting from the surroundings and dehydration of the packaged product.

An alternative to EMA that may have potential is to package the prepared produce in high O_2 MAP (80–90% O_2). Statistically valid evidence of the effectiveness of high O_2 MAP on iceberg lettuce was shown by Day (1998). This so-called 'oxygen shock' or 'gas shock' treatment has been found to be particularly effective at inhibiting enzymatic browning, preventing anaerobic fermentation reactions and inhibiting aerobic and anaerobic microbial growth, thereby extending shelf-life and circumventing the disadvantages of microperforated films, as previously mentioned (Day, 1998). The positive effect of high O_2 packaging of prepared produce is hypothesised to be caused by substrate inhibition of polyphenol oxidase (PPO) or alternatively high levels of colourless quinones, which, subsequently formed, may cause feedback production of PPO.

Another method to inhibit enzymatic browning is MAP (20% CO_2 and 80% N_2) combined with citric and ascorbic acids, as browning inhibitors. This gas mixture and pre-treatment gave the best sensory quality of sliced potatoes after seven days' storage. O_2 concentrations in the package head-space were less than 1.5% during a seven-day storage period (Laurila *et al.*, 1998). According to Ahvenainen *et al.* (1998), the quality retention of peeled potatoes was as good in a vacuum package as in gas (20% CO_2 + 80% O_2). Gunes and Lee (1997) showed that active modification of the atmosphere inside the package was necessary in order to extend the shelf-life of the potatoes, but MAP alone did not prevent browning. A dipping treatment was essential in packaged potatoes.

Alternative gases such as argon (Ar) and N_2O have been claimed to provide positive effects on the packaging of fresh produce; however, few scientific reports have confirmed these. Ar is used instead of N_2 and is claimed to inhibit enzymatic discoloration, extend the microbial lag phase and delay the onset of textural softening. A possible reason for these effects could be the greater solubility of Ar compared to O_2 and N_2 and its similar atomic size as O_2. In addition, Ar and N_2O are thought to sensitise microorganisms to antimicrobial agents (Day, 1998). Carbon monoxide (CO) at 5–10% (combined with less than

5% O_2) is an effective fungistat, which can be used on commodities that do not tolerate high CO_2 levels (Kader *et al.*, 1989).

Only fruit and vegetables of the highest quality should be used for MAP. Hygienic preparation, disinfection in chilled chlorinated water (if legal), rinsing and dewatering prior to MAP will help to ensure low microbial counts prior to chilled storage and distribution. In addition, knowledge of the intrinsic properties of fresh produce (respiration rate, pH, a_W, biological structure and ethylene production and sensitivity) and which extrinsic factors to optimise (harvesting, handling, hygiene, temperature, relative humidity, MAP materials and machinery, gas/product ratio and light) will help to ensure the safety and extended shelf-life of MA packed fruit and vegetables. Further information about post-harvest technologies for fruits and vegetables can be found in Chapter 9.

4.7.2 Cooked fruit and vegetables

The principal spoilage mechanisms affecting cooked and dressed vegetable products are microbial growth and enzymatic browning. CO_2/N_2 gas mixtures are effective at inhibiting these spoilage mechanisms, thereby significantly extending the shelf-life of such products. A gas/product ratio of 2:1 is recommended. For cooked vegetable products, the heating process should kill vegetative bacterial cells and inactivate degradative enzymes. Consequently, spoilage of cooked vegetable products is primarily due to post-cooking contamination by microorganisms, which can be minimised by good hygiene and handling practices. Similarly, dressed vegetable products can be spoiled by post-packaging contamination.

4.8 The safety of MAP food products

It is evident that MAP can increase the shelf-life of a number of food products, but the MAP technology has limitations. Raw materials should be healthy and measures should be taken to avoid cross-contamination during handling and packaging. All known direct or indirect sources of contamination should be monitored and controlled.

Hygienic working conditions must be applied. To reduce hazards, a quality assurance system including hazard analysis critical control point (HACCP) should be established throughout the production line to ensure an effective and rational approach to the assurance of safety in the MAP of foods.

Temperature control must be strictly maintained within limits throughout the production chain, including the packaging, transportation and storage phases, especially for those food products where microbial safety is not assured by any other means. Chilled stores, distribution vehicles and display cabinets must have sufficient refrigeration capacity to maintain the recommended product temperatures of MA packed chilled foods. This refrigeration capacity must be able to cope with conditions of high ambient temperatures and frequent door openings where applicable. The proper refrigerated temperature of each batch of product must be assured prior to chilled storage, distribution and retail display.

Concerns have been expressed that the increase in shelf-life of MAP products may provide sufficient time for human pathogens to multiply to levels that might render the food unsafe while still edible (Phillips, 1996). It is useful to discriminate between two categories of MA packaged products. The first category includes products eaten without any prior heat treatment, for instance ready-to-eat products like smoked salmon, cured meat, fruit and salad vegetables. The second category includes products that will be subjected to a sufficient heat treatment to kill all vegetative pathogens before serving, like most fresh fish, raw meat and poultry products. Safety concerns regarding pathogenic microorganisms are of primary importance for the first category of products, and deserve first priority during manufacture. This implies the need for strict control of storage temperatures. Typical food pathogens, their growth limits, type of growth, the effect of CO_2 atmosphere and their origins are shown in Table 4.2. Much research has been carried out regarding the safety and the health hazards of MAP on food products, and especially on those pathogens able to multiply at chill temperatures (e.g. *Listeria monocytogenes*) and those able to multiply in anaerobic conditions (e.g. psychrotrophic *Clostridium botulinum*). Packaging under enhanced CO_2 atmospheres at chill temperatures will inhibit the growth of most pathogens (Table 4.2), but MA atmospheres and chilled storage alone are not sufficient to control the growth of *L. monocytogenes* in some products (Sivertsvik *et al.*, 2002). Non-proteolytic *C. botulinum* is a pathogen that is associated with fishery products, which can grow and produce a potent neurotoxin down to 3°C. This generally requires weeks at the low temperature limit (Dodds and Austin, 1997). However, toxin has been detected in MAP fish prior to the products being considered spoiled (Taylor *et al.*, 1990; Garcia and Genigeorgis, 1987; Post *et al.*, 1985), but other challenge tests of *C. botulinum* and fish have shown MA or vacuum-packaged fish to spoil prior to or in coincidence with toxin production (Reddy *et al.*, 1999; Cai *et al.*, 1997; Reddy *et al.*, 1997a; Reddy *et al.*, 1997b; Reddy *et al.*, 1996; Garren *et al.*, 1995; Cann *et al.*, 1984). Even if the results are not conclusive, there is a potential threat for a packaged fish product to become toxic prior to spoilage at storage temperature of 8°C or above. To ensure the safety of food products with extended shelf-life, in cases where the cold chain is not controllable additional hurdles must be applied.

The Advisory Committee on the Microbiological Safety of Food (ACMSF) in the UK published a report in 1992, *Report on Vacuum Packaging and Associated Processes*. Their conclusion was that, if the storage temperature and the shelf-life are less than 10°C and ten days respectively, the risk is low. For products with a shelf-life of longer than ten days, which are not subjected to a heat treatment that will sufficiently inactivate the spores of psychrotrophic *C. botulinum*, combinations of preservative factors such as pH (< 5), water activity (< 0.97), and salt ($> 5\%$) should be introduced to prevent the growth of

Table 4.2 Growth limiting factors, type of growth, the effect of CO_2 atmosphere on growth, and reservoirs for common food pathogens

Microorganism	Minimum growth limits[1] Temp. °C	pH	a_W (or max. salt)	Type of growth	Effect on growth in CO_2 atm[2]	Common foods and reservoirs
Aeromonas spp.	0–4	4.0	>4–5% NaCl	Facultative	Inhibited (weakly)	Freshwater fish, drinking water, produce, poultry, meat
Bacillus cereus	4	4.3	0.95	Facultative	Inhibited	Rice, spices, vegetables, eggs, dairy products
Campylobacter jejuni	32	4.9	0.99	Microaerophile	Inhibited, survival[3]	Water, poultry, cattle, sheep, pets
Clostridium botulinum proteolytic (A, B, F)	10	4.6	0.93	Anaerobic	Unaffected[4]	Meat, fish, vegetables, soil, sediments
C. botulinum non-proteolytic (B, E, F)	3	5.0	0.97 (or ≥ 5.5% NaCl)	Anaerobic	Unaffected[4]	Fish, seafood (type E), meat (type B, F)
C. perfringens	12	5.0	0.95	Anaerobic	Inhibited	Meat, soil, aquatic sediments
Escherichia coli	7	4.4	0.95	Facultative	Inhibited (weakly)	Intestine of warm-blooded animals, faecal contaminated water and soil
E. coli O157:H7	6.5	4.5	0.95	Facultative	Inhibited (weakly)	Intestine of warm-blooded animals, faecal contaminated water and soil
Listeria monocytogenes	0	4.3	0.92	Facultative	Unaffected/inhibited[5]	Vegetables, non-pasteurised dairy products, soil, water, plants, sewage drain
Plesiomonas spp.	8	4.0	> 4–5%-NaCl	Facultative	Inhibited[6]	Fresh and tropical waters, pets, cattle, pigs, shellfish, tropical fish
Salmonella	7	4.0	0.94	Facultative	Inhibited[4]	Poultry, egg, spices, animal feeds, dried ingredients
Staphylococcus aureus	6 (10 for toxin)	4.0 (4.5 for toxin)	0.83 (0.9 for toxin)	Facultative	Inhibited (weakly)	Humans and warm-blooded animals
Vibrio cholerae	10	5.0	0.97	Facultative	Inhibited	Human intestine, faecal polluted water
V. parahaemolyticus	5	4.8	0.94 (Halophile)	Facultative	Inhibited	Fish, seafood, coastal and brackish water
Yersinia enterocolitica	−1	4.2	0.96	Facultative	Inhibited	Swine intestine and pork products

Notes: 1. From European Chilled Food Federation cited from Martens (1997) and Huss *et al.* (1997). 2. Farber (1991), growth and/or survival of pathogen as relative to growth and/or survival in air. 3. The bacteria survive better in CO_2 as compared to air, but growth is (weakly) inhibited. 4. One report of growth stimulation under CO_2. 5. Unaffected by CO_2 atmosphere if at least 5% O_2 present, inhibited under 100% CO_2. 6. Kirov (1997).

psychrotrophic *C. botulinum* (Gibson and Davis, 1995). Since temperature abuse is common throughout the distribution and retail chain and even more frequent among consumer handling post-retail (Reddy *et al.*, 1992), ACMSF introduced even stricter recommendations in 1994, even if there was no evidence of illness caused by vacuum-packaged or MAP chilled foods (Davis, 1998). When chilled storage is the only controlling factor, storage at temperatures between 5°C and 10°C should be limited to five days, and shelf-lives up to ten days could be assigned for storage temperatures at 5°C or below.

It should also be noted that many food poisoning bacteria may survive on dried foods, particularly herbs and spices, and may pose a hazard when subsequently reconstituted or used as an ingredient in high a_W foods. Consequently, strict standards of hygiene and handling should be observed to minimise such food poisoning hazards.

4.9 The future of MAP

Several novel technologies offer the potential of further improvements in safety and shelf-life of MAP products, including the use of active and smart packaging and hurdle technology. Active packaging is an emerging technology in which the food product, packaging and environment interact, resulting in an extended shelf-life. For many food products the most relevant active packaging technologies are O_2 absorbers and CO_2 emitters/producers used either alone to develop a modified atmosphere or in combination with a gas mixture. CO_2 emitters are commercially available (Ageless, Tokyo, Japan) and could be used alone in order to extend the shelf-life of foods (Sivertsvik, 1999). CO_2 could also be produced inside the packages by letting the exudate from the product react with a mixture of sodium carbonate and citric acid inside the drip pad, as described by Bjerkeng *et al.* (1995). The development of 100% CO_2 atmospheres can be achieved by combining dry ice (approx. 1g solid CO_2 per kg product) and vacuum packaging. Care must be taken to avoid direct contact between the dry ice and the product, because of freeze burns. Using this dry ice method on whole salmon in plastic bags with one-way valves to let off excess CO_2 were found to be of superior quality when compared to ordinary MAP and traditional ice packaging (Sivertsvik *et al.*, 1999).

Another possible approach to create modified atmospheres inside a package is to dissolve the CO_2 into the product prior to packaging, using a method called soluble gas stabilisation (SGS). This method can give appropriate packages with smaller gas/product ratio, and thus decrease the package size. The CO_2 is dissolved into the food product at low temperature (~0°C) and elevated pressures (> 2 atm) prior to packaging. This is in contrast to ordinary MAP where CO_2 is introduced into the package atmosphere at the time of packaging. The method can extend shelf-life of different fish products either alone, combined with traditional MAP, or with vacuum packaging (Sivertsvik, 1999, 2000). The additional benefit of this method when compared to MAP is the increased degree of filling that is possible, apart from the inhibition of microorganisms, as achieved by the dissolving CO_2.

Hurdle technology (combined processes) involves the deliberate combination of existing and novel preservation techniques in order to establish a series of preservative factors (hurdles) that any microorganisms present should not be able to overcome. These hurdles may include factors such as the storage temperature, a_W, pH, redox potential, preservatives and novel techniques including MAP, bioconservation, bacteriocins, ultra-high pressure treatment and edible coatings (Leistner and Gorris, 1995). The use of combination processes with MAP is still in its development phase, but in the future this technology could be the way to control pathogen growth and ensure a safe product.

Smart packaging, including time–temperature indicators (TTI), is a technology that appears to have a significant potential, especially with chill-stored MAP products (Labuza *et al.*, 1992). To ensure microbial safety, strict temperature control is needed and temperature abuse should be avoided. TTIs could be applied to monitor the temperature and to detect temperature-abused packages. For further information about active packaging technologies and hurdle technology please see Chapters 5 and 7 respectively.

4.10 References and further reading

4.10.1 References

ACMSF (Advisory Committee on the Microbial Safety of Food) (1992) *Report on Vacuum Packaging and Associated Processes.* HMSO publication. Crown Publishing.

AHVENAINAN, R., SKYTTÄ, E. and KIVIKATAJA, R.-L. (1990) The influence of modified atmosphere packaging on the quality of selected ready-to-eat foods. *Lebens,.-Wiss. u.-Technol.*, **23,** 139–48.

AHVENAINEN, R., HURME, E., HÄGG, M., SKYTTÄ, E. and LAURILA, E. (1998) Shelf life of pre-peeled potato cultivated, stored and processed by various methods. *Journal of Food Protection*, **61**, 591–600.

BARNETT, H.J., STONE, F.E., ROBERTS, G.C., HUNTER, P.J., NELSON, R.W. and KWOK, J. (1982) A study in the use of high concentration of CO_2 in a modified atmosphere to preserve salmon. *Marine Fish Rev.*, **44** (3), 7–11.

BJERKENG, B., SIVERTSVIK, M., ROSNES, J.T and BERGSLIEN, H. (1995) Reducing package deformation and increasing filling degree in packages of cod fillets in CO_2-enriched atmospheres by adding sodium carbonate and citric acid to an exudate absorber. In *Foods and Packaging Material Chemical Interactions.* Ackermann, P., Jägerstad, M. and Ohlsson, T. (eds). The Royal Society of Chemistry, Cambridge.

CAI, P., HARRISON, M.A., HUANG, Y.W. and SILVA, J.L. (1997) Toxin production by *Clostridium botulinum* type E in packaged channel catfish. *Journal of Food Protection*, **60**, 1358–63.

CALLOW, E.H. (1932) Gas storage of pork and bacon. Part I. Preliminary experiments. *Journal of the Society of Chemical Industry*, **51**, 116T–119T.

CANN, D.C., HOUSTON, N.C., TAYLOR, L.Y., SMITH, G.L., THOMSON, A.B. and CRAIG, A. (1984) *Studies of Salmonids Packed and Stored under a Modified Atmosphere.* Torry Research Station, Aberdeen.

CHURCH, I.J. and PARSONS, A.L. (1995) Modified atmosphere packaging technology: a review. *Journal of the Science of Food and Agriculture,* **67**, 143–52.

COULON, P. and LOUIS, P. (1989) Modified atmosphere packaging of precooked foods. In *Controlled/Modified Atmosphere/Vacuum Packaging of Foods.* Brody, A.L. (ed.). Food and Nutrition Press, Trumbull.

COYNE, F.P. (1932) The effect of carbon dioxide on bacterial growth with special reference to the preservation of fish. Part I. *Journal of the Society of Chemical Industry,* **51**, 119T–121T.

COYNE, F.P. (1933) The effect of carbon dioxide on bacterial growth with special reference to the preservation of fish. Part II. *Journal of the Society of Chemical Industry,* **52**, 19T–24T.

DALGAARD, P., GRAM, L. and HUSS, H.H. (1993) Spoilage and shelf-life of cod fillets packed in vacuum or modified atmospheres. *International Journal of Food Microbiology,* **19**, 283–94.

DANIELS, J.A., KRISHNAMURTHI, R. and RIZVI, S.S.H. (1985) A review of effects of carbon dioxide on microbial growth and food quality. *Journal of Food Protection*, **48**, 532–7.

DAVIES, A.R. (1995) Advances in modified atmosphere packaging. In *New Methods of Food Preservation.* Gould, G.W. (ed.) pp. 304–20. Blackie Academic & Professional, Glasgow.

DAVIS, H.K. (1998) Fish and shellfish. In *Principles and Applications of Modified Atmosphere Packaging of Foods.* Blakistone, B.A. (ed.) 2nd edn, pp. 194–239. Blackie Academic & Professional, Glasgow.

DAY, B.P.F. (1992) Guidelines for the good manufacturing and handling of modified atmosphere packed food products. Technical Manual No. 34, Campden Food and Drink Research Association, Chipping Campden, Gloucestershire.

DAY, B.P.F. (1993) Recent developments in MAP. *European Food & Drink Review*, Summer, 87–95.

DAY, B.P.F. (1998) Novel MAP: a brand new approach. *Food Manufacture*, November, 22–4.

DEVLIEGHERE, F., DEBEVERE, J. and VAN IMPE, J. (1998a) Effect of dissolved carbon dioxide and temperature on the growth of *Lactobacillus sake* in modified atmospheres. *International Journal of Food Microbiology*, **41**, 231–8.

DEVLIEGHERE, F., DEBEVERE, J. and VAN IMPE, J. (1998b) Concentration of carbon dioxide in the water-phase as a parameter to model the effect of a modified atmosphere on microorganisms. *International Journal of Food Microbiology*, **43**, 105–13.

DIXON, N.M. and KELL, D.B. (1989) The inhibition of CO_2 of the growth and metabolism of micro-organisms. *Journal of Applied Bacteriology*, **67**, 109–36.

DODDS, K.L. and AUSTIN, J.W. (1997) *Clostridium botulinum*. In *Food Microbiology: Fundamentals and Frontiers*. Doyle, M.E., Beuchat, L.R. and Montville, T.J. (eds) pp. 288–304. ASM Press, Washington DC.

EINARSSON, H. and VALDIMARSSON, G. (1990) Bulk storage of iced fish in modified atmosphere. *Chilling and Freezing of New Fish Products*. IIF/IIR, Paris, pp. 135–40.

EUROPEAN PATENT APPLICATION (2001) Method for packaging mussels and similar shellfish, and thus obtained package with shellfish. EP 1 065 144 A1. Applicant: Roem Van Yerseke BV.

FARBER, J.M. (1991) Microbiological aspects of modified-atmosphere packaging technology: a review. *Journal of Food Protection*, **54**, 58–70.

FIERHELLER, M.G. (1991) Modified atmosphere packaging of miscellaneous products. In *Modified Atmosphere Packaging of Food*. Ooraikul, B. and Stiles, M.E. (eds). Ellis Horwood, New York.

GARCIA, G. and GENIGEORGIS, C. (1987) Quantitative evaluation of *Clostridium botulinum* nonproteolytic types B, E, and F growth risk in fresh salmon tissue homogenates stored under modified atmospheres. *Journal of Food Protection*, **50**, 390–7.

GARREN, D.M., HARRISON, M.A. and HUANG, Y.-W. (1995) Growth and production of toxin of *Clostridium botulinum* type E in rainbow trout under various storage conditions. *Journal of Food Protection*, **58**, 863–6.

GIBSON, D.M. and DAVIS, H.K. (1995) Fish and shellfish products in sous vide and modified atmosphere packs. In *Principles of Modified-atmosphere and Sous-vide Product Packaging*. Farber, J.M. and Dodds, K.L. (eds) pp. 153–74. Technomic Publishing, Lancaster, Penn.

GILL, C.O. (1996) Extending the storage life of raw chilled meats. *Meat Science*, **43**, S99–S109.

GILL, C.O. and MOLIN, G. (1991) Modified atmospheres and vacuum packaging. In *Food Preservatives*. Russel, N.J. and Gould, G.W. (eds) pp. 172–99. Blackie Academic & Professional, Glasgow.

GILL, C.O. and PENNEY, N. (1988) The effect of the initial gas volume to meat weight ratio on the storage life of chilled beef packaged under carbon dioxide. *Meat Science*, **22**, 53–63.

GILL, C.O. and TAN, K.H. (1980) Effect of carbon dioxide on growth of meat spoilage bacteria. *Applied and Environmental Microbiology*, **39**, 317–19.

GRAM, L., TROLLE, G. and HUSS, H.H. (1989) Detection of specific spoilage bacteria from fish stored at low (0°C) and high (20°C) temperatures. *International Journal of Food Microbiology*, **4**, 65–72.

GUNES, G. and LEE, C.Y. (1997) Colour of minimally processed potatoes as affected by modified atmosphere packaging and antibrowning agents. *Journal of Food Science*, **62**, 572–5, 582.

HAINES, R.B. (1933) The influence of carbon dioxide on the rate of multiplication of certain bacteria, as judged by viable counts. *Journal of the Society of*

Chemical Industry, **52**, 13T–17T.

HO, C.S., SMITH, M.D. and SHANAHAN, J.F. (1987) Carbon dioxide transfer in biochemical reactors. *Advances in Biochemical Engineering,* **35**, 83–125.

HURME, E., AHVENAINEN, R. and NIELSEN, T. (1998) Active and smart packaging of food. Technical report in Nordic Network on Minimal Processing, April. SIK, Göteborg.

HUSS, H.H., DALGAARD, P. and GRAM, L. (1997). Microbiology of fish and fish products. In *Seafood from Producer to Consumer: Integrated Approach to Quality.* Luten, J.B., Børresen, T. and Oehlenschläger, J. (eds) pp. 413–30. Elsevier, Amsterdam.

KADER, A.A., ZAGORY, D. and KERBEL, E.L. (1989) Modified atmosphere packaging of fruits and vegetables. *Critical Reviews in Food Science and Technology*, **28** (1), 1–30.

KIROV, S.M. (1997) *Aeromonas* and *Plesiomonas* species. In *Food Microbiology: Fundamentals and Frontiers.* Doyle, M.E., Beuchat, L.R. and Montville, T.J. (eds) pp. 265–87. ASM Press, Washington DC.

KNOCHE, W. (1980) Chemical reactions of CO_2 in water. In *Biophysics and Physiology of Carbon Dioxide.* Bauer, C., Gros, G. and Bartels, H. (eds) pp. 3–11. Springer-Verlag, Berlin.

KOLBE, H. (1882) Antiseptische Eigenschaften der Kohlensäure. *Journal für Praktische Chemie,* **26**, 249–55.

LABUZA, T.P., FU, B. and TAOUKIS, P.S. (1992) Prediction for shelf life and safety of minimally processed CAP/MAP chilled foods: a review. *Journal of Food Protection*, **55**, 741–50.

LAURILA, E., HURME, E. and AHVENAINEN, R. (1998) The shelf life of sliced raw potatoes of various cultivar varieties: substitution of bisulphites. *Journal of Food Protection*, **61** (10), 1363–71.

LEBLANC, R.J. and LEBLANC, E.L. (1992) The effect of superchilling with CO_2 snow on the quality of commercially processed cod (*Gadus morhua*) and winter flounder (*Pseudopleuronectes americanus*) fillets. In *Quality Assurance in the Fish Industry.* Huss, H.H., Jakobsen, M. and Liston, J. (eds) pp. 115–24. Elsevier, Amsterdam.

LEISTNER, L. and GORRIS, L.G.M. (1995) Food preservation by hurdle technology. *Trends in Food Science & Technology*, **6**, 41–6.

MARTENS, T. (1997) Harmonization of safety criteria for minimally processed foods. Inventory report FAIR concerted action FAIR CT96.1020, September.

OGRYDZIAK, D.M. and BROWN, W.D. (1982) Temperature effects in modified-atmosphere storage of seafoods. *Food Technology,* **36** (5), 86–96.

OORAIKUL, B. (1991) Modified atmosphere packaging of bakery products. In *Modified Atmosphere Packaging of Food.* Ooraikul, B. and Stiles, M.E. (eds). Ellis Horwood, New York.

PARKIN, K.L. and BROWN, W.D. (1982) Preservation of seafood with modified atmospheres. In *Chemistry & Biochemistry of Marine Food Products.* Martin, R.E., Flick, G.J., Hebard, C.E. and Ward, D.R. (eds) pp. 453–65.

AVI Publishing, Westport, Conn.

PHILLIPS, C.A. (1996) Review: modified atmosphere packaging and its effects on the microbiological quality and safety of produce. *International Journal of Food Science and Technology*, **31**, 463–79.

POST, L.S., LEE, D.A., SOLBERG, M., FURGANG, D., SPECCHIO, J. and GRAHAM, C. (1985) Development of botulinal toxin and sensory deterioration during storage of vacuum and modified atmosphere packaged fish fillets. *Journal of Food Science,* **50**, 990–6.

RANDELL, K., HATTULA, T., SKYTTÄ, E., SIVERTSVIK, M. and BERGSLIEN, H. (1999) Quality of filleted salmon in various retail packages. *Journal of Food Quality*, **22**, 483–97.

REDDY, N.R., ARMSTRONG, D.J., RHODEHAMEL, E.J. and KAUTTER, D.A. (1992) Shelf life extension and safety concerns about fresh fishery products. *Journal of Food Safety*, **12**, 87–118.

REDDY, N.R., PARADIS, A., ROMAN, M.G., SOLOMON, H.M. and RHODEHAMEL, E.J. (1996) Toxin development by *Clostridium botulinum* in modified atmosphere-packaged fresh tilapia fillets during storage. *Journal of Food Science*, **61**, 632–5.

REDDY, N.R., ROMAN, M.G., VILLANUEVA, M., SOLOMON, H.M., KAUTTER, D.A. and RHODEHAMEL, E.J. (1997a) Shelf life and *Clostridium botulinum* toxin development during storage of modified atmosphere-packaged fresh catfish fillets. *Journal of Food Science,* **62**, 878–84.

REDDY, N.R., SOLOMON, H.M., YEP, H., ROMAN, M.G. and RHODEHAMEL, E.J. (1997b) Shelf life and toxin development by *Clostridium botulinum* during storage of modified-atmosphere-packaged fresh aquacultured salmon fillets. *Journal of Food Protection*, **60**, 1055–63.

REDDY, N.R., SOLOMON, H.M. and RHODEHAMEL, E.J. (1999) Comparison of margin of safety between sensory spoilage and onset of *Clostridium botulinum* toxin development during storage of modified atmosphere (MA)-packaged fresh marine cod fillets with MA-packaged aquacultured fish fillets. *Journal of Food Safety*, **19**, 171–83.

ROSNES, J.T., SIVERTSVIK, M. and BERGSLIEN, H. (1994) Effects of storage temperature on the shelf life of cod (*Gadus morhua*) fillets packaged in modified atmosphere. Paper presented at Western European Fish Technologist Association 24th Annual Meeting, Nantes, 25–29 September.

ROSNES, J.T., SIVERTSVIK, M. and BERGSLIEN, H. (1997) Distribution of modified atmosphere packaged salmon (*Salmo salar*) products. In *Seafood from Producer to Consumer: Integrated Approach to Quality*. Luten, J.B., Børresen, T. and Oehlenschläger, J. (eds) pp. 211–20. Elsevier, Amsterdam.

ROSNES, J.T., SIVERTSVIK, M., SKIPNES, D., NORDTVEDT, T.S., CORNELIUSSEN, C. and JAKOBSEN, Ø. (1998) Transport of superchilled salmon in modified atmosphere. In Proceedings from Hygiene, Quality and Safety in the Cold Chain and Air-Conditioning IIF-IIR-Commission C2/E1. 16–18

September, Nantes, pp. 229–36. International Institute of Refrigeration, Paris.

SIKORSKI, Z.E. and PAN, B.S. (1994) Preservation of seafood quality. In *Seafoods: Chemistry, Processing, Technology and Quality*. Shahidi, F. and Botta, J.R. (eds) pp. 168–95. Blackie Academic & Professional, Glasgow.

SIVERTSVIK, M. (1995) Modified Atmosphere Packaging of Seafood. Proceedings of MAPack'95, the Leading Edge Conference on Modified Atmosphere Packaging, 19–20 October, Anaheim, California. Institute of Packaging Professionals, Herndon, VA.

SIVERTSVIK, M. (1999) Use of O_2-absorbers and soluble gas stabilisation to extend shelf life of fish products, Nordic Foodpack Symposium, 31 May – 2 June, Dansk Teknologisk Institutt, HøjeTåstrup, Denmark.

SIVERTSVIK, M. (2000) Use of soluble gas stabilisation to extend shelf life of fish. In Proceedings of 29th Annual WEFTA Meeting, 10–14 October, Leptocarya – Pieria, Greece. Georgakis, S.A. (ed.) pp. 79–91. Greek Society of Food Hygienists and Technologists, Thessaloniki.

SIVERTSVIK, M., ROSNES, J.T. and BERGSLIEN, H. (1997) Shelf life of whole cooked shrimp (*Pandalus borealis*) in CO_2-enriched atmosphere. In *Seafood from Producer to Consumer: Integrated Approach to Quality*. Luten, J.B., Børresen, T. and Oehlenschläger, J. (eds) pp. 221–30. Elsevier, Amsterdam.

SIVERTSVIK, M., ROSNES, J.T., VORRE, A., RANDELL, K. AHVENAINEN, R. and BERGSLIEN, H. (1999) Quality of whole gutted salmon in various bulk packages. *J. Food Quality*, **22** (4), 387–401.

SIVERTSVIK, M., NORDTVEDT, T.S., AUNE, E.J. and ROSNES, J.T. (2000) Storage quality of superchilled and modified atmosphere packaged whole salmon. In Twentieth International Congress of Refrigeration, Sydney, Australia 19–24 September 1999. Proceedings on CD-ROM. Vol. IV, Paper 490, pp. 2488–95. International Institute of Refrigeration, Paris.

SIVERTSVIK, M., JEKSRUD, W.K. and ROSNES, J.T. (2002) Review article: modified atmosphere packaging of fish and fishery products – significance of microbial growth, activities, and safety. *International Journal of Food Science and Technology*, **37**, 107–27.

SIVERTSVIK, M., KLEIBERG, G.H., SKIPNES, D. and ROSNES, J.T. Unpublished results. Stavanger, Norway.

SØRHEIM, O., AUNE, T. and NESBAKKEN, T. (1997) Technological, hygienic and toxicological aspects of carbon monoxide used in modified-atmosphere packaging of meat. *Trends in Food Science & Technology*, **8**, 307–12.

STAMMEN, K., GERDES, D. and CAPORASO, F. (1990) Modified atmosphere packaging of seafood. *Critical Reviews in Food Science and Technology*, **29**, 301–31.

STILES, M.E. (1991) Modified atmosphere packaging of meat, poultry and their products. In *Modified Atmosphere Packaging of Food*. Ooraikul, B. and Stiles, M.E. (eds). Ellis Horwood, New York.

TAYLOR, L.Y., CANN, D.D. and WELCH, B.J. (1990) Antibotulinal properties of nisin in fresh fish packaged in an atmosphere of carbon dioxide. *Journal of Food Protection,* **53**, 953–7.

VALLEY, G. (1928) The effect of carbon dioxide on bacteria. *The Quarterly Review of Biology*, **3**, 209–24.

4.10.2 Books and proceedings about MAP of foods

BLAKISTONE, B.A. (1998) *Principles and Applications of Modified Atmosphere Packaging of Foods*. 2nd edn. Blackie Academic & Professional, London.

BRODY, A.L. (1989) *Controlled/Modified Atmosphere/Vacuum Packaging of Food*. Food and Nutrition Press, Trumbull.

BRODY, A.L. (1994) *Modified Atmosphere Food Packaging*. IoP Press.

CAMPDEN AND CHORLEYWOOD FOOD RESEARCH ASSOCIATION (1995) Proceedings from Modified Atmosphere Packaging (MAP) and Related Technologies. 6–7 September 1995, Chipping Campden, UK.

CAMPDEN FOOD AND DRINK RESEARCH ASSOCIATION (1990) Proceedings from International Conference on Modified Atmosphere Packaging. Part 1 & 2, 15–17 October 1990, Stratford-upon-Avon, UK.

FARBER, J.M. and DODDS, K.L. (1995) *Principles of Modified Atmosphere Packaging and Sous-vide Packaging*. Technomic, Basel.

INSTITUTE OF PACKAGING PROFESSIONALS (1995) Proceedings from *MAPack'95*: The Leading Edge Conference on Modified Atmosphere Packaging. 19–20 October 1995, Anaheim, Ca.

OORAIKUL, B. and STILES, M.E. (1991) *Modified Atmosphere Packaging of Food.* Ellis Horwood, New York.

5

Active and intelligent packaging

E. Hurme, Thea Sipiläinen-Malm and R. Ahvenainen, VTT Biotechnology, Espoo and T. Nielsen, SIK (Swedish Institute for Food and Biotechnology), Gothenburg

5.1 Introduction

The basic functions of food packaging include the following:

- containment of the foodstuff
- protection of the product, thereby maintaining quality throughout its expected shelf-life
- providing information to the customer
- being easy to use
- selling the product
- being environmentally friendly, e.g. easily disposable or possible to recycle or refill.

Consumers' demands for more environmentally friendly packaging techniques, and for fresh, minimally processed foods with naturally preserved quality, are putting new demands on food packaging (Lund, 1989; Ahvenainen *et al.*, 1994; Ahvenainen *et al.*, 1995a). Food companies have also placed a greater emphasis on increasing product shelf-life and product safety (Ackermann *et al.*, 1995). Modern food packaging technologies include modified atmosphere packaging, active packaging and smart packaging, the main purpose of which is to enhance food safety and quality in as natural a way as possible (Hotchkiss, 1995). Modified atmosphere packaging, including vacuum and gas-flushing techniques, is an important preservation method for fresh, minimally processed foods. The technology is used for a wide range of shelf-stable and ready-to-eat chilled foods. Packaging specialists often consider modified atmosphere packaging as one form of active packaging, because many active packaging methods, like modified atmosphere packaging, modify the composition of the package head-space (the space between the food and the packaging layer).

The idea of using active and smart packaging techniques is not new, but the potential quality and economic advantages of these techniques can be considered to be among the latest advances in the food packaging industry. It must be remembered, however, that these techniques must be inexpensive (relative to the value of the packed product), reliable, environmentally friendly, aesthetically acceptable and compatible with existing distribution systems (Rooney, 1995b; Hurme and Ahvenainen, 1996). This chapter aims to give an overview of different types of active and smart packaging, to describe briefly the chemical and physical principles involved and also to point out a few commercial applications in the marketplace. It updates such reviews of the subject as Nielsen (1998), Ahvenainen and Hurme (1997), Smolander *et al.* (1997) and Hurme and Ahvenainen (1996).

5.2 Definitions

Various terms for new packaging methods can be found in the literature, such as active, smart, interactive, clever or intelligent packaging. These terms are often more or less undefined. In this chapter, **active packaging** is defined as a packaging technique that actively and constantly either:

- changes package permeation properties or the concentration of different volatiles and gases in the package head-space (Abe, 1990; Abe, 1994; Smith *et al.*, 1990; Rooney, 1995b) during storage; or
- actively adds antimicrobial (Abe 1990), antioxidative (Hoojjat *et al.*, 1987) or other quality improving agents, e.g. flavour enhancing substances (Rooney, 1995a), via packaging materials into the packed food in small amounts during storage.

These kinds of packaging materials can also be defined as *interactive*, because they are in active interaction with food. The aim of these active or interactive methods is to maintain a product's desired shelf-life throughout storage. Typical examples of active packaging methods are as follows:

- oxygen (O_2) scavengers or absorbers based on various reactions (organic or inorganic)
- carbon dioxide (CO_2) absorbers or generators
- ethanol emitters
- ethylene absorbers
- moisture absorbers.

Intelligent, **smart** or **clever packaging** is defined as a packaging technique containing an external or internal indicator for the active product history and quality determination. Typical examples of smart packaging methods are as follows:

- time–temperature indicators intended to be fixed onto a package surface (Taoukis *et al.*, 1991)
- O_2 indicators (Krumhar and Karel, 1992; Ahvenainen *et al.*, 1995b)
- CO_2 indicators (Plaut, 1995)
- spoilage or quality indicators, which react with volatile substances from chemical, enzymatic and/or microbial spoilage reactions released from food (Smolander *et al.*, 1998, Mattila and Auvinen, 1990a, 1990b; Mattila *et al.*, 1990).

These kinds of indicators could be called *interactive, smart* indicators, because they are in active interaction with compounds originating from food.

Microwave heating enhancers, such as susceptors, and other temperature regulation methods are also regarded as active or intelligent packaging methods.

5.3 Active packaging techniques

The interest in active and smart packaging systems, and development and research work on them, have increased considerably throughout the world during the last fifteen years. Japan has been a pioneer in the development of commercially significant active packaging systems, mainly O_2 scavengers or various emitters (Rooney, 1995a). Labuza and Breene (1989) and Labuza (1989) were among the first to review the concept of active packaging for a wider international audience. Abe and Kondoh (1989) and Abe (1994) introduced Japanese O_2 absorbers, and Abe (1990) introduced other Japanese active packaging methods. Idol (1991) critically reviewed the use of in-package O_2 scavengers in the USA. Yang (1995) and Rooney (1995a) have compiled broader reviews of the field, including antibacterial packaging materials, O_2 absorbers and moisture absorbers. Taoukis *et al.* (1991) and later Fu and Labuza (1995) have reviewed time–temperature indicators and their potential use as an indicator of temperature abuse. There have also been a number of important international conferences and symposiums devoted specifically to active and intelligent packaging (including Ackermann *et al.*, 1995; Ahvenainen *et al.*, 1994; Ahvenainen *et al.*, 1995a; Ohlsson *et al.*, 1996; CCFRA, 1995).

Many research institutes in Europe, such as PIRA and the Campden and Chorleywood Food Research Association in the UK (CCFRA, 1995; Day, 1994; Barnetson, 1995; Selman, 1996), Technion-Israel Institute of Technology (Miltz *et al.*, 1995), the Royal Veterinary and Agricultural University, Denmark (Andersen and Rasmussen, 1992), Matforsk and Norconserv in Norway (Sørheim *et al.*, 1995; Bjerkeng *et al.*, 1995), TNO in the Netherlands (Rijk *et al.*, 2002; de Kruijf, *et al.*, 2002) and VTT Biotechnology in Finland (Hurme *et al.*, 1994; Ahvenainen *et al.*, 1995b; Hurme *et al.*, 1995, Hurme and Ahvenainen, 1996; Mattila-Sandholm *et al.*, 1995; Randell *et al.*, 1995; Salminen *et al.*, 1996), have been working on shelf-life and quality assurance studies related to active and smart packaging during recent years. Outside

Europe, CSIRO, Australia (Rooney, 1995c) and the University of Minnesota, USA (Labuza, 1989; Fu and Labuza, 1995) have been particularly active.

Despite intensive research and development work on active and smart packaging, there are only a few commercially significant systems on the market. O_2 absorbers added separately as small sachets in the package head-space probably have the most commercial significance currently in active food packaging. Ethanol emitters/generators (Seiler and Russell, 1991; Smith *et al.*, 1995) and ethylene absorbers (Zagory, 1995) are also used, but to a lesser extent than O_2 absorbers. Other commercially significant active techniques include absorbers for moisture and off-odours and absorbers/emitters for CO_2 (Rooney, 1995a).

5.4 Oxygen absorbers

Japan is the most important market area for O_2 absorbers. The first commercial iron-based pouch-type absorbers were launched in Japan in 1977. Since then, the design and applications for new designs have increased sharply with the current annual production of O_2 absorber sachets alone exceeding 7 billion in Japan, several hundred million in the USA and some tens of millions in Europe (Rooney, 1995b). The most common uses for O_2 absorbers are ready-made foods like hamburgers, fresh pastas, noodles, potato sticks, cured meats (sliced ham and sausage), cakes and shelf-stable bread, and confections, nuts, coffee, herbs and spices. Use of O_2 absorber pouches is especially advantageous for O_2 and light-sensitive products, such as bakery products and pizzas, cooked ham, etc., where mould growth or colour changes are a problem (Hurme and Ahvenainen, 1997). The interest in these techniques is increasing in Europe in particular, but more information is needed about consumer attitudes and acceptability in domestic markets, as well as their potential environmental impact, before they can grow significantly.

Active packaging methods of commercial significance have many advantages compared to modified atmosphere packaging using protective gases, but also some disadvantages. The most important advantage of O_2 absorbers is that it is possible to reduce the O_2 concentration to an ultra-low level, which is impossible in commercial gas packaging lines. High O_2 levels promote the growth of many bacteria, yeasts and moulds. Foodstuffs can lose nutrients, flavours and pigments or become rancid through oxidation reactions. Active packaging with O_2 absorbers actively decreases the O_2 concentration in the package head-space, even down to 0.01%, thus retarding the oxidation processes, colour changes, yeast and mould growth, etc., occurring in foods (Smith *et al.*, 1990; Andersen and Rasmussen, 1992; Abe, 1994; Hurme *et al.*, 1995; Randell *et al.*, 1995). Moreover, if the absorber has sufficient capacity, it will also absorb O_2 that leaks into the head-space through pinholes and prolong the shelf-life of the packed foodstuffs (Hurme and Ahvenainen, 1997; Hurme *et al.*, 1995).

Table 5.1 Advantages (+)/disadvantages (−) encountered with gas and vacuum packaging techniques and oxygen absorbers (Hurme and Ahvenainen, 1996)

Feature	Gas packaging	Vacuum packaging	Oxygen absorbers
Investment costs	−	+	++
Packaging costs	+	++	−
Need for food preserving agents	−	−	+
Shelf-life/food quality	+	+	++
Package volume/space saving	−	++	+
Ease of leak detection	−	+	−
Suitable for soft products	+	−	+
Visible/invisible	+	+	+/−[1]
Usable with metal detectors	+	+	+/−[2]
Environmental impacts[3]	+/−	+/−	+/−

Notes: 1. Not studied extensively. 2. Depends on the type of the absorber. 3. No data available.

One big advantage is that the capital investments needed are significantly cheaper with active packaging than with gas packaging. Basically, only sealing systems are necessary. This has a large significance to small and medium sized food enterprises for which packaging machines are often the most expensive investment. On the other hand, the main disadvantage of commercially significant active packaging methods is that they are visible (sachets or labels) whereas, in gas packaging, gases are invisible (Table 5.1; Hurme and Ahvenainen, 1996).

Most of the currently used O_2 absorbers are iron based, using treated iron powder. As a rule of thumb, one gram of iron will react with 300ml O_2. Even if the iron-containing absorbers probably are the most suitable materials (Abe, 1994), they have one disadvantage in that they cannot pass the metal detectors usually installed on the packaging line. This problem can be avoided, e.g. by ascorbic acid or enzyme-based O_2 absorbers. A scavenger of the appropriate size is chosen depending on the O_2 level in the head-space, how much O_2 is trapped in the food initially and the amount of O_2 that will be transported from the surrounding air into the package during storage. For an O_2 absorber to be effective the packaging material used has to be a relatively good barrier to O_2 permeation, otherwise the scavenger will rapidly become saturated and lose its ability to trap O_2. However, the use of O_2 absorber enables the use of more gas-permeable packaging materials than that in similar gas-flushed packages, especially when the adjusted gas-absorbing capacity of the absorbers/emitter was substantial (Hurme and Ahvenainen, 1997).

The most significant manufacturers and trade names for O_2 absorbers are listed in Table 5.2. Various sizes of O_2 absorbers, with the ability to consume 20–2,000 ml O_2, are present on the market. Many of the manufacturers have several types and sizes of absorbers under their trade names. The selection of an

Table 5.2 Manufacturers and trade names of oxygen absorber sachets/labels

Manufacturer	Country	Trade name
Mitsubishi Gas Chemical Co., Ltd	Japan	Ageless
Toppan Printing Co., Ltd	Japan	Freshilizer
Toagosei Chemical Industry Co., Ltd	Japan	Vitalon
Nippon Soda Co., Ltd	Japan	Seaqul
Finetec Co., Ltd	Japan	Sanso-Cut
Multisorb Technologies Co., Ltd	USA	FreshMax
		FreshPax
Standa Industrie	France	ATCO
Bioka Ltd	Finland	Bioka

appropriate sachet has to be based on storage temperature, desired shelf-life, water activity, initial O_2 content and O_2 transmission rate of the packaging material. Most O_2 absorbers are used at ambient temperature, but some scavengers that can act under refrigerated or frozen conditions have been developed. They are applicable to many types of foods including those of high, intermediate or low moisture, as well as greasy foods.

In the USA O_2 scavengers have been used in bottle closures. Beer is a product that is extremely sensitive to O_2 because of flavour deterioration. With modern technology it is possible to fill and close beer bottles in a way that leaves less than 500 ppb oxygen in the bottle. However, even if the pressure inside the bottle is approximately 3 atm, O_2 ingress through the crown has been observed. This permeation is facilitated by the lower partial pressure of O_2 within the bottle. Historically, oxidation of beer flavours has been prevented by addition of antioxidants, such as sulphur dioxide and ascorbic acid, to the product. Recently, however, O_2 scavenging bottle closures have been developed and become an instant success (Teumac, 1995). O_2 absorbing substances, e.g. ascorbates, sulphites and iron compounds, are incorporated in a polymer with suitable permeabilities for water and oxygen. PE and PP exhibit too low water permeability but polyvinyl chloride (PVC) has proved to be the ideal polymer for this application. In addition to having appropriate permeabilities, its properties remain unaffected by incorporation of other materials. Another requirement on this system is that the O_2 scavenging agent must be stable under the process conditions which excludes any type of enzymes. The benefit of this type of packaging is a prolonged shelf-life without addition of preservatives. The only drawback is that the head-space O_2 must not be consumed too rapidly, since some O_2 is required in reactions removing sulphur-containing organic substances that otherwise might cause off-flavour in the beer. This type of closure is currently only used for beer, but might in the future find new applications, for instance in the juice and wine industry.

5.5 Carbon dioxide absorbers and emitters

While most O_2 scavengers have a single objective, i.e. to remove O_2 from the packaging atmosphere, there are a few examples of absorbents with dual functions (Smith *et al.*, 1995). One such product is a sachet containing ascorbic acid and iron carbonate, which produces the same volume of CO_2 as that of consumed O_2. This is necessary in order to prevent package collapse and can be used in products that are shock-sensitive, e.g. potato crisps. Care must be taken that any flavour changes that could be caused by the dissolution of CO_2 in the aqueous or fat phase of the product do not take place. Another type of active packaging, which is utilised for ground coffee, scavenges CO_2 as well as O_2. Fresh roasted coffee can release considerable amounts of CO_2. Unless removed, the generated CO_2 can cause the packaging to burst due to the increasing internal pressure. The reactant commonly used to absorb CO_2 is calcium hydroxide which, at a high enough water activity, reacts with CO_2 to form calcium carbonate.

5.6 Ethylene absorbers

Ethylene is a plant hormone, which is produced during ripening of fruits and vegetables. Ethylene can have both positive and negative effects on fresh produce. It catalyses the ripening process and exposure to high levels of ethylene can be used to induce rapid ripening of products such as bananas and tomatoes. Ethylene can also be used to de-green fruits of the citrus variety. Generally, however, ethylene is seen as an unwanted substance since even in small amounts it can be very detrimental to the quality and longevity of many products. The increased respiration rate, which is an effect of ethylene exposure, leads to softening of fruit tissue and accelerated senescence. Furthermore, ethylene promotes the degradation of chlorophyll and is responsible for a series of post-harvest disorders. Consequently, it is desirable to suppress the effects of ethylene by somehow removing the substance from the horticultural environment.

A wide array of packaging solutions that, according to the manufacturers, are claimed to be efficient ethylene scavengers, can be found on the market. As is often the case with active packaging, the developed products have been patented, and their actual efficiency has not been well documented. Ethylene scavengers are supplied either as sachets, which are placed in the package, or incorporated into some part of the packaging material. Ethylene is a highly reactive molecule which can be absorbed by using a number of substances in the packaging materials, e.g. potassium permanganate, activated carbon and different minerals. Many suppliers in Japan and the USA offer ethylene scavengers based on potassium permanganate, which is embedded in an inert substrate such as silica gel. The substrate adsorbs the ethylene whereafter the permanganate oxidises it to ethanol and acetate. These products contain some 5% $KMnO_4$, and are solely supplied in the form of sachets due to the toxicity of potassium permanganate (Abeles *et al.*, 1992). There are also a number of ethylene scavengers using some form of metallic catalyst, e.g. palladium, placed on activated carbon. Ethylene is adsorbed and then catalytically broken down. Yet another product gains its ethylene-scavenging effect from activated carbon impregnated with a bromine-containing substance. In this case, however, there is some concern that toxic bromine gas can be produced if the sachet comes in contact with water.

Lately, companies in the Far East have brought onto the market several types of packaging containing finely dispersed materials with a putative ability to adsorb ethylene (Matsui, 1989; Someya, 1992). Minerals that are used include zeolites, different kinds of clay and Japanese Oya stone which originates from a cave that, reportedly, has been used for thousands of years for storage of fresh produce. There is no scientific evidence, however, that these minerals actually have ethylene-removing capabilities. There is an alternative explanation to the observation that products stored in polyethylene (PE) bags containing different kinds of minerals have a longer shelf-life than those stored in common PE bags. It could be that the finely dispersed minerals open up pores in the polymeric material, thereby altering the gas-exchange properties, resulting in prolonged shelf-life, as has been discussed in the previous chapter. Finally, integration of electron-deficient dienes and trienes in packaging material have claimed to be an efficient way of scavenging ethylene (Holland, 1992). In one experiment it was shown that incorporation of tetrazine, which is hydrophilic and has to be protected by a hydrophobic polymer such as PE, could reduce the ethylene content by two orders of magnitude within 48 hours. Another beneficial characteristic of tetrazine is that it changes colour when it becomes saturated with ethylene, indicating that it needs replacing.

5.7 Moisture/water absorbers

An unwanted accumulation of liquid water can occur in packages due to transpiration of horticultural produce, drip of tissue fluid from meats or temperature fluctuations in high moisture packages. It is essential to remove excess water since the build-up promotes the growth of moulds and bacteria. Other negative consequences include fogging of films and the appearance of mobile blood and tissue fluid. There is interest not only to absorb the liquid water from the packaging but also to intercept the moisture in the vapour phase in order to lower the water activity at the food surface. As a result unfavourable conditions for mould and bacterial growth are created and an extended shelf-life of the product is obtained.

Drip-absorbent pads can often be found in packaging for cut meat and poultry. They consist of granules of a superabsorbent polymer between two layers of a microporous or non-woven polymer which is sealed at the edges (Rooney, 1995a). The absorbers remove liquid water and prevent discoloration of both the product and the package. The polymers most frequently used for absorbing water are polyacrylate salts and graft copolymers of starch. These superabsorbent polymers are capable of absorbing between 100 and 500 times their own weight of liquid water (Showa Denko KK, 1990). Similar devices, but on a much larger scale, are used in packaging for air transportation of fresh fish and other seafood to absorb melted ice.

By reducing the relative humidity inside a package the water activity at the food surface is lowered. The concomitant result is a positive effect on the longevity of the product. In order to achieve this the water must be absorbed in its vapour phase which is more effectively done using humectants instead of superabsorbing polymers. The Showa Denko Co. in Japan has developed a water vapour absorbing film (Pichit) for use in the household (Labuza and Breene, 1989). The film is a laminate with a core layer of propylene glycol inside a sealed envelope of polyvinyl alcohol (PVA). The PVA film holds the glycol in place but it is very permeable to water. The foodstuff is wrapped in cellophane, placed in a Pichit bag and kept in the refrigerator. The difference in water activity between the food and the glycol means that the water is rapidly drawn out of the food surface and absorbed by the film. The desired effect, i.e. surface dehydration, is normally obtained within four to six hours. This product, which is sold in boxes containing a Pichit bag and ten cellophane sheets, can reportedly provide three to four days' extension of the shelf-life of refrigerated fish. An additional advantage is that, after use, the bag can be washed in water, dried out and reused up to ten times.

During distribution and storage of fresh produce there are great risks for accumulation of water caused by the continuous evaporation from the products and the inevitable temperature fluctuations during handling and transportation. High humidity and condensation of water increase the possibility of microbial growth on the products. This problem can be overcome either by including microporous pads containing water-absorbing inorganic salts or by lining the fibreboard cartons with a protected layer of a solid polymeric humectant (Hudson, 1991; Shirazi and Cameron, 1992; Patterson and Joyce, 1993).

Anti-fog additives can be blended with the polymer resin before extrusion in order to prevent fogging and forming of droplets on the inside of a package. The amphiphilic additives lower the interfacial tension between the polymer and water condensate. As a consequence, the water droplets coalesce and spread as a thin transparent layer across the polymer film. The customer can see the product clearly, but the water remains in close proximity to the food with the potential of causing spoilage. Therefore this type of treatment can merely be regarded as a cosmetic form of active packaging which does not improve the quality or prolong the shelf-life of the packaged product.

5.8 Ethanol emitters

Ethanol has been used as a preservative for centuries. At high concentrations ethanol denatures the proteins of moulds and yeasts and, although the action is not as severe, it exhibits antimicrobial effects even at low levels. Spraying the foodstuff with ethanol prior to packaging can be used to obtain the desired effect, but in some cases a more practical option can be to use sachets generating ethanol vapour. A product called Ethicap consists of an ethanol/water mixture adsorbed onto a silicon dioxide powder, contained in a sachet of a laminate of paper and ethyl vinyl acetate copolymer. The odour of alcohol can be masked by adding traces of flavours, such as vanilla, to the sachet. Ethicap acts by absorbing moisture from the food and releasing ethanol vapour into the packaging head-space. The size of the sachet used depends on the water activity of the food and the desired shelf-life of the product.

Ethanol vapour generators are mainly used in Japan for high moisture bakery goods and for fish products. Considerable shelf-life extensions have been reported for a number of foods (Ethicap Technical Information, 1987; Pafumi and Durham, 1987; Smith *et al.*, 1987; Black *et al.*, 1993). As an example, cakes with high water content remained mould-free for up to 20 times longer when an ethanol-releasing sachet was included in the package. An additional benefit from the use of ethanol vapour generators is the delay of staling which has been observed in bakery products (Black *et al.*, 1993). On the other hand, Salminen *et al.* (1995, 1996) concluded that, although ethanol vapour generators increased the shelf-life of packed bread, they did not retard staling. In addition, they found oxygen absorbers to be at least as effective in shelf-life prolongation as ethanol emitters. Ethanol emitters should advantageously be added into a package just after baking and cooling in sterile conditions.

The main disadvantage of using ethanol vapour for preservation purposes is its off-odour and off-flavour formation to the foodstuff, and absorption from the head-space by the foodstuff. In some cases the ethanol concentration in the product might reach a couple of percent, which might cause regulatory problems. If the product is heated in an oven prior to consumption the accumulated ethanol will evaporate, leaving only traces in the food. Therefore ethanol vapour generators can safely be used in products intended to be heated before use. They can provide a viable alternative or supplement to gas packaging for prolonging shelf-life of foods with regards to microbial spoilage.

5.9 Active packaging materials

Potentially, a great number of unwanted food constituents could be removed by the aid of specially designed food packages which selectively interact with the product. Although some research has been conducted in this area of active packaging, the field remains largely unexplored. Elimination of food components would probably be most applicable to liquid products, where the molecules can move freely and the removal process would not be limited to substances with high vapour pressures at the storage temperature. It is of importance, however, that this technology is not used to mask the development of microbial odours, thereby concealing the marketing of products that are below standard or even dangerous for the consumer.

In contrast to removing unwanted substances from the food, the packaging could presumably be used for depositing desirable compounds in the product. CO_2 and ethanol emitters have previously been described but the packaging material could in theory be used as a carrier of any substance. Consumers have a growing interest in buying preservative-free foods. A possibility for the food manufacturer to satisfy the consumer demands without using additives in the actual product could be to introduce them via the package. Substances that potentially could be released into the food in this manner include antioxidants, stabilisers, flavours and nutrients.

5.10 Oxygen-absorbing packaging materials

Apart from the technical aspects, some consumer resistance to oxygen-absorbing sachets can be foreseen. Even if the contents are safe, there could be concern about unintentional ingestion of sachet contents. Precautions to minimise this risk have been taken by clearly stating 'Do not eat' on the labels and by legislating a minimum size of the sachets. Another concern is that the sachet content could leak out and adulterate the product. Circumvention of these problems could be obtained by affixing it to some part of the package. The oxygen-absorbing substance can be included in polymer films, adhesives, inks or coatings. The oxygen-consuming substrate can either be the polymer itself or some easily oxidisable compound dispersed or dissolved in the packaging material. Metallic sulphites, ascorbic acid and iron compounds are among those substances that have been mixed with polymeric films to obtain oxygen-absorbing effects (Hofeldt and White, 1989; Farrell and Tsai, 1985). In recent years, several Japanese companies have introduced packaging materials containing, for example, a metal compound or ascorbates (Toppan Printing Co. Ltd, Mitsubishi Gas Chemical Co. Inc., Toyobo Kogyo Kaisha).

An additional problem in connection with the use of oxygen-scavenging films is that the films must be stable in the atmosphere prior to use. In other words, the films must not start consuming O_2 until the food has been packaged. This problem has been solved by inclusion of some sort of activation mechanism, which triggers the oxygen-consuming capabilities of the film, in the packaging system. For instance, reagents or catalysts can be supplied at the time of filling or exposure to light might be required for the reactions to take place. Australian researchers have reported that reaction of iron with ground state O_2 is too slow for shelf-life extension (Rooney and Holland, 1979; Rooney *et al.*, 1981). Therefore they developed a photosensitive dye, which was impregnated onto a polymeric film. When the film was irradiated by ultraviolet light the dye activated the O_2 to its singlet state, making the oxygen-removing reaction much faster ($ZERO_2$, Southcorp Australia Ltd). Also plastic oxygen-absorbing concentrates, where the scavenging reaction is triggered by elevated moisture, have been developed (Amosorb, Amoco Chemicals, USA). The use of an enzyme reactor surface would be yet another way of controlling O_2 levels in food packages. A combination of two enzymes, glucose oxidase and catalase, has been applied for O_2 removal (Bioka, Bioka Ltd, Finland). The enzymes can easily be applied to the surface of polyolefins such as PE and PP, which are very good substrates for immobilising enzymes (Scott, 1965). One restriction for the larger implementation of oxygen-absorbing materials has been their relatively low oxygen-absorbing capacity when compared to separate oxygen-absorbing pouches. Therefore it may take some time before such films have been developed, commercialised and adopted by the customers.

5.11 Packaging materials with antioxidants

The packaging industry uses antioxidants for polymer stabilisation and has an interest in finding natural replacements for synthetic antioxidants. In food packages, antioxidants provide both a functional barrier to O_2 diffusion and a transfer of antioxidants to the contained product, thus inhibiting oxidative reactions. Recently, vitamin E has been integrated in polymer films to exert its antioxidative effects and could potentially migrate into foods. The release of vitamin E could eliminate the need for the addition of antioxidants to the actual foodstuff itself. Butylated hydroxytoluene (BHT) has been widely used as an antioxidant to extend the shelf-life of, for example, cereal products (Hoojjat *et al.*, 1987). There is, however, some concern about the safety of BHT for human consumption.

5.12 Enzymatic packaging materials

A wide range of enzymes that actually change the product biochemically can possibly be incorporated into packaging materials, thus making the package a part of the process. Excess cholesterol ingestion can lead to cardiovascular diseases and the introduction of cholesterol reductase into milk cartons could be one way of reducing the cholesterol intake. Consumption of dairy products containing lactose can cause severe discomfort to the relatively large fraction of the population that suffers from lactose intolerance. Incorporation of the lactase enzyme for lactose removal in the food packaging material may rapidly reduce the lactose content in a significant manner (Tetra Pak International, 1977; Brody, 1990).

5.13 Antimicrobial agents in packaging materials

Antimicrobial agents, either mixed directly or applied on the surface of foods, can be used to prolong the shelf-life of packed foods. However, it could also be possible to incorporate antimicrobial agents into packaging materials which then migrate in small amounts into the food. This may be especially effective with vacuum packaging because of the intimate contact between the packaging material and the food surface. Substances with antimicrobial effects can be, for example, nisin produced by *Lactococcus lactis* (Folwer and Gasson, 1991), organic acids, esters and sorbates. Also packaging materials containing chitosan (Toyo Ink Mfg Co., 1992), imazalil (Weng and Hotchkiss, 1992) or horseradish-derived allyl-isothiocyanate enclosed within a cyclic oligosaccharide (Anon., 1993) have been claimed to have an antibacterial property.

The original idea of using certain inorganic materials was because of their claimed capability of emitting infrared radiation with an antimicrobial effect. The mechanism of this growth-inhibiting effect has, however, not been established. Several commercial packaging materials, including ceramic particles, are available. The active compound can consist of, for example, aluminium silicate and silver (Shinagawa Nenryo KK, 1992), dry powder made by substituting antimicrobial copper or silver for calcium atoms in hydroxyapatite (Anon., 1991), synthetic zeolite and silver (Abe, 1990; Kunisaki *et al.*, 1993) or copper, mangane or nickel and silver containing zeolite (Shinanen New Ceramic KK and Chugoku Pearl Hawbai KK, 1989). Also, magnesium oxide and zinc oxide have been proved to be bactericidal and bacteriostatic agents respectively (Sawai *et al.*, 1993).

5.13.1 Adding antimicrobial agents to packaging

The release of antimicrobial substances can be achieved conventionally by adding a permeable or porous sachet containing the substances into the package. Some preservatives can also be incorporated into or onto polymeric packaging materials to provide antimicrobial activity. These agents can be applied by impregnation, mixing or using various coating techniques. The active agents can be placed in intermediate layers or encapsulated to achieve slow release into food. A more sophisticated concept is the use of immobilised enzymes, and materials having chemically bound antimicrobial functional groups on the material surface.

Some functional groups that have antimicrobial activity have been introduced and immobilised on the surface of polymer films by modified chemical methods. Haynie *et al.* (1995) prepared a series of antimicrobial peptides covalently bonded to a water-insoluble resin which proved to have antimicrobial activity. Laser-induced surface functionalisation of polymers has recently been found to be an effective way of imparting functional groups of food packaging polymer surfaces to improve adhesion, to modify barrier properties and to give the polymers antimicrobial activity. Paik *et al.* (1998) recently presented the use of 193 nm UV irradiation using a UV excimer laser to convert amide groups on the surface of polyamide plastic to amines which have an antimicrobial effect.

5.13.2 Potential antimicrobial agents

Substances that can be used as antimicrobial include, for instance, ethanol and other alcohols, organic acids and salts (sorbates, benzoates, propionates), bacteriocins, etc. Hotchkiss (1995) has prepared an inventory of antimicrobial agents of potential use in food packages (Table 7.5). In food contact applications the substances incorporated should be safe and slow to migrate. Generally, it is recognised that metallic ions of silver, copper, zinc and others are safe antimicrobial agents. Ag-substituted zeolite is the most frequently used antimicrobial agent in plastic materials in Japan. The use of silver and Ag-zeolite is described by Ishitani (1995). The purpose of the zeolite is to allow for slow release of antimicrobial metal ions into the surface of the food products. The antimicrobial activity of metals is due to minute quantities of ions formed. Silver nitrate forms silver ions in water solutions which have a strong antimicrobial activity. These ions are considered to have inhibitory activities on metabolic functions of respiratory and electronic transport systems of microbes, and mass transfer across cell membranes. Experiments with yeast revealed that Ag-zeolite shows almost the same degree of activity regardless of the oxygen present and the existence of light.

Ag-zeolite is manufactured from synthetic zeolite by replacing a portion of its natural sodium with silver ions. Zeolite retains Ag-ions in stable conditions. Ag-zeolite content of 1–3% can be laminated as a thin film (3–6 μm) on the food contact surface of the packaging material. Only the ions of the zeolite particles at the surface are active. Sulphates, hydrogen sulphide and certain kinds of sulphur-containing amino acids and proteins are considered to influence the antimicrobial activity of Ag-zeolites. The amount of zeolite may influence the heat-sealing strength and other properties, like the transparency, of the plastic film.

Antimicrobial films can be based on release of inorganic acids, e.g. chlorine dioxide or sulphur dioxide. The materials have the biocidal substance residing in the polymer matrix and the active agents are released upon food contact or in response to environmental changes. For instance, a material capable of releasing chlorine dioxide has been patented by Wellinghoff and Kampa (1977). Systems releasing sulphur dioxide from metabisulfite-incorporated microporous material, effective in controlling mould growth in fruits, may, however, have some toxicological problems (Ozdemir and Sadikoglu, 1998).

Ethanol emitters as films or sachets are particularly effective against mould, but can also inhibit the growth of yeast and bacteria. Several applications of ethanol-emitting films and sachets have been patented. They contain absorbed or encapsulated ethanol in a carrier material which allows for controlled release of ethanol vapour. Sachets containing encapsulated ethanol have prevented microbial spoilage of intermediate moisture foods, cheese products and sweet bread (Labuza and Breene, 1989). To mask the odour of alcohol, some sachets contain small amounts of flavours. In several countries an upper limit for the ethanol content is prescribed.

Smith and Rollin (1954) showed the effectiveness of a sorbic acid wrapper made of moisture-proof cellophane coated by a thermoplastic process on natural and processed cheeses. Common food-grade antimycotic agents have been added to cellulose-based edible film (Vojdani and Torres, 1990). The films were constructed of cellulose derivatives and fatty acids in order to control the release of sorbic acid and potassium sorbate. These films have the greatest application as fruit and vegetable coatings. Propionic acid also has antimicrobial potential (Hotchkiss, 1995). The potential incompatibility of simple antimycotic acids such as propionic, benzoic and sorbic acids to polymers such as LDPE used in packaging films has been overcome by using anhydrides (Weng and Hotchkiss, 1993). The moisture in the food activates the acid by hydrolysis which is migrated into the food. Han and Floros (1997) have studied the incorporation of potassium sorbate into plastic films and their use as antimicrobial release systems. Decomposition of the potassium sorbate is prevented by using as low processing temperatures as possible.

A further advance is biologically derived antimicrobial materials that are bound or incorporated into films and do not need to migrate to the food to be effective. Theoretically, bacteriocins could be attached to the packaging material surface, but it is still unclear whether such bound bacteriocins would remain effective (Hoover and Steenson, 1993). Ming *et al.* (1997) indicated that pediocin-coated bags were effective in inhibiting *Listeria monocytogenes* in fresh or processed meat. Wilhoit (1996) has patented an antimicrobial film comprising nisin or pediocin and a chelating agent which are applied on a range of common packaging films (regenerated cellulose, polyolefins, polyamide, polyester, polyvinyl chloride, polyvinyl acetate, etc.). Natamycin (pimaricin) produced by *Streptomyces natalensis* is an interesting substance for use in packaging of cheeses to prevent the growth of yeast and moulds (Davidson and Doan, 1993).

Antimicrobial enzymes might be bound to the inner surface of food contact films. The enzymes would produce microbial toxins, such as glucose oxidase which forms hydrogen peroxide. Lysozyme applied in cellulose triacetate had a high activity, but applied in polyvinyl alcohol and in polyamide 6,6 yielded low activity (Appendini and Hotchkiss, 1997). Migration by desorption of the lysozyme from the cellulose film was observed. It is still unclear whether significant effects can be achieved in real foods. Padgett *et al.* (1998) compared a heat press method (1 min at 100°C) and a casting method (from boiling solution) to incorporate nisin and lysozyme into films made of soy protein and corn zein. Bactericidal properties were found in both processes. Addition of EDTA increased the inhibitory effect of the films against *Escherichia coli*. The fungicide imazalil was chemically coupled to plastic films (LD polyethylene) to prevent mould growth on cheese surfaces (Weng and Hotchkiss, 1992). Halek and Garg (1989) chemically coupled the antifungal agent benomyl to ionomer film and demonstrated inhibition of microbial growth in defined media. Triclosan is a widely used antibacterial substance in cosmetics and toothpaste, etc. It can be incorporated in plastic articles for kitchen and bathroom. Recently it has also been introduced in applications such as conveyor belts in the food industry (Day, 1998). Another development is incorporation of methyl salicylate into paperboard boxes marketed as an insect repellent, but the substance also has antimicrobial properties (Barlas, 1998).

Naturally occurring preservatives that have been proposed and/or tested for antimicrobial activity in plastics include spice and herb extracts, e.g. from rosemary, cloves, horseradish, mustard, cinnamon and thyme (Day, 1998). Most frequently studied is horseradish-derived allyl isothiocyanate. Chitin is a biopolymer consisting of polysaccharide. Large amounts of chitin are obtained from shellfish and can be used to prepare chitosan by deacetylation. Chitosan has antimicrobial effects and can be used in food packaging materials. A further possibility would be to incorporate radiation-emitting materials into the films. In Japan a material is reported to have been developed that emits long-wavelength infrared radiation, which is thought to be effective against microorganisms (Hotchkiss, 1995).

5.14 Flavour-scalping materials

The possibility of using packaging to eliminate unwanted compounds from the food or the food environment has already been discussed. Naturally, most attention has been given to substances that have a fundamental impact on food quality. O_2, water and ethylene are such compounds. Removal of these substances has been extensively investigated and briefly described in previous sections. The scavenging capabilities of food packages are, however, not limited to the aforementioned compounds. Flavour scalping, i.e. sorption of flavours by food packaging material, is a topic that receives increasing interest in the food industry. This phenomenon proves the ability of polymeric materials to interact with foods by absorbing food components. Generally, flavour scalping is detrimental to food quality in terms of flavour loss, but it could, potentially, be used to good effect in packages that selectively absorb unwanted odours or flavours.

There is a huge potential market for taint removal packaging, but very few applications have been put to commercial use so far. Studies have shown the ability of polyamides and cellulose esters to remove the bitter principle limonene from citrus juices (Chandler and Johnson, 1979; Chandler *et al.*, 1968). By lining a plastic bottle with cellulose triacetate the limonin content in orange juice was reduced to 25% within three days. Although the product has a much greater consumer appeal after elimination of limonin, the process has not been taken up commercially.

Amines, produced during protein degradation in fish muscle, contain malodorous compounds that it would be desirable to remove from fish packaging. The amines have strong basic properties and a Japanese patent based on the interactions between acidic compounds, e.g. citric acid, incorporated in polymers and the off-odours, claimed amine-removing capabilities (Hoshino and Osanai, 1986). Another approach has reached commercial use. A bag made from packaging material containing ferrous salt and citric acid is reported to oxidise amines as they are absorbed by the polymer (Rooney, 1995b). However, this gives cause for another concern, namely the uncertainty of the reaction products.

Aldehydes are formed in the autoxidation of fats and oils and they can render products organoleptically unacceptable at very low levels. This is one of the major problems concerning the shelf-life of fatty foods such as crisps, nuts, biscuits and cereal products. Dupont has developed a food packaging that, reportedly, eliminates aldehydes from the packaging head-space (Dupont Polymers, 1993). A coating formulation composed of zinc compounds and polycarboxylic acids has been patented and is claimed to have taint-removing effects when applied to a polymeric packaging material (Hoshino *et al.*, 1990). The deodorising sheets were shown to eliminate ammonia and hydrogen sulphide from the head-space.

Off-odours and off-flavours in foods can originate from the package itself. Additives and monomers might migrate into the product causing undesirable organoleptic changes. Styrene monomer has an extremely low detection limit which, consequently, could cause taint problems. It has been reported that inclusion of myrcene in styrene-containing packaging material can eliminate this effect by reacting with remaining styrene monomers (Tokas, 1979).

Immobilisation of enzymes for O_2 scavenging has already been described and the same theory can be applied for removal of other unwanted food components. The bitter component naringin, present in citrus juices, can be eliminated by binding in the enzyme naringinase in the packaging material. As a result the juice tastes much sweeter and is valued more highly by the consumer (Hotchkiss, 1997). Incorporation of flavours in packaging material might be used to minimise flavour scalping by removing the driving force from the equilibrium process. Release of flavours might also provide a means to mask off-odours stemming from the food or the packaging material. Further applications of flavour-enriched packaging materials include the possibility of improving the organoleptic quality of the product by emitting desirable flavours into the food and of encapsulating pleasant aromas, that are released upon opening, in the package.

The major problem concerning possible addition of components to foods from packaging material is the need to control the release in such a way that the transfer can match the requirements of the product. One way of providing such control is by employing polymeric films with different diffusion characteristics for the used compounds. Controlled release from packages is most applicable to liquid foodstuffs or solid foods with close-fitting packaging, when the process is solely dependent on diffusion.

5.15 Temperature-sensitive films

As film permeation increases when temperature increases, it is possible to compute the desired temperature-dependent permeations. For example, with fresh respiring products, it would be advantageous for the product shelf-life retention to have film permeability increased at least as much as the respiration rate increases in order to avoid anaerobic conditions. Unfortunately, the permeation rates of most packaging films are only modestly affected by temperature. There are at least two ways of manipulating polymer films, thereby enhancing their permeability and/or altering the ratio of O_2 and CO_2 permeabilities through a polymer film the better to match the respiration behaviour of the fresh respiring product.

Perforations of a few microns' diameter can provide favourable conditions for a variety of fresh produce (Anderson, 1989). The film can also be made from two dissimilar layers, or from two different thicknesses of the same material, both layers containing minute cuts. When the temperature rises or falls, the layers expand at different rates. As the temperature rises, the film at the cut edge retracts and curls upwards to give enlarged holes, thus significantly increasing the film permeability (Anon., 1994).

Another possibility of meeting the requirements of foods with high respiration rates is to incorporate some kind of filler in the polymer resin. The resulting film will contain micropores which facilitate the transport of gases through the packaging material. The gas permeabilities of the polymer can be controlled by filler content, particle size of the filler and degree of stretching of the film (Mizutani, 1993). Filling materials include $CaCO_3$ and SiO_2. Landec Co. has developed a film engineered with an adjustable 'temperature switch' point at which the film's permeation changes rapidly and dramatically. Landec's technology uses long-chain fatty alcohol-based polymeric chains. Under the predetermined temperature-switch point, these chains are in a crystalline state, providing a gas barrier. At the specified temperature, the side chains melt to a gas-permeable amorphous state.

5.16 Temperature control packaging

The sensorial appreciation of a food product is highly dependent on its serving temperature. If the product is to be consumed directly from the packaging it is desirable to use packages that help the product obtain its optimum temperature.

5.16.1 Self-heating

Beverage packages with self-heating functions have been described by Katsura (1989). There is also a growing demand in the outdoor market for food packaging that cooks or prepares the food via built-in heating mechanisms. The principle of heating is based on the theory that heat is generated when certain chemicals are mixed. Ready-meals can be heated, without the aid of a conventional or a microwave oven, by mixing iron, magnesium and salt water (Anon., 1997). The metals are supplied in a PET bag placed on a heat-resistant tray. Salt water, which is supplied in a separate bag, is added to the metal-containing bag and the actual food package is then placed on the tray. Within 15 minutes the food has reached a temperature of 60°C and is ready to eat. Other containers with heating functions are based on the reaction between lime and water.

5.16.2 Accessories in microwave heating

The majority of research on heat transfer in packaging has dealt with microwaveable packaging. The basic problems of microwave-heated foods have been the difficulty in producing browning and crisping of the surface and in obtaining an even temperature in composite foodstuffs containing components with varying heating rates. These problems can now be overcome with the aid of susceptors which convert the microwave energy to heat, generating temperatures that by far exceed those produced by microwave heating alone (Sacharow, 1995; Robertson, 1993). Typically, the heater elements consist of a polyester film onto which a thin layer of a metal film, normally aluminium, has been applied. The microwave energy is absorbed by the coating and re-emitted in the form of heat, leading to localised temperature effects that dry out the food surface, resulting in crisping and eventually browning of the product. Ideally, the temperature of a susceptor rises rapidly to its desired value and is thereafter held at a constant level. The earliest susceptors had some difficulties maintaining the temperature at the same level and concomitant problems with regard to the heating profile arose. Lately, however, considerable research has led to the development of susceptors that can meet this demand as well. Susceptors are used for a range of products including pizzas, pies, pastries, other crust items and for popping popcorn.

Field intensification devices are another recent development that focus microwave energy in order to increase local intensity (Sacharow, 1995). They act as focusing lenses and the degree of intensification depends on the geometrical shape of the system. The benefits include faster heating, the possibility of designing a device that facilitates heating of each part of a multi-component meal to its proper temperature and the ability to produce browning of foods with soft and sticky surfaces which is not feasible with contact susceptors. The negative aspects of field intensification devices are mainly the cost and the extra space required to accommodate the system.

5.16.3 Self-cooling

Although not as widespread as packaging with heating functions, a few examples of cooling cans exist. In Japan this type of package has been used for raw sake. By introducing selected substances into a void in a beverage can, a cooling effect can be obtained. When ammonium chloride and ammonium nitrate are mixed with water, the mixture absorbs heat and thereby lowers the temperature of the product. One drawback of this approach is the need for shaking the packaging before cooling proceeds which means that the technology cannot be used for carbonated beverages such as beer and soft drinks.

5.17 Intelligent packaging techniques

Currently, the intelligent packaging technique with the most commercial value is undoubtedly the external temperature indicators; however, internal O_2 indicators

Table 5.3 Examples of external or internal indicators to be used in smart packaging

Technique	Principle/ reagents	Application
Time–temperature indicators (external)	Mechanical Chemical Enzymatic	Foods stored under chilled and frozen conditions
O_2 indicators (internal)	Redox dyes pH dyes	Foods stored in packages with reduced O_2 concentration
Microbial growth indicators (internal)	pH dyes All dyes reacting with certain metabolites	Perishable foods

Table 5.4 Some manufacturers and trade names of commercial smart indicators

Manufacturer	Country	Trade name
Time–temperature indicators		
Lifelines Technology Inc.	USA	Fresh-Check
Trigon Smartpak Ltd	UK	Smartpak
3M Packaging Systems Division	USA	MonitorMark
Visual Indicator Tag Systems Ab	Sweden	Vitsab
Oxygen indicators		
Mitsubishi Gas Chemical Co., Ltd	Japan	Ageless-Eye
Toppan Printing Co., Ltd	Japan	–
Toagosei Chem. Industry Co., Ltd	Japan	–
Finetec Co., Ltd	Japan	–

Fig. 5.1 A colour changing 'quality label' attached to the package indicating package leakage, temperature abuse and/or product quality.

are also used. One of the indicators' main purposes is to indicate whether the quality of the packaged food has decreased, preferably before the product has deteriorated. Examples of different indicators and manufacturers and trade names of currently available colour indicators are listed in Tables 5.3 and 5.4 and in Fig. 5.1.

5.18 Time–temperature indicators

This appliance should indicate any temperature abuse and, ideally, estimate the remaining shelf-life of the foodstuff. There are two types of temperature indicators: those providing the entire temperature history to indicate the cumulative time–temperature exposure above a critical temperature (time–temperature indicators – TTI) and those providing information as to whether the indicator has been exposed above or below a critical temperature (temperature indicators – TI). Temperature indicators labelled on a package can inform of the heat load to which the package surface has been subjected in the distribution chain, usually expressed as a visible response in the form of mechanical deformation, colour change or colour movement. Hundreds of patents for TIs and for TTIs have been accepted, but they have found little use in the marketplace. However, once the concerns have been addressed there is an enormous potential for the use of TTIs for foods. Their use would advantageously lead to increased quality, safety and integrity of the packaged product.

A number of requirements for TTIs have to be met if they are to be commercially used with any success (Selman, 1996). They must be easy to use and to activate. It is essential that the indicators cannot be tampered with. They must be applied and activated at the time of packaging. Until now, indicators have usually been active all the time, and therefore they have had to be stored below the critical temperature or they have required physical activation. Now there are self-activating indicators available which can be stored at ambient temperature and require no physical activation (Plaut, 1995). TTIs should also accurately respond to the storage temperature and to quick temperature fluctuations. The response must be irreversible and correlate with the actual damage to the food. Further, the indicators need to have the ability to cumulate the time and temperature effects during the storage period. Also, the devices should give a clear read-out which cannot be misinterpreted by the consumer.

The principles that are utilised for TTIs include enzyme reaction, polymerisation, corrosion, melting point temperature and liquid crystals. Generally, the output is either in the form of a colour change, some type of movement or a combination of both effects. Each individual food product reacts differently to the storage conditions, and therefore there is a need for TTIs that respond correctly to the varying critical time–temperature combinations. The basic idea of the TTIs is, nevertheless, identical for all types, i.e. that reactions proceed at different rates as the temperature is altered. The indicator response is changed continuously during storage, or abruptly when a certain time–temperature effect has been exceeded.

The many benefits that can be recognised from the use of TTIs have induced great interest and generated considerable research in this area. The lack of commercial success, however, indicates that more work still needs to be done before the manufacturers and consumers alike can reap the rewards from this technology. Regardless of how well indicators may serve as an excellent ultimate quality assurance of food for both consumers and retailers, they do not necessarily reliably indicate the actual product temperature, but rather the package surface temperature. For example, in the carton folded plastic package, the product temperature can clearly be lower than at the carton surface where the indicator is labelled. This must be absolutely taken into account when determining the indicator type for a specific product. Contradictions between the TTI response and the stated 'use by' date on the packaging can also be foreseen which could lead to confusion among the consumers. Finally, the outputs from the TTIs have to be clear and unambiguous in order to prevent any misunderstanding about the state of the food.

5.19 Oxygen and carbon dioxide indicators

The integrity of a gas package is one of its most essential properties. If the package leaks, the optimised modified atmosphere surrounding the food will deteriorate as the head-space gas merges with the normal atmosphere surrounding the package and the beneficial effects of the modified atmosphere will be lost. The effect of the leakage on the sensory quality of modified atmosphere packaged foods is highly dependent on the type of the food. It is essential to detect a leakage in order to avoid the serious effects of contaminated food on the consumer's health, especially if the sensory quality does not correlate with the microbial quality. In packages with high CO_2 concentration a leakage usually means an increase in the O_2 concentration and a decrease in the CO_2 concentration inside the package. However, it has to be taken into account that the microbial growth likely to follow the leakage will again cause a decrease in the O_2 and an increase in the CO_2 concentration, thus complicating the situation (Ahvenainen *et al.*, 1995b; Smolander *et al.*, 1997).

As the attention paid to the high quality and safety of food is increasing, the need to control the quality of an entire production lot is evident (Gestrelius *et al.*, 1994). Package integrity testing taking place in the food processing plant is not sufficient for the detection of leakage created after the product has left the processing plant. Therefore a leak indicator attached to the package would advantageously ensure the package integrity throughout the whole distribution chain from the manufacturer to the consumer.

Most of the commercial internal O_2 indicators are either colour labels or tablets. O_2 indicators react with O_2 coming from the outside of the package through leakages (leakage indicator) or they are intended to be used with O_2 absorbers to verify that all O_2 has been absorbed from the package head-space (e.g. Yoshikawa *et al.*, 1979; Perlman and Linschitz, 1985; Krumhar and Karel, 1992). Perhaps the most famous O_2 indicator is Ageless-Eye (Mitsubishi Gas Chemical Co., Japan). The indicator can be included with the O_2 scavenger and is pink when there is no O_2 in the environment (0.1% or less) and blue if O_2 does exist (0.5 % or more) (Abe, 1990).

O_2 indicators can be used successfully to confirm that the product has been properly packaged. However, they have their deficiencies in distribution. Most of the O_2 indicators are too sensitive to O_2 from the gas packaging point of view and they are also reversible in colour change reaction. So, in the first case, the indicators may react to the residual O_2 entrapped in the gas packaging process or, in the second case, they may indicate that there is no O_2 in a package (i.e. a package should be intact) even though the product is spoiled by microbial growth which has consumed O_2 entrapped through leakage (Ahvenainen *et al.*, 1995b). The colour change of the indicators should therefore be irreversible.

The first patents describing O_2 indicators to be used in food packages were published in the late 1970s and since then the increase in the number of patented techniques has been between one and ten patents per year. The research and development work on package indicators is very often combined with that on O_2 absorbers and modified atmosphere packaging. Many Japanese companies dealing with O_2 absorbers frequently patent O_2 indicator technology. Research institutes such as CSIRO in Australia (Rooney, 1995a; Rooney, 1994) and VTT Biotechnology in Finland (Ahvenainen *et al.*, 1995b), both having extensive research programmes on modified atmosphere packaging, have also developed O_2 indicator systems. VTT's patented leakage indicator is specially designed for gas packages. The leak indicator does not react with the residual O_2 entrapped in the modified atmosphere package because the oxygen-absorbing component with adjusted capacity for the residual O_2 is included in the indicator and, moreover, the indicator is included in a film composition which protects against oxidation of the indicator during packaging. Package leakage is detected by a clear irreversible colour change (Mattila-Sandholm *et al.*, 1995).

It has been claimed that the colour change of the O_2 indicators used in modified atmosphere packages containing acidic CO_2 gas is not definite enough (Balderson and Whitwood, 1995; Mattila-Sandholm *et al.*, 1995). According to Davies and Garner (1996), the prolonged exposure of their indicator to CO_2 causes a reduced response time for O_2. On the other hand, the indicator of Mattila-Sandholm *et al.* (1995) could be successfully used with modified atmosphere packages containing CO_2. Yamamoto (1992) included a filler-layer and an overprint layer of neutral-alkaline resin in the O_2 indicator to neutralise the CO_2 of the package.

A typical visual O_2 indicator consists of a redox dye, a reducing compound and an alkaline compound. In addition to these main components, compounds such as a solvent (typically water and/or an alcohol) and bulking agent (e.g. zeolite, silica gel, cellulose materials, polymers) are added to the indicator (Yoshikawa *et al.*, 1979). The indicator can be formulated as a tablet (Goto, 1987; Yoshikawa *et al.*, 1979) or a printed layer (Davies and Garner, 1996; Krumhar and Karel, 1992; Yoshikawa *et al.*, 1979) or it can also be laminated in a polymer film (Nakamura *et al.*, 1987).

In addition to the purely visual O_2 indicators, a couple of other systems can also be considered as indicators even if external equipment is needed. These systems possess, however, an internal indicator attached to the package and they can be interpreted non-destructively. The concept of luminescent dyes quenched by O_2 as indicators for food packages was preliminarily introduced by Reininger *et al.* (1996). This optical method can be used for quantitative measurement of O_2 concentration in a non-destructive manner. The advantage in comparison with visual indicators is that misreadings, due to the sensitivity limitations of sight, do not occur. It is, however, clear that the need for optical instruments will increase the costs considerably in comparison to the visual indicator and the most likely application area for this kind of system would be the integrity testing done by the manufacturer at the processing plant. In another method Maurer (1986) suggests a system using the conversion of O_2 to ozone with the aid of UV radiation or an electric field. The presence of ozone is shown with a potassium iodide–starch indicator strip.

CO_2 indicators are mainly intended for gas packages with an elevated CO_2 concentration either to show too low or too high CO_2 concentration. These indicators provide fast colour reaction, demonstrating that the correct CO_2 level has been put in every package. For example, Balderson and Whitwood (1995) describe a reversible CO_2 indicator suitable for modified atmosphere packages. The indicator consists of, for example, five indicator strips. The strips contain CO_2 sensitive indicator material consisting, for example, of an indicator anion and a lipophilic organic quaternary cation (Mills and McMurray, 1991). The colour change of each strip has been designed to take place when the CO_2 concentration is below a certain limit. The concentration of CO_2 is indicated by a change of colour in one or more of the strips.

In principle, CO_2 indicators can be used in package leakage control. However, reversible colour change of the CO_2 indicators may in the worst case result in false information for the consumer: if the packed product is microbiologically spoiled, e.g. due to leakage, the CO_2 produced by microbial metabolism may still keep the head-space CO_2 at a level that indicates good product quality. The colour change of the CO_2 indicators used as leak indicators should therefore be irreversible (Hurme and Ahvenainen, 1996). No such indicators are on the market.

5.20 Freshness and doneness indicators

An ideal freshness indicator which gives information on the integrity, time–temperature history and microbial quality of the packed food would be a great help in assuring the quality and safety of the packaged product. Recently launched FreshTag® indicator labels (COX Recorders, USA) can be used to indicate the freshness of fresh fish. The indicator reacts with a colour change to volatile amines produced in fish during storage.

Even if the indication of microbial growth by CO_2 is difficult in modified atmosphere packages already containing a high concentration of CO_2, it is possible to use the increase in CO_2 concentration as a means of determining microbial contamination in other types of products. The use of the pH-dye bromothymol blue as an indicator for the formation of CO_2 by microbial growth has been suggested in many studies (Wolfbeis and List, 1995; Holte, 1993;

Mattila and Ahvenainen, 1989). The possibilities to use polyphenol oxidase lactase enzyme as an O_2 sensitive component has also been studied (Smolander *et al.*, 1998).

Doneness indicators are another variety of temperature control packaging that deserves mention. They detect and indicate the state of readiness of heated foods. One type of doneness indicators are the ready-buttons commonly placed in poultry. When a certain temperature has been reached the material expands and the button pops out, informing the consumer that the bird is fully cooked. Other packages have labels that change colour when the desired temperature has been reached (Sacharow, 1995). A limitation of doneness indicators is the difficulty in observing a colour change without opening the oven. Therefore an alternative indication system has been developed which, instead of a colour change, generates a whistling sound when the foodstuff has been cooked (Robertson, 1993).

5.21 Consumer and legislative issues

There is a lot of interest in active and intelligent techniques in the food industry. One of the main factors restricting their use is possibly the manufacturers' fear of a negative attitude of consumers towards the separate objects in the head-space of food packages. On the other hand, indicator structures are under development and labels and laminated structures would probably be more easily accepted by the consumer.

Before the food industry can make decisions on the best available active and/or smart packaging technique, research studies are needed both in domestic and foreign markets to evaluate consumer attitudes and behaviour towards these techniques. Even the naming of the different 'absorbers' or 'indicators' in food packages may not sound too familiar to consumers and therefore more common names may be needed. The food industry must be aware of all the consumers' prejudices and fears, as well as expectations and the marketing requirements in order to confirm the successful introduction of new products. One key element is that consumers understand the quality improvement and/or assurance function of active absorbers/emitters and/or colour indicators and also the importance of using them. Consumers must have confidence in the safety of the food they buy (Hurme and Ahvenainen, 1996).

Consumer research studies both in the UK on time–temperature indicator labels (Harris, 1991) and the VTT's study in Finland on O_2 absorber sachets (Mikkola *et al.*, 1997) have, on the other hand, indicated a generally positive attitude towards these techniques. Both the studies revealed that the introduction of these new packaging technologies was recognised to be done alongside a comprehensive publicity and education campaign in the media and information notices in retail outlets. In addition, improved user safety of O_2 absorbers was seen to increase consumer acceptance further. If the manufacturer does not put much trust in consumer education and information and fears an increase in the

Table 5.5 Problems and solutions encountered with introducing new products using active and/or intelligent packaging techniques (Hurme and Ahvenainen, 1996)

Problem/fear		Solution
• Consumer attitude	→	Consumer research: education and information
• Doubts about the performance	→	Storage tests before launching; consumer education and information
• Increased packaging costs	→	Use in selected, high quality products; marketing tool for increased quality and quality assurance
• False sense of security, ignorance of date markings	→	Consumer education and information
• Mishandling and abuse	→	Active compound incorporated into label or packaging film; consumer education and information
• False complaints and returns of packs with colour indicators	→	Colour automatically readable at the point of purchase
• Difficulty of checking every colour indicator at the point of purchase	→	Bar code labels: intended for quality assurance for retailers only

Table 5.6 Advantageous features of intelligent colour indicators (Hurme and Ahvenainen, 1996)

- Correlation with the product quality (not with the environment)
- Irreversible colour change
- Easy to understand
- Easy to read/clear colour changes
- Easy storage and use
- Reasonable price

quantity of mishandling, abuse and false complaints and returns of packs with a colour indicator, it is possible to use bar code label indicators. This enables an automatic checking of the product by a retailer when the product is sold. Table 5.5 outlines some potential problems and solutions that the food industry should take into account before deciding to use active and/or smart packaging techniques. Advantageous features of successful colour indicators are listed in Table 5.6.

Even though technological development of active and smart packaging techniques has been and still is very intensive, the methods have not been regulated in national and international legislation. So, no specific regulations exist on testing their suitability in the direct contact of foods. In practice, this means that, more or less, legislation concerning traditional packaging materials has been applied to these new packaging techniques. On the other hand, combined materials/multistructures into which many active and smart packaging devices can be classified seems to continue to present difficulties from a regulatory point of view, especially as they are not easily tested using protocols intended for pure plastics. Furthermore, it must be kept in mind that the present packaging legislation requires that the migration of substances from the food contact materials is low. For example, the overall migration limit from packaging materials is 60 mg/kg food in the EU. This is more or less contradictory to the idea of interactive packaging techniques (Ahvenainen and Hurme, 1997).

5.21.1 Legislative issues: active packaging techniques

When the acceptability of active packaging techniques for packaging of foods is considered from the migration point of view, techniques can be divided into three groups (Ahvenainen and Hurme, 1997):

- *Group A1*: the systems or techniques from which *no chemical substances are by purpose transferred into packed food*, but they absorb/adsorb compounds from the surroundings of food (oxygen absorbers) or compounds released from foods (carbon dioxide absorbers, ethylene absorbers, moisture absorbers). Mostly, in order to assure their optimal function, these systems are not intended to come into direct contact with packed food; rather they are placed in the free head-space of a package or attached to the inner surface of the package lid. However, in practice most of these systems are more or less in direct contact with packed food.
- *Group A2*: the systems that *emit quality preserving agents*, such as carbon dioxide and ethanol. They are not intended to come into direct contact with foods, either. However, in practice they have the same situation as the systems in the first group.
- *Group A3*: the systems from which *quality preserving agents* (e.g. non-volatile antimicrobial substances or antioxidants) *are by purpose transferred onto the surface of packed food* (interactive packaging materials) and systems that absorb/adsorb substances (e.g. water) from the surface of a food product. It is obvious that in order to have maximal use of these systems, most of these have been intended to come into direct contact with food.

At present, commercially significant systems mainly belong to the groups A1 and A2. If the systems from the groups A1 and A2 are not in direct contact with packed food, migration is not a problem. But, as was mentioned, in practice they often are. In these cases migration can present a problem, particularly with moist or fatty food, if the sachet material is very porous. Thus, the overall migration limit may be exceeded, which is not acceptable even when the substances transferred could be food ingredients according to the food law.

On the other hand, an emitting compound can be such that its use is restricted in food when added directly into food. A good example is ethanol. According to Hotchkiss (1995), there has been some reluctance in some parts of the world to allow the use of ethanol emitters in packed foods which will be consumed without further cooking or processing. Hotchkiss (1995) points out that for those active packaging systems that indirectly add components to foods, the governmental regulatory and health issues will be similar to those related to migration of residual monomers or other plastic components. Laboratory investigations will be required to determine the potential for migration and to quantify the amount of migration. If the amount of additive migration is considered to be of potential significance, toxicological testing may be required (Hotchkiss, 1995).

Migration aspects are the most unclear with the techniques in group A3, where their function is based on active migration of preserving agents and, in addition, other components from the system can also migrate into food. It is probable that the migration limit of 60 mg/kg is exceeded, and thus it can present a hindrance to use these kinds of active packaging materials. It must be noted that responsible manufacturers of interactive packaging materials are aware of food law and incorporate packaging materials only with preserving agents which are in conformity with the food law. However, the present European food packaging regulations do not deal with this particular matter: preserving agents are not on any food packaging material substance list, neither has an official opinion been given on the migration aspects. However, packaging materials in group A3 are very advantageous for consumers, because most probably less preservatives can be used compared to the traditional way where preservatives are added and mixed directly into the food product during the processing. If an interactive packaging material is used, preservatives function precisely there, where most of the spoilage reactions occur: on the surface of the food product.

Therefore, in 1999, a European study was started within the framework of the EU FAIR R&D programme. The study has initiated amendments to European legislation for food contact materials in order to establish and implement active and intelligent systems within the current relevant regulations for packaged food in Europe (Rijk *et al.*, 2002).

5.21.2 Legislative issues: intelligent packaging techniques

When the acceptability of smart packaging techniques in the packaging of foods is considered from the migration point of view, techniques can be divided into two groups (Ahvenainen and Hurme, 1997):

- *Group S1*: *external indicators* which are fixed on an outer surface of a package, such as time–temperature indicators.
- *Group S2*: *internal indicators* intended to be placed in the head-space of a package, such as oxygen and carbon dioxide indicators.

Migration does not pose any problem with the indicators in group S1, because they never come into direct contact with food. Indicators in group S2 are not intended to come into direct contact with packed food, either, but they are placed in the free head-space of a package or fixed to the inner surface of the lid. However, in practice most of these systems are more or less in direct contact with packed food. Then migration can be a problem, particularly with moist or fatty food, if the sachet material is very porous. The situation is very similar to the active package systems in group A2.

5.22 Future trends

Active packaging will probably increase in European countries in the near future, particularly in small and medium-sized enterprises (SMEs) and exporting food companies. This is due to both the consumers' preference for minimally processed and naturally preserved foods and the food industry's eagerness to invest in product quality and safety. The best way to avoid possible negative consumer attitudes towards new packaging techniques would be to incorporate an active absorber and/or emitter into the packaging film or into the label. Development in this area has been very active in recent years. It is possible that in the near future these 'invisible' active absorbers and/or emitters will be launched on the market on a larger scale. One benefit in using active labels rather than separate pouches is that labels can be automatically applied within packages using conventional labelling equipment rather than a specially designed dispensing machine. Oxygen absorbers may then become a potential competitive preservation technique for gas flushing and vacuum packaging.

Smart visual indicators for heat load, oxygen tension or metabolic volatiles might become more important in the near future, because of their capability to ensure product quality in a non-destructive real-time manner throughout the entire distribution system. However, development work is still needed in order to produce colour indicators with all the advantageous features listed in Table 5.6. An alternative solution for smart packaging would be a bar code in the packaging material with an indicator system that reacts with temperature, different volatiles and oxygen, thereby being a real quality indicator for a shopkeeper, but invisible to consumers.

The development and adoption of new active and smart packaging methods are a very multi-dimensional challenge. They require a broad knowledge about the demands of the product to be packed, chemical migration, possible taint problems, environmental aspects and consumer attitudes. More information is needed about the real benefits of various active and smart packaging in regard to food quality and safety. So far, research has mainly concentrated on development of various methods and their testing in a model system, but not so much on function in food preservation with real food products. This is true particularly with interactive packaging materials. Research and development on active and smart packaging areas is and will still be ongoing in several research institutes and companies for many years to come. To be fruitful in this development work, more discussion between researchers, industry, authorities and consumer organisations is needed in order to ensure that really safe and quality-maintaining active and smart packaging systems are developed.

5.23 References

ABE, Y. 1990. Active packaging: a Japanese perspective. *In:* Conference Proceedings: International Conference on Modifed Atmosphere

Packaging. 15–17 October 1991, Alveston Manor Hotel, Stratford-upon-Avon, England. Gloucestershire, The Campden Food and Drink Research Association, Part 1.

ABE, Y. 1994. Active packaging with oxygen absorbers. *In: Minimal Processing of Foods*, VTT Symposium 142. R. Ahvenainen, T. Mattila-Sandholm and T. Ohlsson (eds), pp. 209–23.

ABE, Y. and KONDOH, Y. 1989. Oxygen absorbers. *In: Controlled/Modified Atmosphere/Vacuum Packaging of Foods.* A.L. Brody (ed.). Trumbull, Food & Nutrition Press, pp. 149–74.

ABELES, F.B., MORGAN, P.W. and SALTVEIT, M.E. (1992) *Ethylene in Plant Biology.* Academic Press, San Diego.

ACKERMANN, P., JÄGERSTAD, M. and OHLSSON, T. (eds). 1995. *Foods and Food Packaging Materials: Chemical Interactions.* Special Publication No. 162. Cambridge, The Royal Society of Chemistry.

AHVENAINEN, R. and HURME, E. 1997. Active and smart packaging for meeting consumer demands for quality and safety. *Food Additives and Contaminants*, **14**, 6–7, pp. 753–63.

AHVENAINEN, R., MATTILA-SANDHOLM, T. and OHLSSON, T. (eds) 1994. *Minimal Processing of Foods.* VTT Symposium 142, Espoo.

AHVENAINEN, R., MATTILA-SANDHOLM, T. and OHLSSON, T. (eds) 1995a. *New Shelf-life Technologies and Safety Assessments.* VTT Symposium 148, Espoo.

AHVENAINEN, R., HURME, E., RANDELL, K. and EILAMO, M. 1995b. *The effect of leakage on the quality of gas-packed foodstuffs and the leak detection.* VTT Research Notes 1683, Espoo.

ANDERSEN, H.J. and RASMUSSEN, M.A. 1992. Interactive packaging as protection against photodegradation of the colour of pasteurized, sliced ham. *International Journal of Food Science and Technology*, **27**, pp. 1–8.

ANDERSON, H.S. 1989. Controlled atmosphere package. US Patent 4842875.

ANON. 1991. Anti-microbial additive based on hydroxyapatite and silver. *Food, Cosmetics & Drug Packaging*, **14**, 7, p. 8.

ANON. 1993. Anti-bacterial food packaging made with horseradish. *Food, Cosmetics & Drug Packaging*, **16**, 8, p. 5.

ANON. 1994. Permeable plastics film for respiring food produce. *Food, Cosmetics & Drug Packaging*, **17**, 1, p. 7.

ANON. 1997. *Packnordica*, **1**, 26.

APPENDINI, P. and HOTCHKISS, J.H. 1997. Immobilization of lysozyme on food contact polymers as potential antimicrobial films. *Pack. Tech. Sci.*, **10**, pp. 271–9.

BALDERSON, S.N. and WHITWOOD, R.J. 1995. Gas Indicator for a Package. US Patent 5 439648. Trigon Industries Ltd, New Zealand.

BARLAS, S. 1998. Packagers tell insects: 'Stop bugging us!'. *Packaging World*, **5**, 6, p. 31.

BARNETSON, A. 1995. Intelligent packaging of foods. *In*: Conference Proceedings, Modified Atmosphere Packaging (MAP) and Related

Technologies. Chipping Campden, Campden and Chorleywood Food Research Association.

BJERKENG, B., SIVERTSVIK, M., ROSNES, J.T. and BERGSLIEN, H. 1995. Reducing package deformation and increasing filling degree in packages of cod fillets in CO_2-enriched atmospheres by adding sodium carbonate and citric acid to an exudate absorber. *In: Foods and Food Packaging Materials: Chemical Interactions*. P. Ackermann, M. Jägerstad and T. Ohlsson (eds). Cambridge, The Royal Society of Chemistry, pp. 222–7.

BLACK, R.G., QUAIL, K.J., REYES, M., KUZYK, M. and RUDDICK, L. 1993. Shelf-life extension of pita bread by modified atmosphere packaging. *Food Australia*, **45**, pp. 387–91.

BRODY, A.L. 1990. Active packaging. *Food Engineering*, **62**, 4, pp. 87–92.

CCFRA (CAMPDEN AND CHORLEYWOOD FOOD RESEARCH ASSOCIATION). 1995. Conference Proceedings, Modified Atmosphere Packaging (MAP) and Related Technologies. Chipping Campden, Campden and Chorleywood Food Research Association, 6–7 September.

CHANDLER, B.V. and JOHNSON, R.L. 1979. New sorbent gel forms of cellulose esters for debittering citrus juices. *Journal of the Science of Food and Agriculture*, **30**, pp. 825–32.

CHANDLER, B.V., KEFFORD, J.F. and ZIEMELIS, G. 1968. Removal of limonin from bitter orange juice. *Journal of the Science of Food and Agriculture*, **19**, pp. 83–6.

DAVIDSON, P.M. and DOAN, C.H. 1993. Natamycin. *In*: Davidson, P.M. and Branen, A.L. (eds). *Antimicrobials in Foods*. New York, Marcel Dekker, pp. 395–407.

DAVIES, E.S. and GARNER, C.D. 1996. Oxygen Indicating Composition. UK Patent Application GB 2298273. The Victoria University of Manchester.

DAY, B.P.F. 1994. Modified atmosphere packaging and active packaging of fruits and vegetables. *In: Minimal Processing of Foods*, VTT Symposium 142. R. Ahvenainen, T. Mattila-Sandholm and T. Ohlsson (eds), Espoo, pp. 173–207.

DAY, B.P.F. 1998. Active packaging of foods. *CCFRA New Technologies Bulletin*, **17**, p. 23.

DE KRUIJF, N., VAN BEEST, M., RIJIK, R., SIPILÄINEN-MALM, T., PASEIRO, L., DE MEULEAAER, B. 2002. Active and intelligent packaging: application and regulatory aspects. *Food Additives and Contaminants*, **19**, Suppl., 144–62.

DUPONT POLYMERS. 1993. Bynel IPX101. Interactive packaging resin.

ETHICAP TECHNICAL INFORMATION. 1987. No-mix-type mould inhibitor Ethicap. Freund Industrial Co., Tokyo.

FARRELL, C.J. and TSAI, B.C. 1985. Oxygen scavenger. US Patent 4536409.

FOLWER, G.G. and GASSON, M.J. 1991. Antibiotics-nisin. *In:* Russel, N.J. and Gould, G.W. (eds). *Food Preservatives*. New York, AVI, pp. 135–54.

FU, B. and LABUZA, T.P. 1995. Potential use of time–temperature indicators as an indicator of temperature abuse of MAP products. *In: Principles of Modified-Atmosphere and Sous Vide Product Packaging*. J.M. Farber and

K.L. Dodds (eds). Lancaster, Technomic Publishing Company, pp. 385–423.

GESTRELIUS, H., MATTILA-SANDHOLM, T. and AHVENAINEN, R. 1994. Methods for non-invasive sterility control in aseptically packaged foods. *Trends in Food Science and Technology*, **5**, pp. 379–83.

GOTO, M. 1987. Oxygen Indicator. Japanese Patent JP 62-259059. Mitsubishi Gas Chemical Co., Tokyo, Japan.

HALEK, G.W. and GARG, A. 1989. Fungal inhibition by fungicide coupled to an ionomeric film. *J. Food Safety*, **9**, pp. 215–22.

HAN, J.H. and FLOROS, J.D. 1997. Casting antimicrobial packaging films and measuring their physical properties and antimicrobial activity. *J. Plastic Film Sheet*, **13**, pp. 287–98.

HARRIS, L. 1991. *Time–Temperature Indicators: Research into Consumer Attitudes and Behaviour.* MAFF Publications, National Consumers Council, London.

HAYNIE, S.L., CRUM, G.A. and DOELE, B.A. 1995. Antimicrobial activities of amphiphilic peptides covalently bonded to a water-insoluble resin. *Antimicrobial Agents and Chemotherapy*, **39**, pp. 301–7.

HOFELDT, R.H. and WHITE, S.A. 1989. European Patent Application 89301149.4.

HOLLAND, R.V. 1992. Australian Patent Application PJ6333.

HOLTE, B. 1993. An Apparatus for Indicating the Presence of CO_2 and a Method of Measuring and Indicating Bacterial Activity within a Container or Bag. International Patent Application WO 93/15402.

HOOJJAT, P., HARTE, B.R., HERNANZED, R.J., GIACIN, J.R. and MILTZ, J. 1987. Mass transfer of BHT from high density polyethylene film and its influence on product stability. *Journal of Packaging Technology*, **1**, 3, pp. 78–81.

HOOVER, D.G. and STEENSON, L.R. 1993. *Bacteriocins of Lactic Acid Bacteria.* San Diego, Academic Press.

HOSHINO, A. and Osanai, T. 1986. Japanese Patent 86209612.

HOSHINO, A., SAJI, M. and FUJII, S. 1990. Deodorizing sheet with a deodorizing coating formulation. US Patent 4931360.

HOTCHKISS, J.H. 1995. Safety considerations in active packaging. *In: Active Food Packaging.* Rooney, M.L. (ed.). Glasgow, Blackie Academic & Professional, pp. 238–55.

HOTCHKISS, J.H. 1997. Presentation at Tetra Laval, Lund, Sweden.

HUDSON, A. 1991. Modified Atmosphere Storage of Tomatoes. PhD thesis, University of New South Wales, Sydney, Australia.

HURME, E. and AHVENAINEN, R. 1996. Active and smart packaging of ready-made foods. *In: Minimal Processing and Ready Made Foods.* T. Ohlsson, R. Ahvenainen and T. Mattila-Sandholm (eds). Göteborg, SIK, pp. 169–82.

HURME, E. and AHVENAINEN, R. 1997. Applicability of oxygen absorbers in food packages (in Finnish). PTR Report no. 44.

HURME, E., RANDELL, K. and AHVENAINEN, R. 1995. The effect of leakage and oxygen absorbers on the quality of gas-packed foodstuffs and the detection of leakage. *In:* the IAPRI 9th World Conference on Packaging. Brussels,

Belgian Packaging Institute, pp. 45–52.

HURME, E., VAARI, A. and AHVENAINEN, R. 1994. Active and smart packaging of foods. *In: Minimal Processing of Foods*, VTT Symposium 142. R. Ahvenainen, T. Mattila-Sandholm and T. Ohlsson (eds), Espoo, pp. 149–72.

IDOL, R.C. 1991. A critical review of in-package oxygen scavengers and moisture absorbers. *In:* CAP 91, 6th International Conference on Controlled/Modified Atmosphere/Vacuum Packaging. San Diego, Schotland Business Research Inc., pp. 181–90.

ISHITANI, T. 1995. Active packaging for food quality preservation in Japan. *In*: Ackermann, P., Jägerstad, M. and Ohlsson, T. (eds). *Foods and Food Packaging Materials: Chemical Interactions.* Cambridge, Royal Society of Chemistry, pp. 177–88.

KATSURA, T. 1989. Present state and future trend of functional packaging materials attracting considerable attention. *Packaging Japan*, 21–26 September.

KRUMHAR, K.C. and KAREL, M. 1992. Visual indicator system. US Patent 5,096,813.

KUNISAKI, S., NODA, K., SAEKI, T. and AMACHI, T. 1993. Development and application of aseptic new materials. *In:* Proceedings of 6th International Congress on Engineering and Food, ICEF 6, 23–27 May, Makuhari Messe, Chiba, Japan. Elsevier Science, p. 141.

LABUZA, T.P. 1989. Active food packaging technologies. *In: Engineering and Food, Vol. 2, Preservation Processes and Related Techniques.* W.E.L. Spiess and H. Schubert (eds). London, Elsevier Applied Science, pp. 304–11.

LABUZA, T.P. and BREENE, W.M. 1989. Applications of 'active packaging' for improvement of shelf-life and nutritional quality of fresh and extended shelf-life foods. *Journal of Food Processing & Preservation*, **13**, pp. 1–69.

LUND, D.B. 1989. Food processing from art to engineering. *Food Technology*, **43**, pp. 242–7.

MATSUI, M. 1989. Film for keeping freshness of vegetables and fruit. US Patent 4847145.

MATTILA, T. and AHVENAINEN, R. 1989. Preincubation time and the use of oxygen indicators in determining the microbiological quality of aseptically packed pea and tomato soup. *International Journal of Food Microbiology*, **9**, pp. 205–14.

MATTILA, T. and AUVINEN, M. 1990a. Headspace indicators monitoring the growth of B. cereus and Cl. perfringens in aseptically packed meat soup (part I). *Lebensmittel-Wissenschaft und -Technologie*, **23**, pp. 7–13.

MATTILA, T. and AUVINEN, M. 1990b. Indication of the growth of Cl. perfringens in aseptically packed sausage and meat ball gravy by headspace indicators (part II). *Lebensmittel-Wissenschaft und -Technologie*, **23**, pp. 14–19.

MATTILA, T., TAWAST, J. and AHVENAINEN, R. 1990. New possibilities for quality control of aseptic packages: microbiological spoilage and seal defect detection using headspace indicators. *Lebensmittel-Wissenschaft und -Technologie*, **23**, pp. 246–51.

MATTILA-SANDHOLM, T., AHVENAINEN, R., HURME, E. and JÄRVI-KÄÄRIÄINEN. 1995. Leakage indicator. Pat. FI-94802. (WO 9324820). VTT Biotechnology and Food Research.

MAURER, H. 1986. Verfahren zum Nachweis von Sauerstoff in einer Verpackung. Swiss Patent CH 654 109. Tecan AB, Hombrechtikon, Switzerland.

MIKKOLA, V., LÄHTEENMÄKI, L., HURME, E., HEINIÖ, R-L., JÄRVI-KÄÄRIÄINEN, T. and AHVENAINEN, R. 1997. Consumer attitudes towards oxygen absorbers in food packages. VTT Research Notes 1858, Espoo.

MILLS, A. and MCMURRAY, H.N. 1991. Carbon Dioxide Monitor. International Patent Application WO 91/05252. Abbey Biosystems Ltd, Dyfed, UK.

MILTZ, J., PASSY, N. and MANNHEIM, C.H. 1995. Trends and applications of active packaging systems. *In: Foods and Food Packaging Materials: Chemical Interactions*. P. Ackermann, M. Jägerstad and T. Ohlsson (eds). Cambridge, The Royal Society of Chemistry, pp. 201–10.

MING, X., WEBER, G.H., AYRES, J.W. and SANDINE, W.E. 1997. Bacteriocins applied to food packaging materials to inhibit Listeria monocytogenes on meats. *J. Food Sci.*, **62**, pp. 413–15.

MIZUTANI, Y. 1993. *Ind. Eng. Chem. Res.* **32**, pp. 221–7.

NAKAMURA, H., NAKAZAWA, N. and KAWAMURA, Y. 1987. Food Oxidation Indicating Material. Japanese Patent JP 62-183834. Toppan Printing Co., Tokyo, Japan.

NIELSEN, T. 1998. Active packaging: a literature review. SIK-Rapport No. 631.

OHLSSON, T., AHVENAINEN, R. and MATTILA-SANDHOLM, T. (eds) 1996. *Minimal Processing and Ready Made Foods*. Göteborg, SIK.

OZDEMIR, M. and SADIKOGLU, H. 1998. A new and emerging technology: laser-induced surface modification of polymers. *Trends Food Sci. Tech.*, **9**, pp. 159–67.

PADGETT, T., HAN, I.Y. and DAWSON, P.L. 1998. Incorporation of food-grade antimicrobial compounds into biodegradable packaging films. *J. Food Prot.*, **61**, 1330–5.

PAFUMI, J. and DURHAM, R. 1987. Cake shelf life extension. *Food Technology in Australia*, **39**, pp. 286–7.

PAIK, J.S., DHANASEKHARAN, M. and KELLEY, M.J. 1998. Antimicrobial activity of UV-irradiated nylon film for packaging applications. *Pack. Techn. Sci.*, **11**, pp. 179–87.

PATTERSON, B.D. and JOYCE, D.C. 1993. International Patent Application PCT/AU93/00398.

PERLMAN, D. and LINSCHITZ, H. 1985. Oxygen indicator for packaging. US Patent 4526752.

PLAUT, H. 1995. Brain boxed of simply packed? *Food Processing*, **7**, pp. 23–5.

RANDELL, K., HURME, E., AHVENAINEN, R. and LATVA-KALA, K. 1995. Effect of oxygen absorption and package leaking on the quality of sliced ham. *In: Foods and Packaging Materials: Chemical Interactions*. P. Ackermann,

M. Jägerstad and T. Ohlsson (eds). Cambridge, The Royal Society of Chemistry, pp. 211–16.

REININGER, F., KOLLE, C., TRETTNAK, W. and GRUBER, W. 1996. Quality control of gas-packed food by an optical oxygen sensor. *In: Food Packaging: Ensuring the Safety and Quality of Foods*, 11–13 September, Budapest, ILSI, Hungary.

RIJK, R., VAN BEEST, M., DE KRUIJF, N., BOUMA, K., MARTIN, C., DE MEULENAER, B. and SIPILÄINEN-MALM, T. 2002. Active and intelligent packaging systems and the legislative aspects. *Food Packaging Bulletin*, **10**, 9 & 10, pp. 2–10.

ROBERTSON, G.L. 1993. Packaging of microwavable foods. *In: Food Packaging Principles and Practice*. Robertson, G.L. (ed.). New York, Marcel Dekker, pp. 409–30.

ROONEY, M.L. 1994. Oxygen Scavengers Independent of Transition Metal Catalysts. International Patent Application WO 94/12590. CSIRO.

ROONEY, M.L. (ed). 1995a. *Active Food Packaging*. Glasgow, Blackie Academic & Professional.

ROONEY, M.L. 1995b. Overview of active food packaging. *In*: *Active Food Packaging*. Rooney, M.L. (ed.). Glasgow, Blackie Academic & Professional, pp. 1–38.

ROONEY, M.L. 1995c. Development of active and intelligent packaging systems. *In: New Shelf-life Technologies and Safety Assessments*. VTT Symposium 148. R. Ahvenainen, T. Mattila-Sandholm and T. Ohlsson (eds), Espoo, pp. 75–83.

ROONEY, M.L. and HOLLAND, R.V. 1979. Singlet oxygen: an intermediate in the inhibition of oxygen permeation through polymer films. *Chem. Ind.*, pp. 900–1.

ROONEY, M.L., HOLLAND, R.V. and SHORTER, A.J. 1981. Removal of headspace oxygen by a singlet oxygen reaction in a polymer film. *Journal of the Science of Food and Agriculture*, **32**, pp. 265–72.

SACHAROW, S. 1995. Commercial applications in North America. *In: Active Food Packaging*. Rooney, M.L. (ed.). Glasgow, Blackie Academic & Professional, pp. 203–14.

SALMINEN, A., HURME, E., LATVA-KALA, K., LINKO, P. and AHVENAINEN, R. 1995. Ethanol emitters and oxygen absorbers in the packaging of sliced bread. *In:* the IAPRI 9th World Conference on Packaging. Brussels, Belgian Packaging Institute, pp. 158–64.

SALMINEN, A., LATVA-KALA, K., RANDELL, K., HURME, E., LINKO, P. and AHVENAINEN, R. 1996. The effect of ethanol and oxygen absorption on the shelf-life of packed sliced rye bread. *Packaging Technology and Science*, **9**, 1, pp. 29–42.

SAWAI, J., IGARASHI, H., HASIMOTO, A. and SHIMIZU, M. 1993. Growth-inhibitory effect of ceramics powder slurry on bacteria. 6th International Congress on Engineering and Food, ICEF 6, 23–27 May, Makuhari Messe, Chiba, Japan. Elsevier Science, p. 140.

SCOTT, D. 1965. Oxidoreductase. *In*: *Enzymes in Food Processing*. New York:

Academic Press.

SEILER, D.A.L. and RUSSELL, N.J. 1991. Ethanol as food preservative. *In: Food Preservatives.* N.J. Russell and G.W. Gould (eds). New York, AVI, pp. 153–71.

SELMAN, J.D. 1996. Time–temperature integrators: a survey. *In: Minimal Processing and Ready Made Foods.* T. Ohlsson, R. Ahvenainen and T. Mattila-Sandholm (eds). Göteborg, SIK, pp. 183–91.

SHINAGAWA NENRYO KK and SHINANEN NEW CERAMIC KK. 1992. Freshness maintaining body – has free-oxygen absorber on outside and oxygen penetrating resin material containing silver ion substituted alumino-silicate formed on inside. Japanese Patent JP 04072168-A.

SHINANEN NEW CERAMIC KK and CHUGOKU PEARL HAWBAI KK. 1989. Freshness maintaining packaging material for food – comprises macromolecular film containing silver ion containing zeolite with carbon dioxide absorbed zeolite. Japanese Patent JP 03111264.

SHIRAZI, A. and CAMERON, A.C. 1992. Controlling relative humidity in modified atmosphere packages of tomato fruit. *HortScience*, **27**, 4, pp. 336–9.

SHOWA DENKO KK. 1990. Japanese Patent 4072168.

SMITH, D.P. and ROLLIN, N.J. 1954. Sorbic acid as a fungistatic agent for foods (Part VII). Effectiveness of sorbic acid in protecting cheese. *Food Research*, **19**, pp. 50–65.

SMITH, J.P., HOSHINO, J. and ABE, Y. 1995. Interactive packaging involving sachet technology. *In: Active Food Packaging.* Rooney, M.L. (ed.). Glasgow, Blackie Academic & Professional, pp. 143–73.

SMITH, J.P., OORAIKUL, B., KOERSEN, W.J., VAN DE VOORT, F.R., JACKSON, E.D. and LAWRENCE, R.A. 1987. Shelf life extension of bakery products using ethanol vapor. *Food Microbiology*, **4**, pp. 329–37.

SMITH, J.P., RAMASWAMY, H.S. and SIMPSON, B.K. 1990. Developments in food packaging technology. Part II: Storage aspects. *Trends in Food Science & Technology*, **1**, pp. 111–18.

SMOLANDER, M., HURME, E. and AHVENAINEN, R. 1997. Leak indicators for modified-atmosphere packages. *Trends in Food Science and Technology*, **8**, 4, pp. 101–6.

SMOLANDER, M., HURME, E., SIIKA-AHVO, M. and AHVENAINEN, R. 1998. Biological freshness indicator. *In: Biotechnology in the Food Chain: New Tools and Applications for Future Foods.* Poutanen, K. (ed.). VTT Symposium 177, Espoo, p. 256.

SOMEYA, N. 1992. Packaging sheet for perishable goods. US Patent 5084337.

SØRHEIM, S., LEA, P., ARNESEN, A.K. and HAUGDAL, J. 1995. Colour of beef loins in carbon dioxide with oxygen scavengers. *In: Foods and Food Packaging Materials: Chemical Interactions.* P. Ackermann, M. Jägerstad and T. Ohlsson (eds). Cambridge, The Royal Society of Chemistry, pp. 217–21.

TAOUKIS, P.S., FU, B. and LABUZA, T.P. 1991. Time–temperature indicators. *Food Technology*, **45**, 10, pp. 70–82.

TETRA PAK INTERNATIONAL AB. 1977. German Patent DE2817854A.

TEUMAC, F.N. 1995. The history of oxygen scavenger bottle closures. *In: Active Food Packaging.* Rooney, M.L. (ed.). Glasgow, Blackie Academic & Professional, pp. 193–202.

TOKAS, E.F. 1979. Molding compositions and process for preparing same. US Patent 4180486.

TOYO INK MFG CO. 1992. Freshness-maintaining packaging material – has powders with ethylene absorbing and decomposition properties, antibacterial property powder, and water absorbing resin powder. Japanese Patent JP 04087965-A.

VOJDANI, F. and TORRES, J.A. 1990. Potassium sorbate permeability of methyl cellulose and hydroxylpropyl methyl cellulose coatings: effect of fatty acids. *J. Food Sci.*, **55**, pp. 841–6.

WELLINGHOFF, S.T. and KAMPA, J.J. 1997. Sustained-release biocidal composition, based on chlorine dioxide formation. US Patent 5650446.

WENG, Y.-M. and HOTCHKISS, J.H. 1992. Inhibition of surface molds on cheese by polyethylene film containing the antimycotic imazalil. *Journal of Food Protection*, **55**, 5, pp. 367–9.

WENG, Y.-M. and HOTCHKISS, J.H. 1993. Anhydrides and antimycotic agents added to polyethylene films for food packaging. *Pack. Tech. Sci.*, **6**, pp. 123–8.

WILHOIT, D.L. 1996. Film and method for surface treatment of foodstuffs with antimicrobial compositions. US Patent 5573797.

WOLFBEIS, O.S. and LIST, H. 1995. Method for quality control of packaged organic substances and packaging material for use with this method. US Patent 5407829. AVL Medical Instruments, AG, Switzerland.

YAMAMOTO, H. 1992. Oxygen indicator label for food packaging etc. – comprises filler layer, oxygen indicator layer and over-print layer laminated on surface of sheet-shaped substrate. Japanese Patent JP 04151554. Dainippon Printing Co., Tokyo, Japan.

YANG, T.C.S. 1995. The use of films as suitable packaging materials for minimally processed foods: a review. *In: Food Preservation by Moisture Control: Fundamentals and Applications.* G.V. Barbosa-Cánovas and J. Welti-Chanes (eds). Lancaster, Technomic Publishing Co., pp. 831–48.

YOSHIKAWA, Y., NAWATA, T., GOTO, M. and FUJII, Y. 1979. Oxygen indicator. US Patent 4169811. Mitsubishi Gas Chemical Co., Tokyo, Japan.

ZAGORY, D. 1995. Ethylene-removing packaging. *In: Active Food Packaging.* Rooney, M.L. (ed.). Glasgow, Blackie Academic & Professional, pp. 38–54.

6

Natural food preservatives

A.S. Meyer, K.I. Suhr, P. Nielsen, Technical University of Denmark, Lyngby and F. Holm, FoodGroup Denmark, Aarhus

6.1 Introduction

This chapter focuses on explaining the antimicrobial and antioxidant action of natural compounds to prevent bacterial and fungal spoilage, growth of microbial pathogens and oxidative quality deterioration of processed foods. There is currently a large interest in substituting synthetic food preservatives and synthetic antioxidants for substances that can be marketed as natural. Obviously, the use of new ingredients in foods and beverages is limited by safety and toxicological requirements, and therefore strongly regulated by legislation. In principle, food legislation does not differentiate between natural and synthetic compounds. Additives are only permitted when there is a technological need, they represent no hazard to public health, and they are not misleading for the consumer (Council Directive 89/107/EEC). Nevertheless, especially in Europe, the food industry aims to exploit natural ingredients and additives that can protect against food quality deterioration, but do not have to be labelled as (E-number) additives. This is particularly the case with minimally processed products aimed at health-conscious, critical consumers. Several different natural compounds, a number of extracts and semi-purified mixed fractions have thus been investigated individually and in combination for their antimicrobial, antifungal and antioxidant activity. To understand better how substances of different origin and structure may be workable in protecting processed foods and beverages against different types of quality deterioration, it is important to define what natural preservatives and antioxidants are. It is also important to know their possible mechanisms of protective action. After a more general overview of natural antimicrobials (section 6.2) that are of potential interest for minimal processing focus, the discussion will be oriented towards the different types of plant-derived substances. Application methods, antimicrobial spectra and mode of action will be reviewed. Only a few commercial applications of natural antimicrobials are presently known but there are several examples of new challenging applications. In section 6.3, particular attention is given to the activity mechanisms of natural antioxidants to prevent lipid peroxidation and to possible structure–antioxidant activity relationships. Enzyme-catalysed antioxidant mechanisms are also discussed. A very limited number of the recently discovered natural antimicrobials have been commercialised, while the natural antioxidants already have a strong market position. An overview of the present types of commercially available natural antioxidant preparations, including their sources and suppliers, will be given in section 6.10. This is followed by a brief summary of recent studies of natural substances having potential dual functionality as both antimicrobial food preservatives and antioxidants. Finally, we will give our view on the future needs and challenges in the practical exploitation of natural antioxidants and preservatives in minimally processed foods.

6.2 Antimicrobial agents

In a broad definition, antimicrobial agents are all components that hinder growth of microorganisms. A smaller group of these compounds are what we call food preservatives. According to the definition used by the Commission of the European Communities, 'preservatives are substances which prolong the shelf-life of foodstuffs by protecting them against deterioration caused by micro-organisms' (Directive 95/2/EC). Substances to be used as food preservatives in the EEC have to be applied in concordance with this directive. Somewhat similar rules apply in the USA; here the FDA defines preservatives as 'any chemical that when added to food tends to prevent or retard deterioration thereof, but it does not include common salt, sugars, vinegar, spices or oils extracted from spices, substances added to food by direct exposure thereof to wood smoke, or chemicals applied for their insecticidal or herbicidal properties' (FDA, Code of Federal Regulations: 21 CFR 172, 2000).

This difference in definition will have an impact on the way natural antimicrobial compounds can be used to preserve our food. In the EEC, for example, a substance has to be on the positive list to be used as a preservative. If not, it may in some cases be added as a food ingredient. This could either be as colours or sweeteners, which are regulated separately in specific directives (94/36/EU and 94/35/EU) or as flavourings that are regarded as food additives but regulated separately. In Commission Decision 1999/217/EC there is a list of flavouring substances, which currently are legally accepted in some member states and should benefit from free circulation. This list is currently evaluated according to EU Parliament and Council Regulation EC/2232/96. These issues are discussed in greater detail in a report by the Nordic Council of Ministers (Fabech *et al.*, 2000).

Table 6.1 Enzymes, peptides and other natural antimicrobials with known or suggested applications

Enzymes, proteins, peptides	Origin	Effective against	MIC/allowed dose	Use in (products)	Reference
Lytic enzymes: β-glucanase Chitinase	Plants, fungi, bacteria	Fungi	App. 50 μg/ml	–	12, 14
Chitosan	Shellfish waste	Fungi and yeasts	0.1–5 g/L	Various Asian foods (noodles, soy sauce, cabbage, sardines)	13
Small cationic proteins and peptides/defensines, e.g. magainins (from frog); salmine, clupeine (from fish)	Mammals, insects, fish, amphibians, fungi, plants	Fungi, bacteria, protozoa	10–100 μg/ml (magainins)	Potential pharmaceuticals	10, 14
Lactoferrin, lactoferricin* Other iron-binding glycoproteins: Ovotransferrin Serum transferrins	Milk, animal tissue Eggs Blood	Bacteria Fungi*, bacteria, parasites	*1% for dip *0.1–500 mg/ml	Raw milk, dip for vegetables	3, 11
Lactoperoxidase Other H_2O_2-generating systems: Glucose oxidase Xanthine oxidase	Milk, Salvia	Bacteria, Some yeasts		Milk, milk products, whole egg, fish	4, 6, 8
Lysozyme	Milk and egg	Bacteria	q.s	Ripened cheese, E1105	2, 6
Bacteriocins					
Nisin Other bacteriocins: pediocin, bavaricin, sakacin, etc.	*Lactococcus lactis* ssp. *Lactis*	Gram +, bacterial spores	3–12.5	Dairy products, E234	6

Antibiotics					
Natamycin (pimaricin)	*Streptomyces* spp.	Fungi	1 mg/dm^2 surface	Cheese, dried cured sausage, E235	6
Penicillin	*Penicillium nalgiovence*	Bacteria		Fermentation of sausages	1
Others: bacitracin, streptomycin, tylan, etc.					
Other fermentation products					
Hydrogen peroxide (H_2O_2)	Bacteria, H_2O_2-generating oxidases	Bacteria, yeast, fungi	3% solution for surfaces (in milk: 200–800)	Surface sterilisation, (formerly used in milk), naturally in honey	9
Diacetyl	Hetero fermentative LAB	Bacteria, fungi	300 μg/ml	Butter-flavour (sensoric threshold value: 100 μg/ml)	7
Ethanol	Yeast	Bacteria, fungi	0.5–10%	Drinks, baked goods	9
Reuterin	*Lactobacillus reuteri*	Bacteria, fungi	–	Dairy products, meat, etc.	5

Notes: 1. Andersen and Frisvad, 1994. 2. Beuchat and Golden, 1989. 3. Branen and Davidson, 2000. 4. Conner, 1993. 5. Dillon and Cook, 1994. 6. Directive 95/2/EC. 7. Jay, 1982a, 1982b. 8. Kennedy *et al.*, 2000. 9. Lück and Jager, 1995. 10. Matsuzaki, 1998. 11. Naidu, 2000. 12. Roller, 2000. 13. Roller and Covill, 1999. 14. Selitrennikoff, 2001. * Lactoferricin does not bind iron.

The focus in this section is natural antimicrobials, with special emphasis on substances to be used in minimal processing. Natural antimicrobials can be defined as substances produced by living organisms in their fight with other organisms for space and their competition for nutrients. The main sources of these compounds are plants (e.g. plant secondary metabolites in essential oils and phytoalexins), microorganisms (e.g. bacteriocins and organic acids) and animals (e.g. lysozym from eggs, transferrins from milk). Across the various sources the same types of active compounds can be encountered, e.g. enzymes, peptides and organic acids. An overview of the most important naturally occurring antimicrobials is shown in Table 6.1. It also contains a reference to the sources of the compounds and the concentration range needed for inhibition. The table is not fully comprehensive, as a multitude of specific compounds exist, and the discovery of yet many more is awaited. Several of these compounds are still far from food applications.

At present the most industrially exploited natural antimicrobials are peptides like natamycin and nisin or enzymes such as lysozym (Gould, 1996). They are mainly applied in the dairy industry to prevent mould growth on cheese surfaces (natamycin), germination of *Bacillus* spores in processed cheese (nisin) and *Clostridium tyrobutyricum* in hard cheeses (lysozyme), although they are applicable in a number of non-dairy products as well, such as cured meat and sausages. Other naturally occurring and industrially important compounds like benzoic, propionic and acetic acids are most often chemically synthesised (nature identical) and are not considered as natural antimicrobials, but some organic acids may be extracted and concentrated from natural sources, e.g. acetic, lactic, citric and sorbic acids. Compounds from plants have not had prevalence in commercial use as antimicrobials yet, but the enormous efforts put into exploration of plant compounds, not only in food science but also in pharmacognosy (Nychas, 1995) and in agriculture for crop protection and allelopathic farming (Eckholdt, 2001), reveal a great potential and challenge. Thus, in this chapter emphasis is put on plant-derived compounds. Other relevant compounds, such as animal proteins and peptides, will be mentioned briefly in section 6.3. Biopreservatives from microorganisms (e.g. bacteriocins) are not treated in this chapter but readers are referred to the extensive literature on the subject (Cleveland *et al.* (2001) or Naidu (2000) which contains several chapters on different bacteriocins).

6.3 Antimicrobial proteins and peptides

Lysozyme (E1105) is the only antimicrobial enzyme that has achieved significant commercial application. Lysozyme is present in egg whites, and hydrolyses β-1,4-glucosidal linkages in the peptidoglycan layer of the bacterial cell walls (Beuchat and Golden, 1989). It is thus most effective against Gram-positive bacteria, and is particularly useful against thermophilic spore formers. It can be used freely in ripened cheeses, and is in fact the only enzyme permitted as a food preservative, and thus has an E number in the EU (Table 6.1). Other groups of antimicrobial enzymes are the β-glucanases that hydrolyse the β-glucan structure of fungal walls, and chitinases that cleave cell wall chitin. They have been isolated from tobacco, peas, grains, fruits and other plant sources upon fungal attack (Selitrennikoff, 2001). Other enzymes of interest are the oxidoreductases, e.g. lactoperoxidase, glucose oxidase and catalases, which catalyse reactions producing cytotoxic compounds such as H_2O_2, $OSCN^-$ and OCl^-. Glucose oxidase and catalase may also protect against oxidative food deterioration and are therefore discussed later in section 6.11 of this chapter in relation to enzymatic antioxidant mechanisms.

Peptides with a great range of amino acid sequences, different sizes and structures, exert antimicrobial activity (Barra and Simmaco, 1995). Besides originating from 'traditional' sources, peptides from insects, fish and amphibians have in recent years attracted much attention. Membrane disruption is thought to be central for activity of these peptides (Brul and Coote, 1999).

Peptides with iron-binding properties are found in egg white (ovotransferrin or conalbumin) and in blood (serum transferrin) where they act as the main carrier of iron (Gould, 1996). The antimicrobial effect apparently results from the binding of essential iron needed for growth of microorganisms (Conner, 1993), e.g. many food pathogens can be inhibited such as *Staphylococcus, Clostridium, Listeria, Salmonella, Pseudomonas, Yersinia, Vibrio* and *Aeromonas* (Naidu, 2000). But in real food systems this mechanism has a limited effect. Peptide digestion of bovine lactoferrin has resulted in a very potent peptide (25 amino acids) named lactoferricin. The mode of action is unclear, but it has some similarity to protamines and may act on the cell surface to increase membrane permeability. The much higher activity of lactoferrin hydrolysate compared to the native lactoferrin indicates that ion-chelation is a weak inhibitory mechanism (Branen and Davidson, 2000). The only antimicrobial peptides currently used as food preservatives are the products of lactic acid bacteria: nisin and related compounds like pediocin.

6.4 Plant-derived antimicrobial agents

Plants protect themselves against microorganisms and other predators by synthesising a wide range of compounds. These compounds are also termed secondary metabolites, as they generally are not essential for the basic metabolic processes – the primary metabolism. Secondary metabolites represent a diverse array of chemical compounds mostly derived from the isoprenoid, phenylpropanoid, alkaloid or fatty acid/polyketide pathways (Dixon, 2001).

Antimicrobial compounds from plants can be divided into three groups:

1. Preinfectional agents at the plant surface (constitutive).
2. Agents bound in vacuoles and associated with hydrolytic enzyme activation systems (constitutive).
3. Phytoalexins, which are compounds produced in response to invasion (inducible).

Compounds from the first group are normally always present in the plant as a first line of defence of the plant, in the leaf glands, for example (Harborne, 1988). The second group of compounds are highly reactive but require tissue rupture or cell collapse in order to be activated. They are stored as inactive precursors and their corresponding enzymes appear in other cell compartments. The third group of compounds, phytoalexins, are normally not present in healthy unaffected plants. They are only produced as a response to stress (invasion by microorganism or environmental factors like drought).

It is the compounds from the first two groups that are contained in the essential oil fractions or extracts obtained from healthy plants, and they are the components generally explored for food preservation purposes.

6.4.1 Essential oils

Since ancient times antimicrobial properties of herbs and spices have been acknowledged and exploited for preservation of food and other organic matter as well as for medical treatments (Conner, 1993; Hirasa and Takemasa, 1998). In the Western world emergence and success of synthetic preservatives have practically made 'natural' food preservation extinct, and it is only in the last few decades that scientific interest in this area has re-emerged. The antimicrobial compounds in plant materials are commonly found in the essential oil fractions obtained by steam or supercritical distillation, pressing, or extraction by liquid or volatile solvents. Generally the oils consist of a mixture of esters, aldehydes, ketones, terpenes and phenolic compounds, and harbour the characteristic flavour and aroma of the particular spice or herb. The traditionally most well-known antimicrobial spices and herbs are clove, cinnamon, chilli, garlic, thyme, oregano and rosemary. But also bay, basil, sage, anise, coriander, allspice, marjoram, nutmeg, cardamom, mint, parsley, lemongrass, celery, cumin, fennel and many others have been reported to have an inhibitory effect toward microorganisms (Deans and Ritchie, 1987; Hili *et al.*, 1997; Hammer *et al.*, 1999; Elgayyar *et al.*, 2001; and Table 6.3 below.

Many of the identified active compounds have phenolic structures, and the same compounds can be found ubiquitously in different oils, such as eugenol in cloves and cinnamon leaf, thymol and carvacrol in thyme and oregano, 1,8-cineole in sage, thyme and bay. Structures of some active compounds are presented in Figure 6.1.

6.4.2 Organic acids

The commercially most important preservatives are still the organic acids (Table 6.2). They are all naturally occurring, although the bulk amount of these substances used in foods are synthetically produced. Based on their main antimicrobial effect they can be divided into two groups. One group shows an antimicrobial activity owing primarily to their pH-reducing effect. This group includes acetic acid, citric acid, formic acid, lactic acid, malic acid, oxalic acid

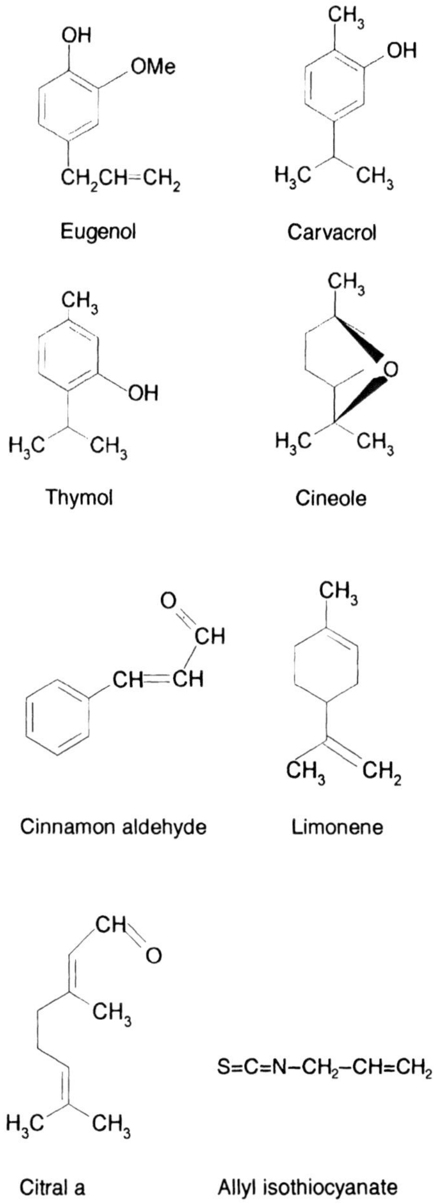

Fig. 6.1 Structures of some active compounds from essential oils.

and tartaric acid. They work either directly by lowering the pH of the food and thus adding stress to the microorganism, or in the undissociated form by migrating through the cell membrane into the cytoplasma of the microorganism where they dissociate and lower the internal pH of the cell. However, several reports have shown that the undissociated form of acetic acid also has

Table 6.2 Natural antimicrobial organic acids

Organic acids	E number[1]	Produced by	pK_a	MIC/allowed dose (ppm)[2]
Acetic acid	E260, 261–3	Bacteria, fungi, animals	4.74	q.s.
Carbonic acid	E290	Bacteria, fungi, plants, animals	6.37; 10.25	q.s.
Citric acid	E330, 331–3	Bacteria, fungi, plants	3.13; 4.77; 6.39	q.s.
Formic acid	E236, 237–8	Plants, animals	3.75	q.s.
Lactic acid	E270, 325–7	Bacteria, fungi, animals	3.08	q.s.
Malic acid	E296	Bacteria, fungi, plants	3.40; 5.11	q.s.
Oxalic acid		Bacteria, fungi, plants	1.23; 4.19	
Tartaric acid	E334, 335–7	Plants	2.98; 4.34	q.s.
Benzoic acid	E210, 211–13	Cranberries, cinnamon, clove a.o.	4.19	200–6,000
Propionic acid	E280, 281–3	Bacteria, plants	4.87	1,000–3,000
Sorbic acid	E200, 201–3	Plants	4.76	200–6,000
Medium chain fatty acids		Plants, animals + microbial fermentation products		*3

Notes: 1. E number as listed in Directive 95/2/EU. 2. The maximum level of use depends on the foodstuff. *3. Kabara, 1993.

antimicrobial action (e.g. Rusul *et al.*, 1987). Other reports have shown that the supplementation of up to 2.5% of citric, malic, tartaric or lactic acid to substrates adjusted to pH 2.5 or 3.5 significantly enhances fungal growth and mycotoxin production (Nielsen *et al.*, 1989; El-Gazzar *et al.*, 1987).

The other group of organic acid preservatives (sorbic acid, benzoic acid and propionic acid) shows antimicrobial activity only when they are present as undissociated acids. The efficiency of these acids therefore depends on the dissociation constant, pK_a. As the pK_a of most acids is between 3 and 5 (Table 6.2), these preservatives are only active at lower pH-values. However, some reports have shown that the dissociated acid may also have some antimicrobial activity (up to 2.5% of the undissociated compound) (Skirdal and Eklund, 1993). Organic acids are, in practice, usually added as the corresponding sodium, potassium or calcium salts because they are more soluble in water.

Sorbic acid is an unsaturated fatty acid (CH_3-CH=CH-CH=CH-COOH) with a high pK_a value (4.76), which makes it active in low acid food and it is therefore applied in a great variety of food products. The compound is active against yeasts, moulds and many bacteria. Microorganisms resistant to sorbate exist and an increasing number of moulds are able to degrade sorbate, producing strong off-flavours by the decarboxylation of sorbic acid into *trans*-1,3-pentadiene (Liewen and Marth, 1985; Kinderlerer and Hatton, 1990).

Benzoic acid (C_6H_5COOH) was first described as a preservative in 1875. The compound occurs naturally in high amounts in cranberries and some other fruits and spices. Sodium benzoate has a pK_a value of 4.19, which gives optimal activity below pH 4.0. Some fungi isolates of *P. roqueforti* have been found to be resistant.

Benzoates have an advantage of low cost compared to other preservatives. An adverse flavour effect is often encountered at higher use levels.

Propionic acid is produced by several lactic acid bacteria and is also found in some fruits and spices. It is effective against moulds and some bacteria but has very limited activity against yeasts. Propionates are able to inhibit *Bacillus subtilis*, which causes 'rope' in bread. Propionates have been used particularly in preservation of grains, bread and other baked products. Levels used are generally higher than with benzoate and sorbate.

6.4.3 Enzyme-released antimicrobial agents

In common food plants two major types of antimicrobial compounds activated upon hydrolytic cleavage exist. In the Allium family (garlic, onion, leek) sulfoxides are converted into pungent and lachrymatory sulfides such as diallyl disulfide upon tissue rupture (Walker, 1994). Most potent is garlic. It contains alliin (propenyl-cysteine sulfoxide), which is hydrolysed by the enzyme alliinase to allicin (2-propenyl-2-propenethiol sulfinate). A recent study concluded that 0.5% (5,000 ppm) of a garlic extract was needed to inhibit important pathogens (Unal *et al.*, 2001). Other studies have shown that these substances inhibit most microorganisms, but only at relatively high concentrations (Beuchat, 1994; Conner, 1993), and the main hindrance of use is sensory effects.

In plants of Cruciferae (cabbage, mustard, horseradish, brussels sprout) glucosinolates are the substrates for hydrolytic enzymes. A number of different isothiocyanates are identified in the Cruciferous family, and some of the compounds are also believed to exert anti-carcinogenic effects (Verhoeven *et al.*, 1997; Mithen *et al.*, 2000). Sinigrin stored in mustard seeds are cleaved by myrosinase to yield allyl isothiocyanate (AITC), which constitutes the main compound of mustard essential oil (95%). *Penicillium expansum*, *Aspergillus flavus* and *Botrytis cinerea* were unable to grow in the presence of 100 μg AITC /L gas-phase, and *Salmonella typhimurium*, *Listeria monocytogenes* and *Escherichia coli* O157:H7 were inhibited by 1000 μg /L (Delaquis and Sholberg, 1997). These authors found *E. coli* O157:H7 to be the most resistant of the bacteria tested, whereas Lin *et al.* (2000) found *L. monocytogenes* to be more resistant than *E. coli* O157 : H7 and *Salmonella montevideo* in broth solution. Applied through the gas phase, the volatile AITC has proved effective in very low doses against food spoilage fungi (Delaquis and Mazza, 1995; Nielsen and Rios, 2000). The gaseous form seems to have higher antimicrobial potential than the liquid (Lin *et al.*, 2000).

6.4.4 Phytoalexins and other plant-derived compounds

An active defence system of plants against invading microorganisms is the *de novo* synthesising of phytoalexins. At least 200 phytoalexins have been characterised from over 20 plant families (Harborne, 1988), and certain chemical types seem to be associated with some plant families. Examples of these are isoflavonoids from Leguminosa as pisatin from garden pea (*Pisum sativum*) and phaseollin from beans (*Phaseolus vulgaris*). Others are rishitin from potatoes and tomatoes (Solanaceae), falcarindiol and 6-methoxy-mellein from carrots (Walker, 1994; Smid and Gorris, 1999). Use of phytoalexins for food preservation has been suggested by many, but practical attempts are lacking or scarce (Smid and Gorris, 1999).

Investigation of plant antimicrobial compounds has been focused on food plants or herbs and spices traditionally known to contain activity. Some authors argue that a number of potentially useful plant types are ignored by this approach (Wilkins and Board, 1989). Also the innate antimicrobial systems of some food products could be brought to play important roles in the preservation of the product itself if minimal processing schemes conserving or enhancing the responsible compounds are used. Examples of these systems are the content of protective polycyclic compounds, the proanthocyanidins, in strawberries. They protect against grey mould, *Botrytis cinerea*, until the berries are ripe and the concentration decreases to non-inhibitory levels (Wilkins and Board, 1989). Similar effects are seen in ripening fruit (peaches and plums) where volatile aldehydes and esters have been shown to inhibit *B. cinerea* and other phytopathogens (Wilson *et al.*, 1987).

6.5 Activity of natural antimicrobials

Inhibition of pathogenic and food spoilage microorganisms by essential oils has been the subject of a great number of studies, ranging from investigation of single compounds towards a few microorganisms to large studies, testing a wide range of essential oils on a broad spectra of microorganisms. An overview of the most important herbs and spices and their active components are presented in Table 6.3. Test methods used in different investigations vary considerably, making comparison between experiments difficult in many cases. Also some of the tests are not relevant for several types of minimally processed foods. There are three main ways to apply natural antimicrobial compounds to minimal processing of food. In processed food like marmalade, juice, soup and sauce, the antimicrobial system can be added directly to the product. Less processed foods like pieces of fruit, salads and other large components can only be preserved by adding the substance to the surface. This can either be done by dipping the product into the substance, spraying it on or applying it in some kind of coat. Alternatively the substances can be added to the packaging atmosphere either during the packaging process or by incorporating the substances into the film, which releases it after packaging. In all cases there is a limit for how much agent it is possible to add to the system. The overall sensory effect has also to be taken into account. The list in Table 6.3 contains plants from which it has been possible to produce extracts or essential oils with an antimicrobial effect at a realistic concentration. In most cases it means that activity has to be present at concentrations much lower than 1,000 ppm.

Table 6.3 Essential oils from spices and herbs, their major active compounds, and examples of minimum inhibitory concentrations (MIC) from different experiments

Spices and herbs	Active components	Effective against	MIC (ppm)	Reference
Allspice	Eugenol	Fungi, yeast	200	5, 11
Basil	Linalool, methyl chavicol	Fungi, yeast, bacteria	> 1,250	8, 21
Bay	Eugenol, linalool, 1,8-cineole, umbellulone	Bacteria (Listeria, Salmonella, Camphylobacter, Staphylococcus, E. coli)	400–5,000	16
Cinnamon	Cinnamon aldehyde, eugenol	Yeasts, bacteria (Staphylococcus, Listeria)	200–750	5, 10, 16
Clove	Eugenol	Fungi, yeasts, bacteria (Bacillus subtilis*)	250–1,000, (*500–600)	2, 5, 10, 11, 16
Garlic (Allium family)	Allicin	Fungi, yeast, bacteria (E.coli)	200– > 1,000	5, 16, 19
Lemon grass	Citral, limonene	Fungi, bacteria	1,000, 5,000	1, 12
Mustard (Cruciferae)	Allyl isothiocyanate	Fungi, yeast, bacteria	0.034–600	7
Oregano	Carvacrol, thymol	Fungi, bacteria	100–1,000, 2.5 (fumigant)	4 , 8, 13, 14, 15, 20
Rosemary	1,8-cineole, camphor, linalool	Bacteria (Bacillus cereus*)	1,000 and above (*600)	6, 16, 17
Sage	Thujone, 1,8-cineole	Gram- bacteria,	750–3,750	18
Thyme	Thymol, carvacrol cymene	Fungi, yeast, bacteria	200–1,000 3 (fumigant)	3, 5, 9, 10, 11, 15, 16

Note: 1. Adegoke and Odesola, 1996. 2. Al-Khayat and Blank, 1985. 3. Arras and Usai, 2001. 4. Basilico and Basilico, 1999. 5. Conner and Beuchat, 1984. 6. Del Campo *et al.*, 2000. 7. Delaquis and Mazza, 1995. 8. Elgayyar *et al.*, 2001. 9. Hammer *et al.*, 1999. 10. Hili *et al.*, 1997. 11. Hitokoto *et al.*, 1980. 12. Mishra and Dubey, 1994. 13. Sivropoulou *et al.*, 1996. 14. Paster *et al.*, 1990. 15. Paster *et al.*, 1995. 16. Smith-Palmer *et al.*, 2001. 17. Soliman *et al.*, 1994. 18. Tzakou *et al.*, 2001, 19. Unal *et al.*, 2001. 20. Ultee *et al.*, 2001. 21. Wan *et al.*, 1998.

The many varieties of species and cultivars of plants give rise to different compositions of essential oils; for instance, in France alone seven different thyme 'chemotypes' have been reported as native (Senatore, 1996). Not only genetic but also climatic factors, as well as time of harvest and the extraction method applied, give huge diversity in the 'same' kind of oils (Tucker and Maciarello, 1986; Perry *et al.*, 1999; Santos-Gomes and Fernandes-Ferreira, 2001). Distribution of active compounds within the plant is also not uniform; for example, in cinnamon bark oil, cinnamaldehyde is the main compound, as opposed to eugenol in cinnamon leaf oil (Jirovetz *et al.*, 2000). Thus, lacking information about essential oil composition raises a problem in comparison of different studies.

Most studies have focused on effect in laboratory systems with little resemblance to actual applications in foods. This renders an overview of effectiveness and a means to compare between substances, but when it comes to actual applications it may result in misleading results. Arras and Usai (2001) found that 250 ppm (v/v) of a thyme oil containing $>80\%$ carvacrol inhibited post-harvest citrus pathogens *Penicillium digitatum, P. italicum, Botrytis cinerea* and *Alternaria citri*. When infected oranges subsequently were stored in chambers with vaporised thyme oil at 250 ppm, pathogen growth was only reduced by 10%. The treatment had to be combined with storage at subatmospheric conditions to prevent growth for an extended period.

Many reports have contradictory conclusions. It has, for example, generally been considered that essential oils are more inhibitory towards fungi than bacteria, and that Gram-positive bacteria are less resistant than Gram-negative (Shelef, 1984; Zaika, 1988). Several authors have, none the less, found this not to be true in their investigations (Tantaoui-Elaraki *et al.*, 1993; Marino *et al.*, 1999; del Campo *et al.*, 2000; Tzakou *et al.*, 2001). The nature of the natural antimicrobial compounds makes it more important (compared to traditional preservation techniques) to test the effect in the real system or in a model system that has a close resemblance to the actual food product. The amount of protein and lipid in the medium, as well as any other substances that can react with the active compounds or alter physical and chemical properties, will affect activity.

Recently experiments conducted in our laboratory have illustrated the impact of methodology on the effect obtained by natural antimicrobial compounds. A range of essential oils were tested for their antifungal activity either by adding 250 ppm of the oil to a rye bread media, which were subsequently inoculated with fungal contaminants, or by releasing a corresponding amount of the oil in a closed container with a piece of rye bread inoculated with the same fungal contaminants. Fig. 6.2 shows the inhibition in percent relative to a control after seven days of growth. Two types of rosemary extract were used. Rosemary-A was a traditional oleoresin with a strong rosemary flavour. The other rosemary type (rosemary-B) had been extracted in a way that reduces the flavour but retains strong antioxidative effect. Fig. 6.2 shows that rosemary-B did not have a good effect on the test fungi in the rye bread media nor as a volatile. The non-modified essential oil had a much stronger effect when used as a volatile.

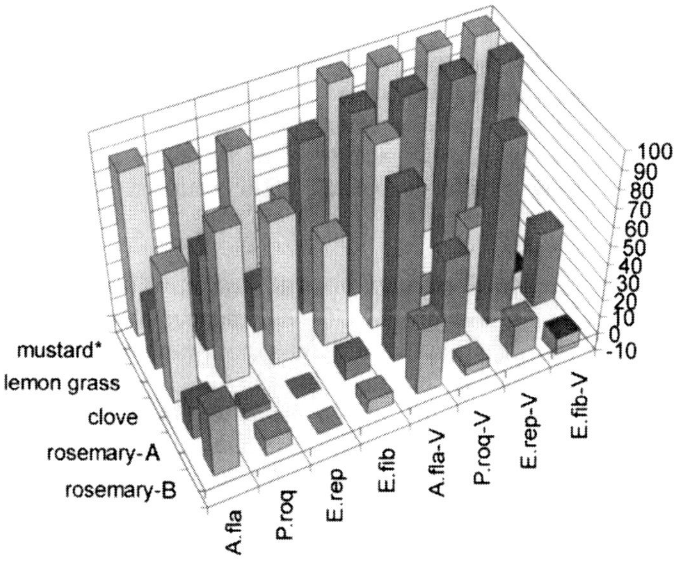

Fig. 6.2 Relative inhibition of some fungi when exposed to essential oils in the media or as a volatile (V). A. fla = *Aspergillus flavous*; P. roq = *Penicillium roqueforti*; E. rep = *Eurotium repens* and E. fib = *Endomyces fibulinger*. Concentrations of essential oils were 250 ppm in all cases, **except* mustard *as volatile* where only 2 ppm was added.

Lemongrass was an efficient antimicrobial compound as a volatile, but not when added to the media. The opposite picture was seen for clove. In particular, mustard oil had a much stronger effect in the gas phase than in the media, as a 100 times higher concentration was added to the media without completely inhibiting growth, contrary to the volatile results. Delaquis and Mazza (1995) have also found large differences depending on the test method applied for mustard oil.

In an early review the correlation between inhibited growth and effect on toxin production has been questioned (Zaika, 1988). For the bacteria *Bacillus cereus* decreased production (80%) of diarrhoeal toxin was found in presence of 60 ppm carvacrol in broth, and the authors concluded that doses below the minimal inhibitory concentration (MIC) can increase food safety by inhibiting toxin production (Ultee and Smid, 2001), although achievement of the same effects in soup demanded a 50-fold increase in carvacrol concentration. Hitokoto and co-workers (1980) and Basilico and Basilico (1999) also showed that inhibition of mycotoxin production by some *Aspergillus* species appeared at lower concentrations than inhibition of growth. However, other authors have shown, in concordance with our own experience, that subinhibitory concentrations can result in an increased mycotoxin production (Nielsen *et al.*, 1989). In fact it should be expected that the production would increase when a small amount of essential oil is added as several of these substances can be partly metabolised and used in the production of secondary metabolites by some fungi.

6.6 Natural food preservatives: mechanisms of action

In order to evaluate the full potential of natural antimicrobials in food preservation, understanding their mode of action on cells is important. Despite the abundant number of studies on antimicrobial activity of natural compounds, only relatively few have investigated their mechanism of actions and detailed knowledge for many specific compounds is still lacking.

6.6.1 Mode of action of essential oils and their constituents

A key cellular event of essential oils affecting microorganisms is hampering of membrane integrity. The plasma or cytoplasma membrane is a hydrophobic barrier, embedding various functional proteins, and crucial for maintaining cellular functions by forming ion gradients to drive energy transducing processes, regulate metabolism, solute transport, etc. By use of different staining methods, it has been shown that increased membrane permeability can be caused by some essential oils and their compounds, e.g. oregano, carvacrol, thymol (Helander *et al.*, 1998; Lambert *et al.*, 2001) and tea tree oil (Cox *et al.*, 2000).

Differences in vulnerability towards active compounds between microorganisms could be explained by differences in cell wall structures, and thus different diffusion rates of the compounds. As Gram-negative bacteria contain an outer membrane structure, some active compounds are believed to be able to cross this into the periplasm through porin proteins providing transmembrane channels (Helander *et al.*, 1998). In contrast to bacteria, fungal membranes contain sterols, which might play a significant role in the differences between fungi and bacteria.

Modelling structure–activity relationships targeted at *Listeria monocytogenes*, lipophilicity and pK_a values were the most important factors for benzoic acids and cinnamic acids (explaining 81% of variance), whereas steric and electronic parameters were more appropriate descriptors for the action of benzaldehydes (Ramos-Nino *et al.*, 1996). It was interpreted that the acids either had to partition through cell membranes to exert effect or were directly involved at the target site itself, whereas aldehydes were reacting with sites more superficially in the membrane. Adjacent hydroxyl groups in the molecules enhanced activity. Lipophilicity and hydroxyl groups were also emphasised as important factors for activity by Lattanzio *et al.* (1994) in another comparative study of benzoic acids, cinnamic acids and flavonoids. Environmental factors are important for interactions of active compounds and microorganisms. At lower pH (5.5 versus 6.5), the inhibitory effect of thyme oil or thymol was found to be greater against *S. typhimurium* as the undissociated form is more lipophilic (Juven *et al.*, 1994). Also, increasing the amount of Tween80, a surface-active agent used to dissolve the active compounds, decreased the effect of thyme oil. Hili *et al.* (1997) supported this with a finding of 50-fold higher activity when no dispersing solvent was used for cinnamon oil against *Saccharomyces cerevisiae*.

(Somewhat contradictory to this, detergents were found to enhance the effect of carvacrol/thymol in a lytic test with lysozyme (Helander *et al.*, 1998). The differences are most probably a matter of dosing.) Detergents increase solubility in the water phase of the hydrophobic compounds and lessen affinity to membranes and proteins (Juven *et al.*, 1994).

6.6.2 Organic and fatty acids

The antimicrobial effect of weak organic acids as sorbic, propionic and benzoic acid has traditionally been explained by acidification intracellularly due to undissociated acid crossing the plasma membrane and dissociating inside the cell at the higher pH. Detrimental effects are believed to be: dissipation of pH gradient across the membrane; diminishing the proton motive force; intracellular accumulation of toxic anions; stress on pH homeostasis; and inhibition of essential metabolic enzymes due to the altered pH (Brul and Coote, 1999). However, the inhibitory mechanism seem to be more complex; for example, poor correlation has been found between measured intracellular pH values and reduced growth upon exposure to sorbic acid (Bracey *et al.*, 1998) and weak acids with identical pK_a values exert different effects (Eklund, 1980). Correlation between inhibitory effect and carbon chain length of organic acids, alcohols and aldehydes suggests that membrane-active properties play a role for activity (Stratford and Anslow, 1998). Antifungal effect of fatty acids has also been demonstrated to cause a dissipation of proton gradient (Breeuwer *et al.*, 1997), which supports a membrane hampering mechanism.

6.6.3 S-containing compounds

The active compound allyl isothiocyanate, from, for example, horseradish and mustard, exerts its toxic effect by reacting with protein. Denaturation is caused by oxidative cleavage of disulfide bonds and also non S-containing amino acids (lysine and arginine) can be attacked (Kawakishi and Kaneko, 1987). Metabolic enzymes as well as other proteins will thus be unable to perform their normal functions and eventually this will kill the microorganism (Delaquis and Mazza, 1995; Delaquis and Sholberg, 1997).

Allicin, the active compound from onions, is believed to have a somewhat similar effect to allyl isothiocyanate altering protein and enzymes by oxidation of thiols to disulfides. According to Unal *et al.* (2001), Wilis showed in 1956 that enzyme inhibition was associated with -SO-S- and not -SO-, -S-S- or -S- groups.

6.7 Application in food products

Despite the vast evidence of antimicrobial effect of natural compounds on food pathogens and spoilage organisms, commercial use is still very limited. One reason for this could be due to the fact that food matrixes usually are much more complex than laboratory media, and higher dosing is often required to achieve the same effect. Thus, the necessity of using real food systems in evaluation of natural preservatives has been more evident in recent years. In soft cheese models essential oils of bay, clove, cinnamon and thyme have been tested as preservatives against *Salmonella enteritidis* and *Listeria monocytogenes* (Smith-Palmer *et al.*, 2001). The fat content of the cheese had a significant effect on activity with decreased effect at higher fat content. Clove oil was the most effective overall, but all oils had an effect at 1% doses in low fat cheese. In opposition to the 0.5–1% oil required for inhibitory effect in cheese, a concentration of less than 0.1% had earlier been reported for the same oils and bacteria in *in vitro* studies.

Mayonnaise made with extra virgin olive oil was more bactericidal against *Salmonella enteritidis* compared to other oils, and this correlated to higher acidity and the range of phenolic compounds in the extra virgin oil (Radford *et al.*, 1991). In a model of the effect of oregano essential oil on *Esherichia coli* O157:H7 in eggplant salad, pH and storage temperature acted synergistically, and a concentration of 0.7% oregano oil proved to be effective and organoleptically acceptable (Skandamis and Nychas, 2000).

Extended shelf-life of meats sprayed with spice extracts has been reported (Ziauddin *et al.*, 1996). Ready-to-eat chicken surface-treated with plant extracts showed that clove and pimento leaf were the most effective of nine extracts tested against *Aeromonas hydrophila* and *L. monocytogenes* (Hao *et al.*, 1998). On roasted beef 20 μl/L horseradish essential oil vapour inhibited *L. monocytogenes*, *S. aureus*, *E. coli* and *S. typhimurium* – a fivefold higher dose compared to findings on agar media (Ward *et al.*, 1998). Off-flavours from oxidation of fat were found to be delayed by horseradish oil preservation, but flavour from the oil itself was, on the other hand, detectable in the roasted beef (Delaquis *et al.*, 1999). In Japan commercial use of allyl isothiocyanate (the active compound in horseradish/mustard) has been developed as a preservative agent for bread, pasta, lettuce, grains, dried mushrooms and fruits (Hirasa and Takemasa, 1998). Besides direct use as a preservative, an alternative is, for example, use of basil oil (0.1% or 1%) for washing lettuces, which was shown to be comparable with 125 ppm chlorine in a test by Wan *et al.* (1998).

Volatile antimicrobial substances from spices and herbs have proved efficient in the control of fungal spoilage by common bread-spoiling fungi (Nielsen and Rios, 2000). Mustard essential oil had the strongest inhibitory effect on all tested fungi for more than two weeks, when applied at 1 μl per 250 ml culture flask. Cinnamon and garlic essential oils required slightly higher concentrations to inhibit growth of all fungi completely. The sensory threshold was slightly higher than the minimal inhibitory concentration (MIC) for allyl isothiocyanate (AITC) on rye bread, thus the required shelf-life of rye bread could be achieved by active packaging with AITC, whereas wheat bread (hot-dog bread) may require the additional effect of other preserving factors to avoid off-flavour formation, due to a slightly lower sensory threshold.

Essential oils of spices and herbs are attractive as potential preservatives, because they have been applied to food for centuries, are generally regarded as safe (GRAS) and will most likely be more acceptable for consumers than synthetic preservatives. The drawbacks are sensory effects, which limits their use to food that already has a strong flavour. Another possibility is use of the most active compounds rather than the whole oil in order to improve efficiency and reduce redundant flavour compounds, e.g. oregano can be substituted by carvacrol and thymol (Skandamis and Nychas, 2000). But the most acclaimed possibility is combining the active compounds with other preservative factors in order to get highest effect at lowest dosage. A combination strategy will also diminish risk of resistance development by the microorganisms. Examples of combinatory preservation strategies are, for example, modified atmosphere packaging combined with essential oregano oil vapour (Tsigarida *et al.*, 2000) or thyme oil (Juven *et al.*, 1994), synergistic effect of pulsed electric field, nisin and carvacrol (Pol *et al.*, 2001), lysozyme and nisin with pulsed high pressure treatment (Garcia-Graells *et al.*, 1999), and as mentioned previously in the salad models factors like pH/acidity. A combination of salt, low pH and low temperature has been shown to act synergistically with rosemary oil (del Campo *et al.*, 2000), and eugenol, monolaurin and sodium citrate is another combination with inhibitory effects (Blaszyk and Holley, 1998).

6.8 Natural antioxidants in food systems

Antioxidants can be very generally defined as 'substances that in small quantities are capable of preventing or retarding oxidation processes' (Schuler, 1990). In food legislation, antioxidant additives are defined as 'substances which prolong the shelf life of foodstuffs by protecting them against deterioration caused by oxidation, such as fat rancidity and color changes' (Directive 95/2/EC). Natural antioxidants only signify that the substances occur in nature, and that the antioxidant compounds in question are obtained from their natural source (Yanishlieva-Maslarova, 2001). Hence there are no specific chemical properties that distinguish natural antioxidants from the contrasting term 'synthetic' antioxidants. The latter term relates to those compounds that are chemically synthesised and not found in nature. With respect to terminology, a grey area principally exists for those compounds that occur in nature, but are synthesised for their use as antioxidant food additives for economical reasons. However, usually when natural compounds such as ascorbic acid are chemically synthesised they are included among natural antioxidants, or at least designated as 'nature-identical' antioxidants. Natural antioxidants may be present in high amounts in certain minimally processed foods if the raw material is rich in antioxidants, and if no significant depletion occurs during processing: examples include certain gently processed fruit juices, e.g. orange juice which is high in ascorbic acid, and extra virgin olive oil, which is rich in tocopherols and olive phenols, notably tyrosol and hydroxytyrosol which have potent antioxidant activities. In principle, natural antioxidants may also be formed during food preparation processes such as fermentation, curing and smoking or in certain cases when the processing involves heating steps. Thus, although heat generally induces degradation of antioxidants, notably of antioxidative enzymes, and – as discussed elsewhere in this book – may cause other significant quality changes in food and beverage processes – heat may in some cases induce formation of maillard reaction products or other types of degradation compounds having high antioxidant activity (Lingnert and Lundgren, 1980; Pokorny and Schmidt, 2001). In the following, however, mainly those natural antioxidative agents intended to be used as food *additives* or as novel ingredients in modern, gentle food processing regimes are covered. The role of natural antioxidants in nutrition is a subject in itself, and in this chapter the focus is on use of natural antioxidant additives as food quality protectants, notably as inhibitors of lipid oxidation to prevent off-flavour development and rancidity in processed foods. The two most widely and longest used natural food antioxidants used to inhibit lipid oxidation, namely ascorbic acid (vitamin C) and α-tocopherol (vitamin E), are nevertheless well recognised also for their nutritional effects as vitamins and biological antioxidants. In agreement with the definition of an antioxidant, natural antioxidants may also be used to protect quality characteristics such as colour, notably by preventing enzyme-catalysed oxidative browning. Antioxidant additives may also preserve structural and textural properties in conventionally as well as minimally processed foodstuffs. In the following, however, mechanisms of natural antioxidants to retard lipid peroxidation in foods are at the centre stage, as addition of antioxidants to foods to prevent lipids from oxidative deterioration is presently of highest practical importance. Apart from α-tocopherol, other tocopherols and ascorbic acid, surprisingly few natural antioxidant compounds are at present of commercial importance as food additives. The main ones are citric acid and constituents extracted from the spices rosemary, sage and oregano. Several other naturally occurring substances may, however, be applicable as food antioxidants provided they work efficiently, are tasteless and odourless, non-toxic, and if economy and legislation permits. Potential antioxidant substances for food use have been extracted from many different sources, and the scientific literature is abundant with reports on the antioxidant effects of natural antioxidants, where the molecular composition and characteristics of the active constituents may or may not be known. Compounds derived from spices and edible plants, i.e. phytochemicals, have received particular research attention during the last 10–15 years, and – as discussed later in this chapter – several new types of plant-based antioxidant preparations are now commercialised. However, antioxidant substances, including antioxidative enzymes, obtained from other natural sources, e.g. from milk, other animal or microbial sources, are also being actively investigated.

6.9 Activity mechanisms of natural antioxidants

Antioxidant action in real foods is very complex as antioxidants may work by different mechanisms and many natural antioxidants are multifunctional and can exert their antioxidant activity by several different mechanisms. Both the chemical structure of the antioxidant as well as a number of physico-chemical factors of the food system determine which type of antioxidant mechanism will prevail and in turn define the antioxidant efficiency. Hence, the main mechanism of oxidation and consequently the most relevant antioxidant strategy differs in different food and beverage products depending on their physical and chemical composition as well as on their exposure to oxidation accelerators such as air oxygen, heat, light and transition metals during processing and storage. Apart from chemical structure and polarity of the antioxidant, antioxidant performance is obviously dependent on the type of lipid to be protected. Also the pH in the aqueous phase, the heterophasic composition of the system, including the interfacial properties and type of emulsifier, strongly influence the oxidation rate and oxidation mechanisms as well as localisation and effectiveness of antioxidants (Frankel and Meyer, 2000). In addition, proteins, including those employed as emulsifiers, may directly interact with antioxidant performance presumably via several different interaction mechanisms (Frankel and Meyer, 2000). Detailed knowledge on oxidation mechanisms and antioxidant stability and action in true food processes as well as in real multiphasial food systems is therefore essential for successful, innovative exploitation of natural antioxidants in minimally as well as in conventionally processed foods and beverages. There is still much to learn!

Metal chelation and radical chain breaking by primary radical interception and alkoxyl breaking reactions are presently regarded as the most important antioxidant mechanisms of natural antioxidants in food systems (Frankel, 1998). However, quenching of singlet oxygen and enzymatic removal of oxygen are also effective in preventing lipid peroxidation in foods. All these mechanisms are therefore discussed in the following. Table 6.4 lists various antioxidant mechanisms that can be workable in conventionally and minimally processed foods and corresponding examples of antioxidants that work by various interception mechanisms.

6.9.1 Metal chelation

Metal-catalysed initiation by copper or iron ions is a prevalent promotion mechanism of lipid peroxidation in food and beverage systems. Metal-catalysed initiation encompasses two different types of reaction (Frankel, 1998):

1. Where a transition metal ion in its high valency state ($Me^{(n+1)+}$) directly abstracts hydrogen from the lipid substrate (LH) to form a lipid radical (L•):

$$LH + Me^{(n+1)+} \longrightarrow L\bullet + H^+ + Me^{n+} \qquad (6.1)$$

Table 6.4 Theoretical antioxidant mechanisms and examples of both synthetic and natural antioxidant compounds listed according to where they interrupt the lipid autoxidation reaction chain. Not all of the antioxidant examples listed are currently used in foods

Antioxidant mechanism	Examples of antioxidants
Metal chelation	Citric acid, EDTA,[1] lactoferrin, phytic acid (phosphates)
Oxygen scavenging	Ascorbic acid, glucose oxidase-catalase
Singlet oxygen quenching	Carotenoids, tocopherols
Active oxygen scavenging	Superoxide dismutase, catalase, mannitol
Primary radical chain-breaking	Tocopherols, ascorbic acid and derivatives, gallic acid and gallates,[1] BHA,[1] BHT,[1] several natural polyphenols, rosemary and sage antioxidants
Alkoxyl radical interruption	Tocopherols (some rosemary compounds)
Secondary chain-breaking	Glutathione peroxidase and glutathione–S-transferase, thiodipropionic acid and its derivatives[1]

Note: EDTA: ethylene diamine tetraacetic acid; antioxidants marked[1] are all 'synthetic' antioxidants. Thiodipropionic acid and its derivatives are permitted as food additives in the USA, but not in the EU.

2. Where metal ions catalyse production of radicals by decomposition of lipid hydroperoxides (LOOH) as in reaction (6.2), below. Both peroxyl radicals (LOO•) and alkoxyl radicals (LO•) can be formed by metal catalysis with the decomposition reaction to produce alkoxyl radicals generally considered as the most rapid and important in food systems (Frankel, 1998):

$$LOOH + Me^{n+} \longrightarrow LO\bullet + OH^{-} + Me^{(n+1)+} \tag{6.2}$$

Previously, ligands able to deactivate prooxidative metal ions by chelation were termed 'synergists'. However, according to the definitions of antioxidants given above, agents that are able to complex with transition metal ions to hinder metal-catalysed initiation reactions and decomposition of lipid hydroperoxides ought to be included as antioxidants. Metal chelators can be classified as 'preventive antioxidants' defined either as 'preventing introduction of initiating radicals' (Scott, 1993) or as 'reducing the rate at which new chains are started' (Burton and Ingold, 1981). Several natural antioxidants, especially polyphenolic, phytochemical compounds, possess potential metal-binding sites in their structure. Part of their antioxidant function may therefore be elicited through metal chelation (Fig. 6.3). For anthocyanins, i.e. the flavonoids responsible for red and purple colours in many fruits and vegetables, metal chelation was thus suggested to be one of the mechanisms in play in the antioxidant action of anthocyanins towards oxidising lecithin liposomes in a food model system (Satué-Gracia *et al.*, 1997). Citric acid is a natural compound widely added during processing of vegetable oils to prevent metal-catalysed initiation and decomposition of lipid hydroperoxides.

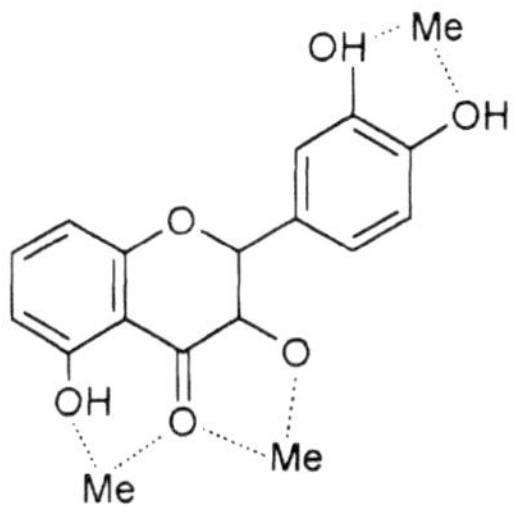

Fig. 6.3 Possible metal chelation sites in natural, flavonoid type antioxidants. The possible binding site at the di-hydroxy moiety in the B ring is assumed to apply also to phenolic acids containing this catechol moiety. Binding sites adapted from Hudson and Lewis, 1988; Nardini *et al.*, 1995.

Its effect is increased by synergistic interaction with the naturally occurring tocopherols in the oils (Frankel, 1998). Other metal chelating compounds such as phytic acid and lactoferrin have been shown in various, heterophasic food model systems to be able to inhibit lipid peroxidation significantly (Empson *et al.*, 1991; Huang *et al.*, 1999; Satué-Gracia *et al.*, 2000). These natural metal chelators may be applicable as antioxidants in real foods if they are economical and if legislation permits. Phytic acid is included among food antioxidants in the *Codex Alimentarius* (Mikova, 2001), but is not generally permitted to be used as a food additive in the European Union nor the USA, because it is presumed to possess potential antinutritive effects due to its ability to bind cationic mineral nutrients. The same concern may be invoked for the use of other natural metal chelators as antioxidant agents. Nevertheless, a main application advantage of targeting metal chelation is that very low usage levels of the metal chelating compounds are generally required for efficient oxidative protection compared to the doses of chain-breaking antioxidants (Dziezak, 1986). This is because only trace amounts of metals may accelerate oxidation, and therefore it is usually sufficient that only these trace amounts are chelated. In many newer, rapid antioxidant assays only the ability of natural compounds to trap free radicals is tested. There is therefore limited knowledge on the metal chelation properties of many recently investigated natural antioxidant compounds, including metal chelating abilities of many phytochemicals.

6.9.2 Radical chain-breaking

An early mechanistic definition of antioxidants was 'compounds able to accept radicals' (Uri, 1961). In fact, antioxidants that work as chain-breakers do not necessarily 'accept' radicals, but rather reduce radicals by hydrogen donation. Thus, radical chain-breaking antioxidants (AH) prevent propagation of lipid autoxidation by donation of hydrogen atoms to peroxyl radicals (LOO•) thereby producing lipid hydroperoxides (LOOH):

$$LOO\bullet + AH \longrightarrow LOOH + A\bullet \tag{6.3}$$

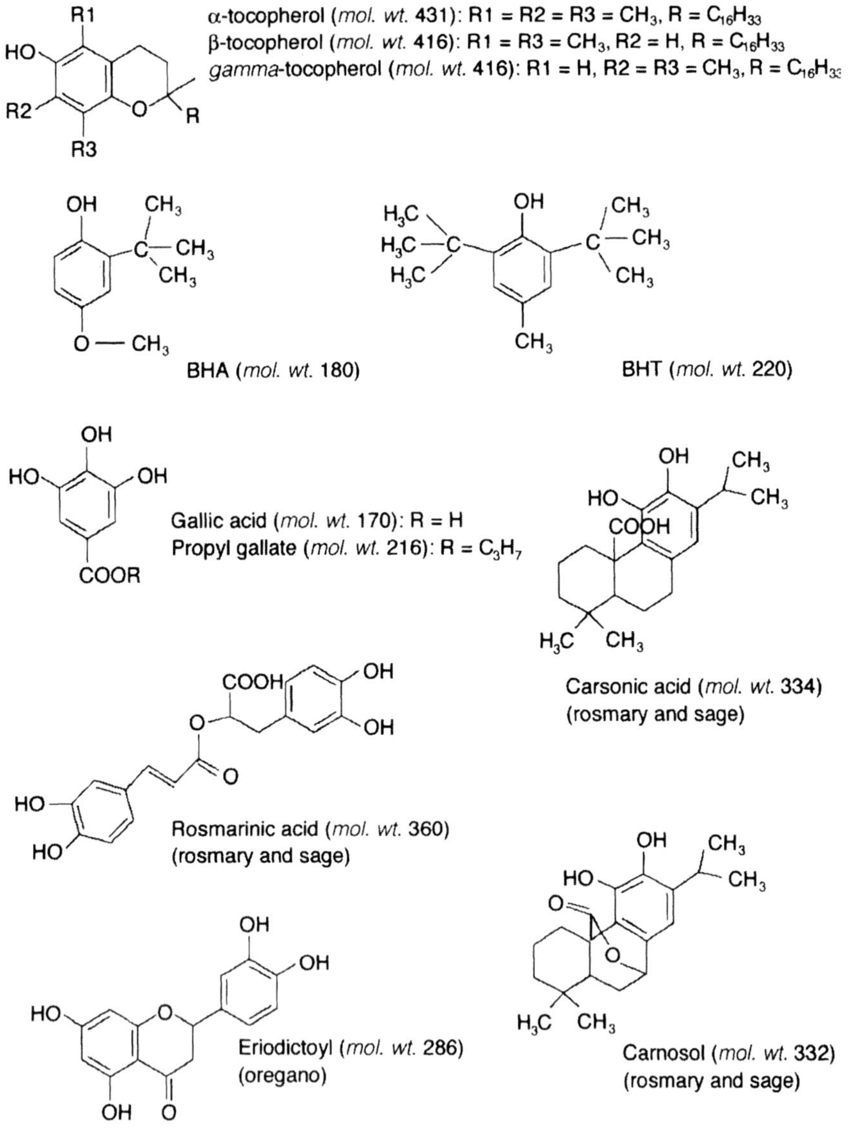

Fig. 6.4 Examples of phenolic structures in synthetic antioxidants (BHA and BHT) and in natural antioxidants originating from plant materials (oil seeds) and from the spices rosemary, sage, and oregano.

Hindered phenols are the most common antioxidant compounds to readily scavenge lipid peroxyl radicals by hydrogen donation from the hydroxyl group(s) (Scott, 1993). The donation of hydrogen may happen directly or by donation of an electron followed by deprotonation of the phenolic antioxidant. Recently it was suggested for catechins that donation of non-phenolic hydrogen atoms might be involved in antioxidant action (Kondo *et al.*, 1999). Under conditions of limited oxygen, antioxidants may also break lipid peroxidation by directly interrupting lipid radicals (L•), reducing them back to the original lipid:

$$L\bullet + AH \longrightarrow LH + A\bullet \quad (6.4)$$

As mentioned above, ascorbic acid and tocopherols are common natural antioxidants used in foods. Both ascorbic acid and tocopherols work as primary chain-breaking antioxidants. Incidentally, ascorbic acid may also work as an oxygen scavenger (see Table 6.4) or via other mechanisms: ascorbic acid is thus able to regenerate the tocopheroxyl radical back to tocopherol resulting in a synergistic effect with tocopherol (Lambelet *et al.*, 1985). In certain systems, notably when iron is present, ascorbic acid may act as a prooxidant (Kanner and Mendel, 1977; Jacobsen *et al.*, 1999). Synthetically synthesised antioxidants such as BHT (butylated hydroxy toluene: 2,6-di-t-butyl-4-methylphenol), BHA (butylated hydroxy anisole: a mixture of t-butyl-p-hydroxyanisole isomers), TBHQ (tertiary butylhydroquinone – which is permitted for food use in the USA, but not in the EU), as well as ascorbyl palmitate, the gallic acid esters propyl-, octyl-, and dodecyl gallate in addition to gallic acid itself, also work by the radical chain-breaking mechanism. Except for ascorbic acid and its derivatives, most chain-breaking antioxidant compounds are hindered phenols that are good scavengers of lipid peroxyl radicals by donation of the phenolic hydrogen atoms. Many of the natural phytochemical antioxidants that have been found to exert potent antioxidant activity in various food model systems also possess phenolic structures (Fig. 6.4). These compounds include the polyphenolic flavanols, various flavonoids and phenolic acids as well as the active molecules of the spices rosemary (*Rosmarinus officinalis L.*) and sage (*Salvia officinalis L.*), notably rosmarinic acid, carnosol, and carnosic acid, as well as eriodictoyl in oregano (*Oreganum vulgare L.*) (Fig. 6.4). Relations between the molecular structure of antioxidants and their antioxidant efficiency is discussed further in a later section of this chapter.

6.9.3 Alkoxyl radical interruption

In foods lipid hydroperoxides are readily decomposed into a complex mixture of compounds, some of which impact the objectionable odour and flavour characteristics of rancid foods. The first step in the lipid hydroperoxide decomposition is usually the homolytic cleavage of the O-O bond, as this bond is relatively weak with a low activation energy (~184 kJ/mole) for cleavage (Kaur and Perkins, 1991):

$$LOOH \longrightarrow LO\bullet + \bullet OH \quad (6.5)$$

Decomposition of lipid hydroperoxides is accelerated by heat and transition metal ions (reaction (6.2), above). Therefore natural antioxidants with metal chelating properties may also inhibit this reaction. It is important to note that this decomposition (reaction (6.5)) produces the highly reactive hydroxyl radical, •OH, which is able to abstract hydrogen directly from a number of sites along a fatty acid carbon chain, and thereby fuel the formation of new lipid radicals to start new chains of oxidation events:

$$\bullet OH + LH \longrightarrow L\bullet + OH \qquad (6.6)$$

Mannitol and formate are known to be able to scavenge hydroxyl radicals (Gutteridge, 1982), but there is a scarcity of knowledge on potential hydroxyl radical scavenging properties of natural antioxidants, and no natural antioxidants are presently exploited in foods primarily because of their hydroxyl scavenging properties. One reason for this may be the extremely high reactivity of the hydroxyl radical reacting non-selectively with most organic molecules with high rates (Larson, 1997, ch. 2). This makes it difficult to target the hydroxyl radical with antioxidant additives added only in low amounts compared to the substrate (lipid) in need of protection. Another reason may be that in the majority of foods, it is not the number of hydroxyl radical-driven initiations, but rather the rate of propagation and decomposition of LO• (see below) that determines the oxidative quality deterioration rate.

From reaction (6.5) above, the resulting peroxyl radicals, LO•, may undergo further breakdown by a number of complex pathways to form aldehydes and other decomposition products that contribute to oxidative flavour deterioration of lipid-bearing foods:

$$LO\bullet \longrightarrow \text{Secondary decomposition products} \qquad (6.7)$$

Antioxidants, including those that occur naturally, can inhibit these decomposition reactions (6.7) by reducing the alkoxyl radicals directly to form lipid hydroxy compounds (Frankel, 1998):

$$LO\bullet + AH \longrightarrow LOH + A\bullet \qquad (6.8)$$

The activity of tocopherols to inhibit aldehyde formation by alkoxyl radical interception is, for example, an important part of their antioxidant activity in food systems (Frankel, 1998). An alternative mechanism has been suggested whereby antioxidants work as electron acceptors intercepting the alkoxyl radicals in a termination reaction that results in formation of antioxidant–lipid complexes (Scott, 1993):

$$LO\bullet + A\bullet \longrightarrow LOA \qquad (6.9)$$

It is not known to what degree this antioxidant radical interception reaction (6.9) occurs in real food and beverage systems.

6.9.4 Quenching of singlet oxygen

Due to its electronic configuration (with two unpaired electrons with parallel spins in the π^* (2p) orbital), oxygen in its ground state, i.e. ordinary triplet oxygen, $^3O_2$, cannot react directly with fatty acid molecules because this would require expenditure of energy. Nevertheless, ordinary triplet oxygen is a very reactive molecule which reacts readily with free lipid *radicals* to propagate lipid peroxidation as implied above. Oxygen itself can also be activated to radicals and various excited forms: such activated oxygen species include the hydroxyl radical, •OH (see reactions (6.5) and (6.6) above), hydrogen peroxide, H_2O_2, the superoxide radical, $O_2{\cdot}^-$, and singlet oxygen, $^1O_2$ (or $^1\Delta O_2$).

The singlet oxygen has a paired set of electrons in one molecular orbital, and is thus a highly electrophilic oxygen species seeking electrons to fill the vacant molecular orbital in the π^* (2p) orbital (Halliwell and Gutteridge, 1989). $^1O_2$ therefore combines rapidly with molecules containing high densities of electrons, including unsaturated fatty acids and other compounds with many double bonds, such as carotenoids and tocopherols. In foods as well as during food processing, the singlet oxygen can be generated by photosensitisation of various molecules such as chlorophyll and synthetic colorants that absorb visible or near UV light to become electronically excited and subsequently, in a series of reaction events, transfer the excitation energy to oxygen (Gordon, 2001). Tocopherols and carotenoids can inhibit photosensitised $^1O_2$ oxidation by different quenching mechanisms: (a) by deactivation of the excited sensitiser to prevent generation of $^1O_2$ (Krinsky, 1979); (b) by physically quenching $^1O_2$ without chemical reaction; (c) by chemically reacting with $^1O_2$ (Burton and Ingold, 1984; Fragata and Bellemare, 1980; Stahl and Sies, 1993). The latter mechanism causes oxidative destruction of the antioxidant. The rate of physical quenching is several orders of magnitude higher than the chemical quenching (Fragata and Bellemare, 1980), but the ratio between physical and chemical quenching mechanisms depends on the polarity of the medium. Moreover, it differs among natural antioxidants depending on their structural traits as discussed later in this chapter. In real food applications $^1O_2$ scavenging by physical quenching is confounded by the chemical quenching mechanism that depletes the antioxidants and, in the case of carotenoids, results in bleaching of the colour.

6.9.5 Enzymic antioxidant mechanisms

Certain enzyme activities can be added to foods to act as antioxidants by promoting three different types of reaction: (a) removal of oxygen; (b) removal of active oxygen species (notably hydrogen peroxide and superoxide radicals); and (c) reduction of lipid hydroperoxides (Meyer and Isaksen, 1995). Removal of oxygen by glucose oxidase coupled with catalase is both the oldest and most studied oxygen scavenging enzyme system in food applications. The overall net reaction is:

Glucose oxidase:	$2\,\beta\text{-D-glucose} + 2\,O_2 \longrightarrow 2\,\delta\text{-D-gluconolactone} + 2\,H_2O_2$
Catalase:	$2\,H_2O_2 \longrightarrow 2\,H_2O + O_2$
Spontaneous reaction:	$\delta\text{-D-gluconolactone} + H_2O \longrightarrow \text{D-gluconic acid}$
Net reaction catalysed:	$2\,\beta\text{-D-glucose} + O_2 \longrightarrow 2\,\text{D-gluconic acid}$

This principle has been shown to be workable in a number of food products including products rich in lipids such as mayonnaise and salad dressing (Mistry and Min, 1992; Isaksen and Adler-Nissen, 1997). The glucose oxidase-catalase enzyme system purified from *Aspergillus niger* is legally permitted as an antioxidative food additive in the USA, but is only permitted as a food processing aid, not a food additive, in the European Union. Other oxidases or enzyme combinations catalysing removal of oxygen via reduction of other substrates than glucose have only been little studied: therefore, although some oxidases may in fact seem more widely applicable than glucose oxidase, their usefulness in real food systems has not been unequivocally demonstrated (as discussed in Meyer and Isaksen, 1995). Superoxide dismutases, which belong to the class of enzymes able to remove active oxygen species, catalyse the removal of superoxide radicals during production of oxygen and hydrogen peroxide. Thus, in conjunction with catalase, to remove hydrogen peroxide, this system can catalyse the removal of reactive oxygen radicals in food systems during production of oxygen and water:

$$\begin{array}{ll} \text{Superoxide dismutase:} & 4\,O_2\cdot^- + 4\,H^+ \longrightarrow 2\,H_2O_2 + 2\,O_2 \\ \text{Catalase:} & 2\,H_2O_2 \longrightarrow 2\,H_2O + O_2 \\ \text{Net reaction catalysed:} & 4\,O_2\cdot^- + 4\,H^+ \longrightarrow 2\,H_2O + 3\,O_2 \end{array}$$

Although this antioxidant principle was patented already in the early 1970s, and was demonstrated as workable in various lipid oxidation *model* systems with free fatty acids, it had no antioxidant effect in food systems containing real food oils. We proposed previously that a likely explanation for this observation is that superoxide catalysed oxidation is not a significant oxidation mechanism in more realistic food systems containing oils and not only free fatty acids (Meyer *et al.*, 1994). Furthermore, the superoxide dismutase enzyme purified from *Saccharomyces cerevisiae* yeast was shown to denature in emulsions (Refsgaard *et al.*, 1992). The last mechanism is based on the abilities of glutathione peroxidases and glutathione S-transferases to catalyse the reduction of lipid hydroperoxides to hydroxides during oxidation of glutathione (G-SH). Glutathione peroxidase has a higher specific activity on lipid hydroperoxides (LOOH) than the transferase that can use different substrates (indicated RX) in addition to LOOH in the reaction:

$$\begin{array}{ll} \text{Glutathione peroxidase:} & 2\,\text{G-SH} + \text{LOOH} \longrightarrow \text{G-S-S-G} + \text{LOH} + H_2O \\ \text{Glutathione S-transferase:} & \text{G-SH} + \text{RX} \longrightarrow \text{G-SX} + \text{RH} \\ & (\text{G-SH} + \text{LOOH} \longrightarrow \text{G-S-OH} + \text{LOH}) \end{array}$$

Both enzymes have been shown to retard lipid oxidation in model systems with linoleic acid as the lipid substrate (Williamson and Ball, 1988; Williamson, 1989; Meyer, 1996). Especially glutathione peroxidase purified from bovine blood was very efficient (Meyer, 1996). But the workability and robustness of these enzymes in real food systems has yet to be demonstrated. Since enzymes, being protein catalysts, are generally sensitive to heat, they may denature during conventional food and beverage processing operations involving relatively long holding times at elevated temperatures. Enzymes may also denature at interfaces, including lipid–water and air–water interfaces (Meyer and Isaksen, 1995). The relevance and application of enzymes as antioxidative food additives may therefore increase in minimally processed foods, notably those types of process not involving harsh heat treatment or, for example, vigorous mixing procedures.

6.9.6 Relation between molecular structure and antioxidant activity

Phenolic compounds possessing multiple hydroxyl groups, especially 3′,4′*o*-hydroxy groups, are generally the most efficient chain-breaking antioxidants as this structure permits electron delocalisation. The better antioxidant efficiencies of phenols having two adjacent hydroxy groups may also be related to metal chelating capacity. Good free radical-scavenging activity of natural (as well as of synthetic) antioxidants involves the ability to easily donate a hydrogen as well as stabilisation of the resulting antioxidant radical by electron delocalisation. A major factor governing the antioxidative efficacy of primary chain-breaking antioxidants has thus long been said to be due to the stability and lack of further oxidative reactivity of the primary antioxidant radical (A•) that is formed after the hydrogen abstraction (Schuler, 1990). With flavonoids, notably three structural traits have been highlighted as important determinants for hydrogen donation and of antioxidant efficiency (Shi *et al.*, 2001):

1. The presence of a 3′,4′o-hydroxyl group, i.e. a catechol structure in the B ring (for phenolic acids simply the presence of this dihydroxy structure).
2. Presence of hydroxyl groups at the 5 position in the A ring and at the 3 position in the middle, C ring.
3. The 4-oxo groups in conjugation with a 2,3 double bond in the C ring.

The high antioxidant activity of natural phenolic compounds having a catechol structure on the phenolic ring may also be related to the ability to form a transient aryloxyl radical having high antioxidant activity: hence, from data obtained recently in studies of solvent effects on phenolic antioxidants (Foti and Ruberto, 2001), it was proposed that the strong antioxidant activity of catechol structures in natural antioxidants could be due to the presumed low activation energy of the transfer of the second phenolic H atom in the reaction between the antioxidant hydroquinone radical and a lipid peroxyl group to yield the *o*-quinone (Fig. 6.5).

Carotenoids with 9 or more double bonds are particularly efficient as $^1O_2$ quenchers. Thus, a β-carotene molecule, having 11 conjugated double bonds, is capable of quenching on average ~250 molecules of $^1O_2$ before being destroyed by chemically reacting with $^1O_2$ (Korycka-Dahl and Richardson, 1978); estimates of quenching of up to 1,000 $^1O_2$ molecules by β-carotene have also been published (Foote *et al.*, 1970). In comparison, a molecule of α-tocopherol may, at best, physically quench around 40 molecules of $^1O_2$ before being oxidatively degraded (Korycka-Dahl and Richardson, 1978).

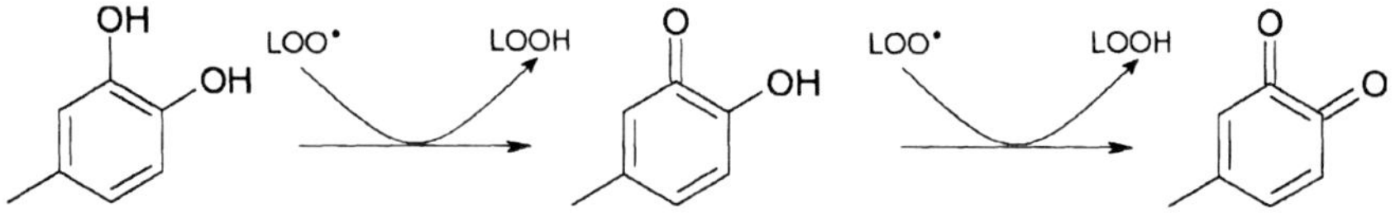

Fig. 6.5 Proposed reaction scheme of the catechol moiety in natural phenolic antioxidants eliciting their activity through the radical chain-breaking mechanism. The rate constant of the second reaction, shown here as a reaction via the transient aryloxyl radical, may be higher than the first hydrogen transfer reaction from the ortho-hydroxyl group (adapted from Foti and Ruberto, 2001).

6.10 Commercial natural antioxidants: sources and suppliers

As mentioned previously, nature is rich in natural antioxidants protecting against oxidative degradation. Of these, antioxidants from plants have increasingly been utilised for stabilisation of foods and they exist in either pure forms or as concentrated preparations produced from, in particular, spices and herbs, seeds and beans, berries and fruits or from tea leaves. There is a scarcity of published knowledge available on the molecular composition of these antioxidant concentrates. As a result, the precise action mechanisms of the antioxidants present and their individual and combined efficiency is at present not elucidated in detail. Thus, despite our rather detailed understanding of the various mechanisms by which natural antioxidants may act – as discussed in the previous sections – it is currently difficult to predict the activities and efficiencies of various plant extracts of mixed composition without knowing the compositional profiles of the preparations. In addition, certain natural antioxidant phenols may act synergistically or even antagonistically (Meyer *et al.*, 1998), which further complicates predictions of antioxidant effectiveness of mixed concentrates. At present, therefore, the application of most natural antioxidant concentrates is based on empirical knowledge from tests in model systems and a limited number of trials in real products.

In 1996 it was estimated by Frost and Sullivan that the natural antioxidants made up 26% of the whole food antioxidant market, growing by 6–7% per year. The forecast for the natural antioxidants is a 38% share of the market in 2004. To that comes the steady growing market for health products believed to contribute positively to the health of human beings.

In Table 6.5 is seen an overview of the major commercial or pre-commercial and natural antioxidants for food preservation. These may be pure or, more often, crude preparations formulated to be used in foods with no or minimal contribution to taste, flavour or colour at 0.01–1% addition but some natural antioxidants, e.g. the flavonoids, the carotenoids and some spice or herb extracts, may contribute to colour or flavour.

In the table, some of the major active compounds and suggested mechanisms are introduced but very often the mechanisms are unclear and strongly depend on the actual food matrix.

Table 6.5 Examples of some commercial and natural antioxidant preparations

Antioxidant	Some active compounds	Antioxidant mechanism	Comments
Tocopherol/ tocotrienol	$\alpha, \beta, \gamma, \delta$ tocopherols and tocotrienols	Chain breaker, singlet oxygene quencher	From oil seeds and cereal oils. Synergists: ascorbic and citric acids, phospholipids Ref. 1, 2
Ascorbic acid	Ascorbic acid and salts	Chain breaker, oxygene scavenger	From acerola cherries, rose hips, paprika and citrus. Ref. 1, 2
Rosemary extracts	Rosmarinic acid, carnosic acid, carnosol	Chain breaker, metal chelator	Water- or oil-soluble preparations. Slight taste and colour. Synergists: ascorbic acid. Ref. 1, 3, 4
Oregano extracts	Rosmarinic acid, protocatechinic acid, caffeic acid, tocopherols, thymol, carvacrol	Chain breaker, metal chelator	Water- or oil-soluble preparations. Slight colour and taste. Synergists: ascorbic acid. Ref. 1, 3, 4, 5
Sage extracts	Rosmarinic acid, flavon glycosides	Chain breaker, superoxide radical scavenger, reducer of transition metals	Slight colour and taste. Ref. 6
Green tea extracts	Catechin, epigallocatechin, gallic acid	Chain breaker, metal chelator	Some colour and taste. Ref. 7, 8, 9, 10
Olive extracts	Biophenols, e.g. hydroxytyrosol, vanillic acid, caffeic acid, oleuropein	Chain breaker	From olive seeds or leaves. Ref. 11, 12, 13, 19
Organic acids	Citric, tartaric, malic acid and salts	Metal chelator	From many vegetable sources. Synergists: tocopherol. Ref. 14
Phytic acid	Phytic acid derivatives	Metal chelator, chain breaker	From many seeds and cereals. Ref. 15
Lecithin	Phospholipids, e.g. phosphatidylethanolamine, phosphatidylcholine, phosphatidylserine	Decompose hydro- peroxides, synergist to chain breakers, metal chelator	Fat- or water-dispersable preparations. Synergists to tocopherols and flavonoids. Ref. 16, 17
Natural carotenoids	α-carotene, β-carotene, lycopene, lutein, zeaxanthin, cryptoxanthin	Singlet oxygene quencher, some are chain breakers, secondary stabiliser of many primary antioxidants	From many fruits and vegetables. Often coloured. Ref. 11
Natural flavonoids	Anthocyanins, cyanidins, quercetins, catechins	Chain breaker, metal chelator, radical scavenger, superoxide scavenger	From many fruits and vegetables. Often coloured. Ref 4, 11, 18

Notes: 1. Schuler, 1990. 2. Frankel, 1998. 3. Nakatani, 1994. 4. Larson, 1997. 5. Tisimidou *et al.*, 1994. 6. McCarthy *et al.*, 2000. 7. Cao *et al.*, 1996. 8. He, 1997. 9. Manzocco *et al.*, 2000. 10. Roedig-Penman *et al.*, 1997. 11. Wanasundara *et al.*, 1997. 12. Le Tutour, 1992. 13. Paiva-Martins and Gordon, 2001. 14. Gordon, 1990. 15. Graf *et al.*, 1984. 16. Jung *et al.*, 1997. 17. Saito, 1997. 18. Wang *et al.*, 1997. 19. Uccella, 2001.

6.10.1 Extracts from seeds and beans

Most unrefined plant oils are naturally protected against oxidation by a number of antioxidants, in particular the tocopherols, tocotrienols, simple phenolics, flavonols, isoflavonoids, lecithin and many others.

Most of these are lost during the refinement of the oils but can be recovered as natural antioxidants. Among these, the tocopherols are the most important. They are often mixtures of α, β, γ, δ tocopherols and tocotrienols, but contain mostly γ and δ which often are more efficient antioxidants in foods at elevated temperatures than vitamin E (α-tocopherol). The tocotrienols are claimed to have a 40–60 times higher antioxidant activity compared to the tocopherols (Palm Oil Research Institute of Malaysia, private communication).

At low concentrations (100 ppm) in purified fish oil triacylglycerols, the retardation of hydroperoxides decreases in the order α-tocopherol $>$ γ-tocopherol $>$ δ-tocopherol but the reverse order is seen at 1,000 ppm. This is due to a concentration dependence of the antioxidant capacity, showing a maximum at 100 ppm and 500 ppm for α- and γ-tocopherols, respectively (Kulas and Ackman, 2001), while the δ form increase by concentration up to 1,500–2,000 ppm.

Some of the largest producers of natural tocopherols are Eisai (Japan), Henkel/Cognis (Germany), ADM (USA), Eastman Chemical (USA) and Cordia, but Daminco Inc. (USA), Jan Dekker International (the Netherlands) and Kemin Foods (USA) also produce or market natural tocopherols. Important synergists to the tocopherols are citric acid and other organic carboxylic acids, phosphoric acid and phospholipids, vitamin C, monoglycerides, alanin, cystein and rosemary extracts (e.g. Schuler, 1990).

Another important antioxidant from oil seeds is lecithin, which is a phospholipid also contributing to stabilising food emulsions and is approved as a natural emulsifier. Lecithin may complex metals, e.g. iron and copper, thereby reducing the metal-catalysed oxidation in foods. It has been shown that the high antioxidant effects of the patented antioxidant mixture consisting of ca. 8% ascorbic acid, ca. 86% lecithin and ca. 5% γ-tocopherol, i.e. the A/L/T system (Löliger and Saucy, 1989) rests on a synergistic action involving the lecithin. Thus, in this system, lecithin not only functions as an emulsifier to keep ascorbic acid and gamma-tocopherol in close proximity, but also appears to participate in a synergistic radical exchange process between these other antioxidants (Lambelet *et al.*, 1994). This ability of lecithin may partly explain why in some studies, lecithin protected oxidation sensible oils better than tocopherols (Jung, 2001). Producers of lecithin are the main refiners of vegetable oils, e.g. Lucas Meyer (Germany) or Riceland (USA).

Phytic acid is found in many seeds, e.g. canola, corn germ, seam, wheat and soy. It is extracted and may be used as a food or non-food antioxidant, mainly chelating metal ions, but its strong iron-binding effects may be detrimental for food uses because of the strongly depressed iron availability for nutritional purposes.

Isoflavonoids are isolated and concentrated from the soy production in particular. The products are rich in genistein and daidzein but in foods their antioxidant activity is not very high (Larson, 1997). However, they may have synergistic effects to other antioxidants. There are many producers of this product, e.g. Central Soya.

Furthermore, it has been claimed that extracts from oat are good antioxidants containing dihydrocaffeic acid and phospholipids and that sesame seed/sesame oil contains several active antioxidants including sesamol, sesamolin and sesamolinol (Wanasundara *et al.*, 1997). Sesame fractions have also been claimed to be used commercially as food antioxidants but it has not been possible to identify producers. Ferulate esters from rice bran are also claimed to be efficient antioxidants protecting linoleic acid better than α-tocopherol (Xu and Godber, 2001).

6.10.2 Extracts of berries and fruits, etc.

Many fruits and berries are rich in antioxidants, which are or can be extracted and concentrated. When added to foods these ingredients are able to contribute significantly to the stability of the food in question by improving the oxidation stability.

The antioxidant components are mostly vitamin C, organic acids and a large number of phenolics, e.g. the hydroxycinnamates such as caffeic acid and ferulic acid, as well as various flavonoids, some of which are strongly coloured, e.g. anthocyanins, which are often utilised as natural food colorants. Some fruits are also very rich in carotenoids, including β-carotene, which may protect against oxidation via singlet oxygen quenching as discussed in section 6.9.4.

A large number of companies produce extracts or concentrates from fruits and berries ranging from the traditional juice producers to companies that specialise in natural flavours and colours. For example, the following companies can be mentioned: Optipure (USA), Chr. Hansen (Denmark), Overseal Natural Ingredients (GB), Quim Dis (France), Inheda (France), Folexco (USA) and many of the flavour houses mentioned earlier.

Some other interesting natural antioxidants are extracts and concentrates from the acerola cherry or rose-hips, containing up to 20 g ascorbic acid per kg berry. The concentrates contain between 17% and 26% ascorbic acid, even up to 50%. These products are produced and marketed by Anidro Do Brasil, but also by Quim Dis (France). Combined with a neutralisation of the acid, this may be a strong new natural antioxidant in many foods.

Other good future candidates may be extracts of bilberries containing 25% anthocyanidins (Optipure), prune extracts with high antioxidant score (California Prune Board), blackcurrant extracts containing anthocyanins (Ocean Spray, USA) and extracts of cranberries, elderberries and strawberries, all strongly coloured.

Grape skin and seeds are also excellent sources of many polyphenols and the byproducts from the wine-making process are a basis for interesting commercial activities (Tomera, 1999). The antioxidants in grape products are resveratrol, quercetin, anthocyanins, cinnamates and other flavonoids. Some of the grape seed extracts are claimed by the producers to possess up to 50 times stronger antioxidant activity than C or E vitamins. This type of antioxidant may be of interest when food protection, colours and a health image are wanted. Producers of grape or wine extracts are Overseal Natural Ingredients (GB), Inheda (France), Quim Dis (France), Seppic Inc (USA) and Polyphenolics (USA).

Crude olive oil, in particular extra virgin olive oil, contains 50–800 mg/kg polyphenols, in particular hydroxytyrosol, tyrosol, vanillic acid, caffeic acid and oleuropein (Visioli and Galli, 1998). These antioxidants are shown to have an equal or larger antioxidant capacity than vitamin C, vitamin E and BHT. Several olive oil producers have investigated the possibility of producing commercial natural antioxidants from olives, e.g. CreAgri Inc. (USA).

Finally, Nutrinova has developed a new food ingredient claimed to have good food antioxidant function. The product is produced from carob fruit vast and contains lignins, simple phenols and tannins.

6.10.3 Extracts from spices and herbs

Spices and herbs have been used for many centuries to enhance flavour and extend the shelf-life of various foods. More than 32 spices have been shown to behave as antioxidants in lard and in particular plants from the Lamiaceae family appear to be effective and often even more effective than BHA. Among the most effective are rosemary, marjoram, sage, oregano, basil, savoury and hyssop (Risch and Ho, 1994) as well as allspice, clove, thyme, turmeric, mace, nutmeg, ginger, cassia and cinnamon (Charalambous, 1994).

Today, rosemary, sage and oregano are exploited on a commercial scale and are produced in several forms, in particular reduced in colour and flavour and often mixed with other natural antioxidants and synergists. In fact, by extraction or molecular distillation it is often possible to isolate the antioxidant principles without any colour or flavour. They are produced to protect fats and oils, emulsions, meat, fish, etc., in the level of 0.1% and are often found to be more efficient than synthetic antioxidants (e.g. Formanek *et al.*, 2001; McCarthy *et al.*, 2000).

Rosemary extracts are produced by a large number of producers of flavours and essential oils, in particular Kalsec (USA), Naturex (France), Chr. Hansen (Denmark), SKW Chemicals (Germany), Ecoms (Canada), Evesa (Spain), R.C. Treatt (GB), Frutarom Meer (USA), Kanegrade (GB), Adrian (France) and Bush Broake Allen Ltd (GB).

Only some of the producers are offering products with reduced flavour and colour while others are sole producers of essential oils, oleoresins or chemical isolated active compounds, e.g. carvone, eugenol or thymol.

Naturex, Chr. Hansen and Kalsec produce some of the most developed natural antioxidants for food shelf-life. These companies deliver flavour- and colour-reduced products and mixes with other different natural antioxidants, e.g. natural tocopherols, natural vitamin C, natural citric acid, natural carotenoids or other spice extracts.

Naturex produces a range of products called 'OxyLess', which are either water soluble or oil soluble to be used in concentrations of 0.03–0.5%. Application examples are chicken fat, frying oils, snacks, sausages, pork, fish oil and mayonnaise and the effects are often claimed to be better than the synthetic antioxidants, e.g. BHA and BHT. The company also delivers mixed products with other natural antioxidants, flavours and colours.

Chr. Hansen produces a product range called 'FlavorGuard', which is also reduced in flavour and colour and combined with other natural antioxidants. Recommended usage in foods is 0.02–0.1% based on fat content and a large number of application examples are available.

Kalsec is a third large producer of natural antioxidants for foods. 'Herbalox' and 'Duralox' are delivered in several forms and mixes, and Kalsec includes extracts from rosemary and sage as well as natural vitamin C, tocopherols and citric acid. The antioxidative effects are often claimed to be higher than the synthetic antioxidants.

The most active components in rosemary extracts are carnosic acid and carnosol (see Fig. 6.4), which is said to account for 90% of the activity but also rosmanol, epirosmanol, isorosmanol, rosmarinic acid, rosmaridiphenol, rosmadial and rosmariquinone are expected to contribute.

Oregano extracts are new on the market and are produced by the Israeli company RAD International. The product 'Origanox' is delivered in water-soluble or oil-soluble forms; the former is claimed to have strong antimicrobial activities. Flavours and colours have been reduced for food use. Just as in thyme and ginger, oregano extracted from *Origanum vulgare* contains thymol, carvacrol, zingerone and 6-gingerol, but also rosmarinic acid, protocatechuic acid and caffeic acid as the antioxidant principles.

The recommended usage levels for most applications are 0.01–0.2% and the product may show synergistic effects to other natural antioxidants. The products have a well-documented antioxidant effect in many foods and are claimed by the producer to be more efficient than BHA, vitamin C, tocopherols and rosemary oil (not extracts), using the Rancimat or TBA tests. The products are effective in meat, poultry, fish, oils, emulsions, flavours, soups, bakery products, dairy products, etc.

6.10.4 Tea extracts

Existing new natural antioxidants are the extracts and concentrates from green tea, which often contain 30% polyphenolics or more. Green tea antioxidants are different from other natural sources and consequently they may show interesting synergistic effects. The major antioxidants in green tea are the catechins (flavonoids), e.g. epicatechin, epigallocatechin, epicatechin gallate and epigallocatechin gallate.

Recently, the products have been modified to meet the needs in the food industry and some producers claim a low influence on colour and taste of the foods in question and a higher antioxidant capacity compared to other natural antioxidants. Tea antioxidants are typically added to 800 ppm and often mixed with tocopherols and citric acid.

The green tea antioxidants are claimed to be effective in oils, fish and emulsions. The tea catechins may be added directly to the food or to animal feeds; 200–300 mg tea catechins in chicken feed improved significantly the oxidation stability of frozen or refrigerated meat and was comparable to the same quantity of tocopherol (Tang *et al.*, 2001) and several studies show an effective protection of pork, beef and fish after addition of tea catechins to the foods at the 300 mg/kg level (McCarthy *et al.*, 2001; Shuze Tang *et al.*, 2000), often with a better protection than by BHA/BHT or tocopherols.

The largest producers are Japanese or Chinese companies, e.g. Maruzen (Japan), Takeda (Japan), Sunphenon/Taiyo Kagaku (Japan), Naturex (China) but also AIM (USA), Inheda (France) and Optipure (USA).

6.10.5 Others

Antioxidative enzymes, in particular the glucose oxidase-catalase systems (see section 6.9) or L-ascorbate oxidase, are produced by most enzyme producers.

Also of interest are the strong iron complexing proteins, in particular the lactoferrins isolated from milk. By complexing iron, metal-catalysed oxidation can be reduced. Lactate-rich fractions from cheese whey do have the same effect and are claimed to have the same antioxidant capacity as BHT in shelf-life extension of pork, for example. Producers are among others DMV (the Netherlands), Armor Protéines (France) and Arla Foods Ingredients (Denmark).

Interesting natural antioxidants can also be found from the sea. Some sea algae and shellfishes are excellent sources of α-carotene, β-carotene, zeaxanthin, cryptoxanthin, astaxanthin and lutein. Producers are Henkel Corporation and Gist-Brocades. It has also been claimed that extracts from shrimp vast containing, for example, chitin/chitosan and rockfish have good antioxidant efficacy.

6.11 Natural compounds with dual protective functionality as preservatives and antioxidants

6.11.1 Metal chelators: organic acids

Citric acid is quantitatively the most widely used organic acid additive in foods, and is utilised in many different products. The production of citric acid is based on fermentation of *Aspergillus niger*. Citric acid is a weak organic acid ($pK1 = 3.1$; $pK2 = 4.7$; $pK3 = 5.4$), and is thus not completely dissociated at the pH range of 5–7 of many food commodities (soft drinks and fruit juices have a lower pH, usually $pH < 4$). As a metal chelator, citric acid is widely used as an antioxidative agent, and is widely recognised to act as an antioxidant synergist to both ascorbic acid and tocopherols. In addition, combinations of citric acid with various radical chain-breaking antioxidants have been documented to prevent oxidative deterioration and rancidity in a wide range of products (Dziezak, 1986). Due to its ability to chelate metals and decrease pH, citric acid is also widely used as a browning inhibitor. Thus, the chelation of copper retards enzyme-catalysed browning reactions in foods. Citric acid may also affect the growth and survival of microorganisms in foods due to its acidic properties. Weak organic acids exert an inhibitory effect against microorganisms, as described in section 6.4. More systematic studies, on whether citric acid added to foods as a metal chelator to retard lipid oxidation (resulting in a final concentration in the food of 0.005–0.02% by weight of citric acid) also exerts significant antimicrobial activities, could fill an important gap in our understanding of such dual functionality of weak organic acids.

6.11.2 Metal chelators: lactoferrin

Long before the recently demonstrated antioxidant potency of lactoferrin was discovered, lactoferrin was known to exert bacteriostatic effects towards a range of microorganisms in a variety of food and food model systems. Like its antioxidant activity, the antimicrobial activity is ascribed to the ability of lactoferrin to sequester free iron required for microbial growth (Oram and Reiter, 1968). Thus, both the antibacterial as well as the antioxidant activity of lactoferrin from bovine milk is decreased by added iron ions (Bullen *et al.*, 1972; Satué-Gracia *et al.*, 2000). Nevertheless, lactoferrin seems well suited to be used as a dual purpose additive for its antimicrobial as well as its antioxidant properties in iron supplemented foods, e.g. in commercial infant formulas (Satué-Gracia *et al.*, 2000). Further research to elucidate the applicability of lactoferrin and other natural metal chelators of dual functionality as antimicrobial preservatives and antioxidative agents in real 'low metal' foods is highly warranted.

6.11.3 Plant extracts

Very recently, a number of aqueous extracts of black teas, sage and linden flowers, as well as especially eucalyptus wood hydrolysates, were demonstrated to exert both antioxidant and antimicrobial effects in a number of model systems (Cruz *et al.*, 2001; Yildirim *et al.*, 2000, 2001). However, further assessment of the applicability and safety of these mixed, natural constituents in food applications is required before their real applicability potential can be assessed. Likewise, the detailed elucidation of the active constituents and their activity mechanisms is required. A number of herb and spice essential oils have also been successfully tested for both antimicrobial and antioxidant effects (Baratta *et al.* 1998), which is obvious as the same phenolic compounds (carvacrol, thymol, etc.) play a key role for both. Sensory issues seem to be the main limiting factor for the use of these compounds.

6.11.4 Enzymes

In addition to its antioxidative potency by catalysing oxygen removal, glucose oxidase has been shown to produce a bactericidal effect when added to a biofilm model system consisting of *Pseudomonas aeruginosa* attached to soft contact lenses (Johansen *et al.*, 1997). It was speculated that this antibacterial activity of glucose oxidase was due to the cytotoxicity of the H_2O_2 or simply an effect of O_2 depletion (Fuglsang *et al.*, 1995), but the possible presence and role of catalase in the tested glucose oxidase preparation was not investigated (see glucose oxidase-catalase reactions above). At present the potential use of enzymes as food protectants, including oxidoreductase systems such as glucose-oxidase coupled with catalase, to inhibit both bacterial growth and oxidative deterioration are far from elucidated nor realised at present. Notably, it seems important to evaluate possible negative effects of retarding the growth of strictly aerobic microorganisms in minimally processed foods, as this might, in certain cases, give a competitive advantage to undesirable anaerobic bacteria, including pathogenic *Clostridia*.

6.12 Conclusion and future trends

Minimal processing is a preservation concept undergoing strong development because of consumer preferences for fresh or fresh-like foods, high sensory quality, adequate convenience and high safety but without chemical additives.

The technological and scientific challenge is obvious but possibly can never be met using a single preservation method. Several hurdles to quality degradation will be necessary to minimise enzymatic degradation, oxidation of lipids, proteins, colours and nutrients, to reduce non-enzymatic colour and texture changes and to prevent growth of pathogenic as well as food spoilage microorganisms.

The needs of good preservation hurdles are especially strong for non-thermal minimally processed foods. In these, a much more selective microbial growth inhibition compared to heated foods is often seen, as well as a considerable hydrolytic and oxidative enzymatic activity and in some minimal processing treatments, even enhanced enzymatic activity may result, e.g. after high pressure treatments.

Some natural preservatives are today used as excellent hurdles in food deterioration but more systematic studies on multi-synergistic effects are scarce in real food systems, e.g. combining lactoferrin, organic acids and oregano extracts with modified atmosphere packaging and pulsed electric field technology to prevent microbial growth. Further, more fundamental knowledge of the mechanism of action of the natural preservatives is needed in order to set up an effective preservation strategy for each product type. Very often it is seen that natural preservatives have strong bactericidal effect on many pathogens in laboratory test systems but only weak to moderate effect in real foods. For this reason, much more research on shielding and complexing factors from the food matrix is needed.

While the natural preservatives are still in their nascent state, the antioxidants are much more advanced and do have a large market share. We do understand many fundamental mechanisms of action and, empirically, many synergists are identified. But our understanding of the intermolecular interferences behind multi-hurdle preservation is still limited, e.g. how flavonoids or carotenoids may synergistically enhance the effect of tocopherols.

The natural concept is not only limited to preservatives and antioxidants. New challenging research areas are natural enzyme inhibitors to delay the enzymatic deterioration and improved selective and natural complexing agents, e.g. influencing vital functions of microorganisms or stabilising food texture.

6.13 References and further reading

ADEGOKE G O and ODESOLA B A (1996), Storage of maize and cowpea and inhibition of microbial agents of biodeterioration using the powder and essential oil of lemon grass (*Cymbopogon citrates*), *Int. Biodet. Biodeg.*, **37**, 81–4.

ANDERSEN S J and FRISVAD J C (1994), Penicillin production by Penicillium nalgiovense, *Letters in Applied Microbiology*, **19**, 486–8.

AL-KHAYAT M A and BLANK G (1985), Phenolic spice components sporostatic to *Bacillus subtilis*, *Journal of Food Science*, **50**, 971–4.

ARRAS G and USAI M (2001), Fungitoxic activity of 12 essential oils against four postharvest citrus pathogens: chemical analysis of *Thymus capitatus* oil and its effect in subatmospheric pressure conditions, *Journal of Food Protection*, **64**, 1025–9.

ATKINSON P and BLAKEMAN J P (1982), Seasonal occurrence of an antimicrobial flavanone, sakuranetin, associated with glands on leaves of *Ribes nigrum*, *The New Phytologist*, **92**, 63–74.

BABIC I, NGUYEN-THE C, AMIOT M J and AUBERT S (1994), Antimicrobial activity of shredded carrot extracts on food-borne bacteria and yeast, *Journal of Applied Bacteriology*, **76**, 135–41.

BARATTA M T, DORMAN H J D, DEANS S G, FIGUEIREDO A C, BARROSO J G and RUBERTO G (1998), Antimicrobial and antioxidant properties of some commercial essential oils, *Flavour Fragr. J.*, **13**, 235–44.

BARRA D and SIMMACO M (1995), Amphibian skin: a promising resource for antimicrobial peptides, *TIBTECH*, **13**, 205–9.

BARTOLOME B, ESTRELLA I and HERNANDEZ M T (2000), Interaction of low molecular weight phenolics with proteins (BSA), *Journal of Food Science*, **65**, 617–21.

BASILICO M Z and BASILICO J C (1999), Inhibitory effects of some spice essential oils on *Aspergillus ochraceus* NRRL 3174 growth and ochratoxin A production, *Letters in Applied Microbiology*, **29**, 241.

BEUCHAT LR (1994) Antimicrobial properties of spices and their essential oils. In Dillon V M and Board R G (eds) *Natural Antimicrobial Systems and Food Preservation*, pp. 167–79. Wallingford: CAB International.

BEUCHAT LR and GOLDEN DA (1989), Antimicrobials occurring naturally in foods, *Food Technology*, **43**, 134–42.

BISIGNANO G, LAGANA MG, TROMBETTA D, ARENA S, NOSTRO A, UCCELLA N *et al.* (2001), In vitro antibacterial activity of some aliphatic aldehydes from *Olea europaea* L. FEMS *Microbiology Letters*, **198**, 9–13.

BLASZYK M and HOLLEY RA (1998), Interaction of monolaurin, eugenol and sodium citrate on growth of common meat spoilage and pathogenic organisms. *International Journal of Food Microbiology*, **39**, 175–83.

BOGDANOV S (1985), Characterisation of antibacterial substances in honey, *Lebensm.-Wiss. u.-Technol.*, **17**, 74–6.

BRACEY D, HOLYOAK CD, NEBE-VON CARON G and COOTE PJ (1998), Determination of the intracellular pH (pH_i) of growing cells of *Saccharomyces cerevisiae*: the effect of reduced-expression of the membrane H+-ATPase. *Journal of Microbiological Methods*, **31**, 113–25.

BRANEN J and DAVIDSON PM (2000), Activity of hydrolysed lactoferrin against foodborne pathogenic bacteria in growth media: the effect of EDTA. *Letters in Applied Microbiology*, **30**, 233–7.

BREEUWER P, DE REU JC, DROCOURT J-L, ROMBOUTS F and ABEE T (1997), Nonanoic acid, a fungal self-inhibitor, prevents germination of *Rhizopus oligosporus* sporangiospores by dissipation of the pH gradient, *Applied and Environmental Microbiology*, **63**, 178–85.

BRUL S and COOTE P (1999), Preservative agents in foods mode of action and microbial resistance mechanisms, *International Journal of Food Microbiology*, **50**, 1–17.

BULLEN J J, ROGERS H J and LEIGH L (1972), Iron-binding proteins in milk and resistance to Esherichia coli infection in infants, *Br Med J*, **1**, 69–75.

BURTON GW and INGOLD KU (1981), Autoxidation of biological molecules. I. The antioxidant activity of vitamin E and related chain-breaking phenolic antioxidants in vitro, *J Am Chem Soc*, **103**, 6472–7.

BURTON G W and INGOLD KU (1984), β-carotene: an unusual type of lipid antioxidant, *Science*, **224**, 569–73.

CAO, G, SOFIC E and PRIOR RL (1996), Antioxidant capacity of tea and common vegetables, *J Agric Food Chem*, **44**, 3426–31.

CHARALAMBOUS G (1994), *Spices, Herbs and Edible Fungi*, Amsterdam: Elsevier Science.

COMMISSION DECISION 1999/217/EC: *http://europa.eu.int/comm/food/fs/sfp/flav02_en.pdf*

CONNER DE (1993), Naturally occurring compounds. In Davidson PM and Branen A L (eds) *Antimicrobials in Food*, pp. 441–68. New York: Marcel Dekker.

CONNER DE and BEUCHAT LR (1984), Effects of essential oils from plants on growth of food spoilage yeasts, *Journal of Food Science*, **49**, 429–34.

COUNCIL DIRECTIVE 89/107/EEC: *http://europa.eu.int/comm/food/fs/sfp/flav07_en.pdf*

COX S D, GUSTAFSON J E, MANN C M, MARKHAM J L, LIEW Y C, HARTLAND R P *et al.* (1998), Tea tree oil causes K^+ leakage and inhibits respiration in *Escherichia coli*, *Letters in Applied Microbiology*, **26**, 355–8.

COX S D, MANN C M, MARKHAM J L, BELL H C, GUSTAFSON J E, WARMINGTON J R *et al.* (2000), The mode of antimicrobial action of the essential oil of Melaleuca alternifolia (tea tree oil), *Journal of Applied Microbiology*, **88**, 170–5.

CRUZ J M, DOMINGUEZ J M, DOMINGUEZ H and PARAJO J C (2001), Antioxidant and antimicrobial effects of extracts from hydrolysates of lignocellulosic materials, *J Agric Food Chem*, **49**, 2459–64.

DE LUCCA A J, BLAND J M, JACKS T J, GRIMM C and WALSH T J (1998), Fungicidal and binding properties of the natural peptides cecropin B and dermaseptin, *Medical Mycology*, **36**, 291–8.

DEANS S G (1991) Evaluation of antimicrobial activity of essential (volatile) oils. In Linskens H F and Jackson J F (eds) *Essential Oils and Waxes*, pp. 309–20. Berlin Heidelberg: Springer-Verlag.

DEANS S G and RITCHIE G (1987), Antibacterial properties of plant essential oils, *International Journal of Food Microbiology*, **5**, 165–80.

DEL CAMPO J D, AMIOT M J and NGUYEN T C (2000), Antimicrobial effect of rosemary extracts, *Journal of Food Protection*, **63**, 1359–68.

DELAQUIS P J and MAZZA G (1995), Antimicrobial properties of isothiocyanates in food preservation, *Food Technology*, **49**, 73–84.

DELAQUIS P J and SHOLBERG P L (1997), Antimicrobial activity of gaseous allyl isothiocyanate, *Journal of Food Protection*, **60**, 943–7.

DELAQUIS P J, WARD S M, HOLLEY R A, CLIFF M C and MAZZA G (1999), Microbiological, chemical and sensory properties of pre-cooked roast beef preserved with horseradish essential oil, *Journal of Food Science*, **64**, 519–24.

DENYER S P and HUGO W B (1991), *Mechanisms of Action of Chemical Biocides: Their Study and Exploitation.* Oxford: Blackwell Scientific Publications.

DIKER K S, AKAN M, GULSEN H and YURDAKOK M (1991), The bactericidal activity of tea against *Campylobacter jejuni* and *Campylobacter coli*, *Letters in Applied Microbiology*, **12**, 34–5.

DILLON V M and COOK P E (1994), Biocontrol of undesirable microorganisms in food. In Dillon V M and Board R G (eds) *Natural Antimicrobial Systems and Food Preservation*, pp. 255–96. Wallingford: CAB International.

DIRECTIVE 94/35/EU: *http://europa.eu.int/comm/food/fs/sfp/flav10_en.pdf*

DIRECTIVE 94/36/EU: *http://europa.eu.int/comm/food/fs/sfp/flav08_en.pdf*

DIRECTIVE 95/2/EC: *http://europa.eu.int/comm/food/fs/sfp/flav11_en.pdf*

DIXON R A (2001), Natural products and plant disease resistance, *Nature*, **411**, 843–7.

DZIEZAK J D (1986), Preservatives: antioxidants – the ultimate answer to oxidation, *Food Technol*, **40**, 94–102.

ECKHOLDT, A (2001). Allelopati, *JordbrugsForskning*, **7**, 2–3.

EJECHI B O, NWAFOR O E and OKOKO F J (1999), Growth inhibition of tomato-rot fungi by phenolic acids and essential oil extracts of pepperfruit (*Dennetia tripetala*), *Food Research International*, **32**, 395–9.

EKLUND T (1980), Inhibition of growth and uptake processes in bacteria by some chemical food preservatives, *Journal of Applied Bacteriology*, **48**, 423–32.

ELGAYYAR M, DRAUGHON F A, GOLDEN D A and MOUNT J R (2001), Antimicrobial activity of essential oils from plants against selected pathogenic and saprophytic microorganisms. *Journal of Food Protection*, **64**, 1019–24.

EL-GAZZAR, F E, RUSEL G and MARTH E M (1987), Growth and aflatoxin production by *Aspergillus parasiticus* NRRL 2999 in the presence of lactic acid and at different initial pH values, *Journal of Food Protection*, **50**, 940–4.

EMPSON K L, LABUZA T P and GRAF E (1991), Phytic acid as a food antioxidant, *J Food Sci*, **56**, 560–3.

EUROPEAN PARLIAMENT AND COUNCIL REGULATION EC/2232/96: *http://europa.eu.int/comm/food/fs/sfp/flav04_en.pdf*

FABECH B, HELLSTRØM T, HENRYSDOTTER G, HJULMAND-LASSEN M, NILSSON J, RÜDINGER L, SIPILÄINEN-MALM T, SOLLI E, SVENSSON K, THORKELSSON A E and TUOMAALA V (2000), Active and intelligent food packaging: a Nordic report on the legislative aspects, TemaNord 2000, 584, Nordic Council of Ministers, Copenhagen.

FDA, CODE OF FEDERAL REGULATIONS: 21 CFR 172, 2000: *http://www.access.gpo.gov/gao/images/PDF.gif*

FOOTE C S, DENNY R W, WEAVER L, CHANG Y C and PETERS J (1970), Quenching of singlet oxygen, *Ann N Y Acad Sci*, **171**, 139–48.

FORMANEK Z, KERRY J P, HIGGINS F M, BUCKLEY D L, MORRISEY F A and FARKAS J (2001), Addition of synthetic and natural antioxidants to alfa tocopheryl-acetate supplemented beef patties: effects of antioxidants and packaging on lipid oxidation, *Meat Sci.*, **58** (4), 337–41.

FOTI M and RUBERTO G (2001), Kinetic solvent effects on phenolic antioxidants determined by spectrophotometric measurements, *J Agric Food Chem*, **49**, 342–8.

FRAGATA M and BELLEMARE F (1980), Model of singlet oxygen scavenging by α-tocopherol in biomembranes, *Chem. Phys. Lipids*, **27**, 93–9.

FRANKEL E N (1998), *Lipid Oxidation*, Dundee: The Oily Press.

FRANKEL E N, HUANG S-W, AESCHBACH R and PRIOR E (1996), Antioxidant activity of a rosemary extract and its constituents, carnosic acid, carnosol, and rosmarinic acid, in bulk oil and oil-in-water emulsions, *J Agric Food Chem*, **44**, 131–5.

FRANKEL E N and MEYER A S (2000), The problems of using one-dimensional methods to evaluate multifunctional food and biological antioxidants, *J Sci Food Agric*, **80**, 1925–41.

FUGLSANG C C, JOHANSEN C, CHRISTGAU S and ADLER-NISSEN J (1995), Antimicrobial enzymes: applications and future potential in the food industry, *Trends Food Sci Technol*, **6**, 390–6.

FYFE L, ARMSTRONG F and STEWART J (1998), Inhibition of *Listeria monocytogenes* and *Salmonella enteriditis* by combinations of plant oils and derivatives of benzoic acid: the development of synergistic antimicrobial combinations, *International Journal of Antimicrobial Agents*, **9**, 195–9.

GARCIA-GRAELLS C, MASSCHALCK B and MICHIELS CW (1999), Inactivation of *Escherichia coli* in milk by high-hydrostatic-pressure treatment in combination with antimicrobial peptides, *Journal of Food Protection*, **62**, 1248–54.

GORDON M (2001), The development of oxidative rancidity in foods, in Pokorny J, Yanishlieva N and Gordon M (eds) *Antioxidants in Food: Practical Applications*, pp. 7–21. Cambridge: Woodhead.

GORDON M (1990), The mechanism of antioxidant action in vitro. In Hudson B J F (ed.) *Food Antioxidants.* London: Elsevier Applied Science.

GOULD GW (1996), Industry perspectives on the use of natural antimicrobials and inhibitors for food applications. *Journal of Food Protection*, Suppl. S, 82–6.

GRAF E, MAHONEY J R, BRYANT R G and EATON J W (1984), Iron-catalyzed hydroxl radical formation: stringent requirements for free iron coordination site, *J Biol Chem*, **259**, 3620.

GROHS BM and KUNZ B (1999), Antimicrobial effect of spices on sausage-spoiling microorganisms using a model medium for sausage type Frankfurter, *Advances in Food Sciences*, **21**, 128–35.

GUTTERIDGE JMC (1982), The role of superoxide and hydroxyl radicals in phospholipid peroxidation catalysed by iron salts, *FEBS Lett*, **150**, 454–8.

HA JU, KIM YM and LEE DS (2001), Multilayered antimicrobial polyethylene films applied to the packaging of ground beef, *Packaging Technology and Science*, **14**, 55–62.

HALLIWELL B and GUTTERIDGE JMC (1989), *Free Radicals in Biology and Medicine*, 2nd edn, Oxford, Clarendon Press.

HAMMER KA, CARSON CF and RILEY TV (1999), Antimicrobial activity of essential oils and other plant extracts. *Journal of Applied Microbiology*, **86**, 985–90.

HAO YY, BRACKETT RE and DOYLE MP (1998), Efficacy of plant extracts in inhibiting *Aeromonas hydrophila* and *Listeria monocytogenes* in refrigerated, cooked poultry, *Food Microbiology*, **15**, 367–78.

HARBORNE JB (1988), Natural fungitoxins. In Hostettemann L and Lea PJ (eds) *Biologically Active Natural Products*, pp. 195–211. Oxford: Oxford University Press.

HE Y (1997), Antioxidant activity of green tea and its catechins in a fish meat model system, *Journal of Agricultural and Food Chemistry*, **45**, 4260–6.

HELANDER IM, ALAKOMI HL, LATVA KK, MATTILA ST, POL I, SMID EJ *et al.* (1998), Characterization of the action of selected essential oil components on Gram-negative bacteria, *Journal of Agricultural and Food Chemistry*, **46**,

3590–5.

HILI P, EVANS C S and VENESS R G (1997), Antimicrobial action of essential oils: the effect of dimethylsulphoxide on the activity of cinnamon oil, *Letters in Applied Microbiology*, **24**, 269–75.

HIRASA K and TAKEMASA M (1998), *Spice Science and Technology*. New York: Marcel Dekker.

HITOKOTO H, MOROZUMI S, WAUKE T, SAKAI S and KURATA H (1980), Inhibitory effects of spices on growth and toxin production of toxigenic fungi, *Applied and Environmental Microbiology*, **39**, 818–22.

HOLYOAK C D, STRATFORD M, MCMULLIN Z, COLE M B, CRIMMINS K, BROWN A J P *et al.* (1996), Activity of the plasma membrane H^+-ATPase and optimal glycolytic flux are required for rapid adaption and growth of *Saccharomyces cerevisiae* in the presence of the weak-acid preservative sorbic acid, *Applied and Environmental Microbiology*, **62**, 3158–64.

HSIEH PC, MAU JL and HUANG SH (2001), Antimicrobial effect of various combinations of plant extracts, *Food Microbiology*, **18**, 35–43.

HUANG S-W, FRANKEL E N, AESCHBACH R and GERMAN J B (1997), Partition of selected antioxidants in corn oil–water model systems, *J Agric Food Chem*, **45**, 1991–4.

HUANG S-W, SATUÉ-GRACIA M T, FRANKEL E N and GERMAN J B (1999), Effect of lactoferrin on oxidative stability of corn oil emulsions and liposomes, *J Agric Food Chem*, **47**, 1356–61.

HUDSON B J F and LEWIS J I (1983), Polyhydroxy flavonoid antioxidants for edible oils: structural criteria for activity, *Food Chem*, **10**, 47–55.

ISAKSEN A and ADLER-NISSEN J (1997), Antioxidative effect of glucose oxidase and catalase in mayonnaises of different oxidative susceptibility. I. Product trials, *Lebensm.-Wiss.u.-Technol.*, **30**, 841–6.

JACOBSEN C, ADLER-NISSEN J and MEYER A S (1999), Effect of ascorbic acid on iron release from the emulsifier interface and on the oxidative flavor deterioration in fish oil enriched mayonnaise, *J Agric Food Chem*, **47**, 4917–26.

JACOBSEN C, SCHWARZ K, STÖCKMANN H, MEYER A S and ADLER-NISSEN J (1999), Partitioning of selected antioxidants in mayonnaise, *J Agric Food Chem*, **47**, 3601–10.

JAY J M (1982a), Antimicrobial properties of diacetyl, *Appl. Environ. Microbiol.*, **44**, 525–32.

JAY J M (1982b), Effect of diacetyl on foodborne microorganisms, *J. Food Sci.*, **47**, 1829–31.

JIROVETZ L, BUCHBAUER G and EBERHARDT R (2000), Analysis and quality control of essential cinnamon oils (bark and leaf oils) of different origin using GC, GC-MS and olfactometry: determination of the coumarin and safrole content, *Ernaehrung*, **24**(9), 366–9.

JOHANSEN C, FALHOLT P and GRAM L (1997), Enzymatic removal and disinfection of bacterial biofilms, *Appl Environ Microbiol*, **63**, 3724–8.

JUNG U K (2001), Antioxidant effects of natural lecithin on borage oil, *Food Sci and Biotech*, **10**(4), 354–9.

JUVEN B J, KANNER J, SCHVED F and WEISSLOWICZ H (1994), Factors that interact with the antibacterial action of thyme essential oil and its active constituents, *Journal of Applied Bacteriology*, **76**, 626–31.

KABARA J J (1993), Medium-chain fatty acids and esters. In Davidson P M and Branen A L (eds) *Antimicrobials in Foods*. New York: Marcel Dekker.

KANNER J and MENDEL H (1977), Prooxidant and antioxidant effects of ascorbic acid and metal salts in a β-carotene-linoleate model system, *J Food Sci*, **42**, 60–4.

KAUR H and PERKINS M J (1991), The free-radical chemistry of food additives. In Aruoma O I and Halliwell B (eds) *Free Radicals and Food Additives*, pp. 17–35. London and New York: Taylor & Francis.

KAWAKISHI S and KANEKO T (1987), Interaction of proteins with allyl isothiocyanate, *Journal of Agricultural and Food Chemistry*, **35**, 85-8.

KENNEDY M, O'ROURKE A, MCLAY J and SIMMOUNDS R (2000), Use of a ground beef model to assess the effect of the lactoperoxidase system on the growth of Escherichia coli O157:H7, Listeria monocytogenes and Staphylococcus aureus in red meat, *Int. J. Food Microbiol.*, **57**, 147–58.

KINDERLERER J and HATTON P V (1990), Fungal metabolites of sorbic acid. *Food Additives and Contaminants*, **7**, 657–69.

KONDO K, KURIHARA M, MIYATA N, SUZUKI T and TOYODA M (1999), Mechanistic studies of catechins as antioxidants against radical oxidation, *Arch Biochem Biophys*, **362**, 79–86.

KORYCKA-DAHL M B and RICHARDSON T (1978), Activated oxygen species and oxidation of food constituents, *Crit Rev Food Sci Nutr*, **9**, 209–41.

KOUTSOUMANIS K, LAMBROPOULOU K and NYCHAS G J E (1999), A predictive model for the non-thermal inactivation of Salmonella enteritidis in a food model system supplemented with a natural antimicrobial, *International Journal of Food Microbiology*, **49**, 63–74.

KRINSKY N I (1979), Carotenoid protection against oxidation, *Pure Appl Chem*, **51**, 649–60.

KULAS, E and ACKMAN R G (2001), Properties of alfa, gamma and delta tocopherol in purified fish oil triacylglycerols, *J of Am Oil Chem Soc*, **78**(4), 361–7.

LAMBELET P, SAUCY F and LÖLIGER J (1985), Chemical evidence for interactions between vitamins E and C, *Experientia*, **41**, 1384–8.

LAMBELET P, SAUCY F and LÖLIGER J (1994), Radical exchange reactions between vitamin E, vitamin C and phospholipids in autoxidizing polyunsaturated lipids, *Free Rad. Res.*, **20**, 1–10.

LAMBERT R J W, SKANDAMIS P N, COOTE P J and NYCHAS G-J E (2001), A study of the minimum inhibitory concentration and mode of action of oregano essential oil, thympl and carvacrol, *Journal of Applied Microbiology*, **91**, 453–62.

LARSON R A (1997), *Naturally Occurring Antioxidants*, New York: Lewis Publishers.

LATTANZIO V, DE CICCO V, DI VENERE D and LIMA GaSM (1994) Antifungal activity of phenelics against fungi commonly encountered during storage, Italian *J. Food Sci.*, **1**, 23–30.

LE TUTOUR B (1992), Antioxidative activities of Olea-Europaea leaves and related phenolic compounds, *Phytochem*, **31**, 1173–8.

LIEWEN M B and MARTH E H (1985), Growth and inhibition of micro-organisms in the presence of sorbic acid: a review, *Journal of Food Protection*, **48**, 364–75.

LIN C-M, PRESTON III J F and WEI C-I (2000), Antibacterial mechanism of allyl isothiocyanate, *Journal of Food Protection*, **63**, 727–34.

LINGNERT H and LUNDGREN B (1980), Antioxidative Maillard reaction products. IV Application in sausages, *J Food Proc Pres*, **3**, 87–103.

LÖLIGER J and SAUCY F (1989), Melange antioxydant synergique. European Patent EP 0 326 829 B1.

LOZIENE K, VAICIUNIENE J and VENSKUTONIS P R (1998), Chemical composition of the essential oil of creeping thyme (Thymus serpyllum s.l.) growing wild in Lithuania, *Planta Medica*, **64**(8), 772–3.

LÜCK E and JAGER M (1995), *Antimicrobial Food Additives: Characteristics, Uses, Effects*, 2nd edn, Berlin: Springer-Verlag.

MANZOCCO L, ANESE M and NICOLI M C (2000), Antioxidant properties of tea extracts as affected by processing, *Lebensm.-Wiss. u.-Technol*, **31**, 694–8.

MARINO M, BERSANI C and COMI G (1999), Antimicrobial activity of the essential oils of Thymus vulgaris L. measured using a bioimpedometric method, *Journal of Food Protection*, **62**, 1017–23.

MATSUZAKI K (1998), Magainins as paradigm for the mode of action of pore forming polypeptides, *Biochimica et Biophysica Acta*, **1376**, 391–400.

MAYERHAUSER C M (2001), Survival of enterohemorrhagic *Escherichia coli* O157 : H7 in retail mustard, *Journal of Food Protection*, **64**, 783–7.

MCCARTHY T L, KERRY J P, KERRY J F, LYNCH P B and BUCKLEY D J (2001), Assessment of the antioxidant potential of natural food and plant extracts in fresh and previously frozen pork patties, *Meat Science*, **57**(2), 177–84.

MEYER A S (1996), Oxidoreductases as food antioxidants, in Proceedings 21st World Congress of the International Society for Fat Research (ISF), Vol. 2, England, P J Barnes & Associates, pp. 343–5.

MEYER A S, HEINONEN M and FRANKEL E (1998), Antioxidant interactions of catechin, cyanidin, caffeic acid, quercetin, and ellagic acid on human LDL oxidation, *Food Chem.*, **61**, 71–5.

MEYER A S and ISAKSEN A (1995), Application of enzymes as food antioxidants, *Trends Food Sci Technol*, **6**, 300–4.

MEYER A S, RØRBÆK K and ADLER-NISSEN J (1994), Critical assessment of the applicability of superoxide dismutase as an antioxidant in lipid foods, *Food Chem.*, **51**, 171–5.

MIKOVA K (2001), The regulation of antioxidants in food, in Pokorny J, Yanishlieva N and Gordon M (eds) *Antioxidants in Food: Practical Applications*, pp. 267–84. Cambridge: Woodhead.

MISTRY B and MIN D B (1992), Reduction of dissolved oxygen in model salad dressing by glucose oxidase-catalase dependent on pH and temperature, *J Food Sci*, **57**, 196–9.

MISHRA A K and DUBEY N K (1994), Evaluation of some essential oils for their toxicity against fungi causing deterioration of stored food commodities, *Appl. Environ. Microbiol.*, **60**, 1101–5.

MITHEN R F, DEKKER M, VERKERK R, RABOT S and JOHNSON IT (2000), The nutritional significance, biosynthesis and bioavailability of glucosinolates in human foods, *Journal of the Science of Food and Agriculture*, **80**, 967–84.

MONTVILLE T J and CHEN Y (1998), Mechanistic action of pediocin and nisin: recent progress and unresolved questions, *Applied Microbiology and Biotechnology*, **50**, 511–19.

NAIDU, A S (2000), Lactoferrin. In Naidu A S (ed.) *Natural Food Antimicrobial Systems*. Boca Raton: CRC Press.

NAKATANI, N (1994), Antioxidative and antimicrobial constituents of herbs and spices. In Charalambous G (ed.) *Spices, Herbs and Edible Fungi*. Amsterdam: Elsevier Science BV.

NARDINI M, D'AQUINO M, TOMASSI G, GENTILI V, DI FELICE M and SCACCINI C (1995), Inhibition of human low-density lipoprotein oxidation by caffeic acid and other hydroxycinnamic acid derivatives, *Free Radic Biol Med*, **19**, 541–52.

NIELSEN P V, BEUCHAT L R and FRISVAD J C (1989), Growth and fumitremorgin production by *Neosartorya ficheri* as affected by food preservatives and organic acids, *J. Appl. Bacteriol.*, **66**, 197–207.

NIELSEN P V and RIOS R (2000), Inhibition of fungal growth on bread by volatile components from spices and herbs, and the possible application in active packaging, with special emphasis on mustard essential oil. *International Journal of Food Microbiology*, **60**, 219–29.

NYCHAS G J E (1995), Natural antimicrobials from plants. In Gould G W (ed.) *New Methods of Food Preservation*, pp. 58–89. Glasgow: Chapman & Hall.

ORAM J D and REITER B (1968), Inhibition of bacteria by lactoferrin and other iron-chelating agents, *Biochim Biophys Acta*, **170**, 351–65.

ÖZCAN M (1998), Inhibitory effects of spice extracts on the growth of Aspergillus parasiticus NRRL2999 strain, *Zeitschrift für Lebensmittel Untersuchung und Forschung A/Food Research and Technology*, **207**, 253–5.

ÖZCAN M and BOYRAZ N (2000), Antifungal properties of some herb decoction, *Eur Food Res Technol*, **212**, 86–8.

PAIVA-MARTINS and GORDON M H (2001), Isolation and characterization of the antioxidant component 3,4-dihydroxyphenylethyl 4-formyl-3-formylmethyl-4-hexenoate from olive leaves, *J Agric Food Chem*, **49**, 4214–19.

PASTER N, JUVEN B J, SHAAYA E, MENASHEROV M, NITZAN R, WEISSLOWICZ H *et al.* (1990), Inhibitory effect of oregano and thyme essential oils on moulds and foodborne bacteria, *Letters in Applied Microbiology*, **11**, 33–7.

PASTER N, MENASHEROV M, RAVID U and JUVEN B (1995), Antifungal activity of oregano and thyme essential oils applied as fumigants against fungi attacking stored grain, *Journal of Food Protection*, **58**, 81–5.

PERRY N B, ANDERSON R E, BRENNAN N J, DOUGLAS M H, HEANEY A J, MCGIMPSEY J A and SMALLFIELD, B M (1999), Essential oils from Dalmatian sage (Salvia officinalis L.): variations among individuals, plant parts, seasons, and sites, *Journal of Agricultural and Food Chemistry*, **47**(5) 2048–54.

POKORNY J and SCHMIDT S (2001), Natural antioxidant functionality during food processing. In Pokorny J, Yanishlieva N and Gordon M (eds) *Antioxidants in Food: Practical Applications*, pp. 331–54. Cambridge: Woodhead.

POL I E, MASTWIJK H C, SLUMP R A, POPA M E and SMID E J (2001), Influence of food matrix on inactivation of Bacillus cereus by combinations of nisin, pulsed electric field treatment, and carvacrol, *Journal of Food Protection*, **64**, 1012–18.

RADFORD S A, TASSOU C C, NYCHAS G-J E and BOARD R G (1991), The influence of different oils on the death rate of *Salmonella enteritidis* in homemade mayonnaise, *Letters in Applied Microbiology*, **12**, 125–8.

RAMOS-NINO M E, CLIFFORD M N and ADAMS M R (1996), Quantitative structure activity relationship for the effect of benzoic acids, cinnamic acids and benzaldehydes on *Listeria monocytogenes*, *Journal of Applied Bacteriology*, **80**, 303–10.

REFSGAARD H H F, MEYER A and ADLER-NISSEN J (1992), Inactivation of copper, zinc superoxide dismutase from Saccharomyces cerevisiae in lipid food model systems, *Lebensm.-Wiss. u.-Technol.*, **25**, 564–8.

RHODES M J C (1996), Physiologically-active compounds in plant foods: an overview. *Proceedings of the Nutrition Society*, **55**, 371–84.

RISCH, S J and HO C T (1994), *Spices, flavour chemistry and antioxidant properties*, American Chemical Society Symposium Service 660, Washington DC.

ROEDIG-PENMAN, A and GORDON M H (1997), Antioxidant properties of catechins and green tea extracts in model food emulsions, *J Agric Food Chem*, **45**, 4267–70.

ROLLER S (2000), The quest for natural antimicrobials as novel means of food preservation: status report on a European research project, *Int. Biodet. Biodeg.*, **36**, 333–45.

ROLLER S and COVILL N (1999), The antifungal properties of chitosan in laboratory media and apple juice, *International Journal of Food Microbiology*, **47**, 67–77.

RUSUL G, EL-GAZZAR F E and MARTH E M (1987), Growth and aflatoxin production by *Aspergillus parasiticus* NRRL 2999 in the presence of acetic or propionic acid and at different initial pH values. *Journal of Food Protection*, **50**, 909–14.

SAITO H (1997), Antioxidant activity and active sites of phospholipids as antioxidants, *JAOCS*, **74**(12), 1531–6.

SANTOS-GOMES P C and FERNANDES-FERREIRA M (2001), Organ- and season-dependent variation in the essential oil composition of Salvia officinalis L. cultivated at two different sites, *Journal of Agricultural and Food Chemistry*, **49**, 2908–16.

SATUÉ-GRACIA M T, FRANKEL E N, RANGAVAJHYALA N and GERMAN J B (2000), Lactoferrin in infant formulas: effect on oxidation, *J Agric Food Chem*, **48**, 4984–90.

SATUÉ-GRACIA M T, HEINONEN M and FRANKEL E N (1997), Anthocyanins as antioxidants on human low-density lipoprotein and lecithin-liposome systems, *J Agric Food Chem*, **45**, 3362–7.

SCHULER P (1990), Natural antioxidants exploited commercially. In Hudson B J F (ed.) *Food Antioxidants*, pp. 99–170. London and New York: Elsevier Applied Science.

SCHWARZ K, FRANKEL E N and GERMAN J B (1996), Partition behaviour of antioxidative phenolic compounds in heterophasic systems, *Fett/Lipid*, **98**, 115–21.

SCOTT G (1993), Initiators, prooxidants and sensitizers. In Scott G (ed.) *Atmospheric Oxidation and Antioxidants*, Vol. 1, pp. 83–119. Amsterdam, Elsevier.

SELITRENNIKOFF C P (2001), Antifungal proteins, *Applied and Environmental Microbiology*, **67**, 2883–94.

SENATORE F (1996), Influence of harvesting time on yield and composition of the essential oil of a thyme (Thymus pulegioides L.) growing wild in Campania (Southern Italy), *Journal of Agricultural and Food Chemistry*, **44** (5), 1327–32.

SHELEF L A (1984), Antimicrobial effects of spices, *Journal of Food Safety*, **6**, 29–44.

SHI H, NOGUCHI N and NIKI E (2001), Introducing natural antioxidants. In Pokorny J, Yanishlieva N and Gordon M (eds) *Antioxidants in Food: Practical Applications*, pp. 147–58. Cambridge: Woodhead.

SHUZE TANG S Z, SHEEHAN D and BUCKLEY D J (2000), Antioxidant activity of added tea catechins on lipid oxidation of raw minced red meat, poultry and fish muscle, *International J of Food and Tech*, **36** (6), 685–92.

SIKKEMA J, DE BONT J A M and POOLMAN B (1995), Mechanisms of membrane toxicity of hydrocarbons, *Microbiological Reviews*, **59**, 201–22.

SIVROPOULOU A, PAPANIKOLAOU E, NIKOLAOU C, KOKKINI S, LANARAS T and ARSENAKIS M (1996), Antimicrobial and cytotoxic activities of origanum essential oils, *J. Agric. Food Chem.*, **44**, 1202–5.

SMID E J and GORRIS L G M (1999), Natural antimicrobials for food preservation. In Rahman M S (ed.) *Handbook of Food Preservation*, pp. 285–308. New York: Marcel Dekker.

SKANDAMIS P N and NYCHAS G-J E (2000), Development and evaluation of a model predicting the survival of Escherichia coli O157:H7 NCTC 12900 in homemade eggplant salad at various temperatures, pHs, and oregano essential oil concentrations, *Applied and Environmental Microbiology*, **66**,

1646–53.
SKIRDAL I M and EKLUND T (1993), Microculture model studies on the effect of sorbic acid on *Penicillium chrysogenum, Cladosporium cladosporioides* and *Ulocladium atrum* at different pH levels, *J. Appl. Bacteriol.*, **74**, 191–5.
SMITH-PALMER A, STEWART J and FYFE L (2001), The potential application of plant essential oils as natural food preservatives in soft cheese, *Food Microbiology*, **18**, 463–70.
SOLAR-RIVAS C, ESPIN J C and WICHERS H J (2000), Review: Oleuropein and related compounds, *Journal of the Science of Food and Agriculture*, **80**, 1013–23.
SOLIMAN F M, EL KASHOURY E A, FATHY M M and GONAID M H (1994), Analysis and biological activity of the essential oil of Rosmarinus officinalis L. from Egypt, *Flavour and Fragrance Journal*, **9**, 29–33.
STAHL W and SIES H (1993), Physical quenching of singlet oxygen in cis-trans isomerization of carotenoids. In Canfield L M, Krinsky N I and Olson J A (eds) *Carotenoids in Human Health*, pp. 10–19. New York: New York Academy of Sciences.
STRATFORD M and ANSLOW P A (1998), Evidence that sorbic acid does not inhibit yeast as a classic 'weak acid preservative'. *Letters in Applied Microbiology*, **27**, 206.
TANG S Z, KERRY J P, SHEEHAN D, BUCKLEY D J and MORRISSEY P A (2001), Antioxidative effect of dietary green tea catechins on lipid oxidation of long-term frozen stored chicken meat, *Meat Sci.*, **57**(3), 331–6.
TANTAOUI-ELARAKI A, LATTAOUI N and ERRIFI A (1993), Composition and antimicrobial activity of the essential oils of *Thymus broussonettii, T. zygis* and *T. satureioides, Journal of Essential Oil Research*, **5**, 45–53.
THOMPSON D P (1989), Fungitoxic activity of essential oil components on food storage fungi, *Mycologia*, **81**, 151–3.
TOMERA J F (1999), Current knowledge of the health benefits and disadvantages of wine consumption, *Trends in Food Sci and Tech*, **10**, 129–38.
TSIGARIDA E, SKANDAMIS P and NYCHAS G-J E (2000), Behaviour of Listeria monocytogenes and autochthonous flora on meat stored under aerobic, vacuum and modified atmosphere conditions with or without the presence of oregano essential oil at 5°C, *Journal of Applied Microbiology*, **89**, 901–9.
TUCKER A O and MACIARELLO M J (1994), The essential oils of some rosemary cultivars, *Flavour and Fragrance Journal*, **1**(4–5), 137–42.
TZAKOU O, PITAROKILI D, CHINOU I B and HARVALA C (2001), Letters: composition and antimicrobial activity of the essential oil of Salvia ringens, *Planta Medica: Natural Products and Medicinal Plant Research*, **67**(1), 81–6.
UCCELLA N (2001), Olive biophenols: novel ethnic and technological approach, *Trends in Food Sci and Tech*, **11**, 328–39.
ULTEE A, KETS E P W and SMID E J (1999), Mechanisms of action of carvacrol on the food-borne pathogen Bacillus cereus, *Applied and Environmental Microbiology*, **65**, 4606–10.

ULTEE A and SMID E J (2001), Influence of carvacrol on growth and toxin production by Bacillus cereus, *International Journal of Food Microbiology*, **64**, 373–8.

UNAL R, FLEMING H P, MCFEETERS R F, THOMPSON R L, BREIDT F and GIESBRECHT F G (2001), Novel quantitative assays for estimating the antimicrobial activity of fresh garlic juice, *Journal of Food Protection*, **64**, 189–94.

URI N (1961), Mechanisms of antioxidation. In Lundberg W O (ed.) *Autooxidation and Antioxidants*, pp. 133–69. New York: Interscience.

VERHOEVEN D T H, VERHAGEN H, GOLDBOHM R A, VAN DEN BRANDT P A and VAN POPPEL G (1997), A review of mechanisms underlying anticarcinogenicity by brassica vegetables, *Chemico-Biological Interactions*, **103**, 79–129.

VISIOLI, F and GALLI C (1998), Effects of minor constituents of olive oil on cardiovascular disease: new findings, *Nutrition Rev*, **56**(5), 142–7.

WALKER J R L (1994) Antimicrobial compounds in food plants. In Dillon V M and Board R G (eds) *Natural Antimicrobial Systems and Food Preservation*, pp. 181–204. Wallingford: CAB International.

WAN J, WILCOCK A and COVENTRY M J (1998), The effect of essential oils of basil on the growth of Aeromonas hydrophila and Pseudomonas fluorescens, *Journal of Applied Microbiology*, **84**, 152–8.

WANASUNDARA P K J P D, SHAHIDI F and SHUKLA V K S (1997), Endogenous antioxidants from oilseeds and edible oils, *Food Rev Int*, **13**(2), 225–92.

WANG, H, CAO G and PRIOR R L (1997), Oxygen radical absorbing capacity of anthocyanins, *J Agri Food Chem*, **45**, 304–9.

WARD S M, DELAQUIS P J, HOLLEY R A and MAZZA G (1998), Inhibition of spoilage and pathogenic bacteria on agar and pre-cooked roast beef by volatile horseradish distillates, *Food Research International*, **31**, 19–26.

WILKINS K M and BOARD R G (1989), Natural Antimicrobial Systems. In Gould G W (ed.) *Mechanisms of Action of Food Preservation Procedures*, pp. 285–362. Essex: Elsevier Science Publishers.

WILLIAMSON G (1989), Purification of glutathione S-transferase from lamb muscle and its effect on lipid peroxidation, *J Sci Food Agric*, **48**, 347–60.

WILLIAMSON G and BALL S K M (1988), Purification of glutathione S-transferase from lean pork muscle and its reactivity with some lipid oxidation products, *J Sci Food Agric*, **44**, 363–74.

WILSON C L, FRANKLIN J D and OTTO B E (1987), Fruit volatiles inhibitory to *Monilinia fructicola* and *Botrytis cinerea*, *Plant Disease*, **71**, 316–19.

XU Z M and GODBER J S (2001), Antioxidant activities of major components of gamma-oryzanol from rice bran using a linoleic acid model, *J of Am Oil Chem Soc*, **78**(6), 645–9.

YANISHLIEVA-MASLAROVA N V (2001), Inhibiting oxidation. In Pokorny J, Yanishlieva N and Gordon M (eds) *Antioxidants in Food: Practical Applications*, pp. 22–70. Cambridge, Woodhead.

YILDIRIM A, MAVI A and KARA A A (2001), Determination of antioxidant and antimicrobial activities of Rumex crispus L. extracts, *J Agric Food Chem*, **49**, 4083–9.

YILDIRIM A, MAVI A, OKTAY M, KARA A A, ALGUR Ö F and BILALOGLU V (2000), Comparison of antioxidant and antimicrobial activities of tilia (Tilia Argentea Desf Ex DC), sage (Salvia Trilob L.), and black tea (Camellia Sinensis) extracts, *J Agric Food Chem*, **48**, 5030–4.

ZAIKA L L (1988), Spices and herbs: their antimicrobial activity and its determination, *Journal of Food Safety*, **9**, 97–118.

ZIAUDDIN K S, RAO H S and FAIROZE N (1996), Effect of organic acids and spices on quality and shelf-life of meats at ambient temperature, *Journal of Food Science and Technology*, **33**, 255–8.

7

The hurdle concept

H. Alakomi, E. Skyttä, I. Helander and R. Ahvenainen, VTT Biotechnology, Espoo

7.1 Introduction

The demand by consumers for high quality foods having 'fresh' or 'natural' characteristics has led to the development of foods that are preserved using mild technologies. The most important preservation technique for many foods is refrigeration. However, because of the difficulty in maintaining sufficiently low temperatures throughout the production, distribution and storage chain, additional barriers (or 'hurdles') are required to control the growth of spoilage or pathogenic microorganisms. The concept of combining several factors to preserve foods has been developed by Leistner (Leistner, 1995) and others into the *hurdle effect* (each factor is a hurdle that microorganisms must overcome). This in turn has led to the application of hurdle technology, where an understanding of the complex interactions of temperature, water activity, pH and other factors is used to design a series of hurdles that ensure microbiological safety of processed foods.

Hurdle technology has arisen in response to a number of developments:

- consumer demands for healthier foods that retain their original nutritional properties
- the shift to ready-to-eat and convenience foods which require little further processing by consumers
- consumer preference for more 'natural' foods which require less processing and fewer chemical preservatives.

The goal of food preservation is to put pathogenic and spoilage microorganisms in a hostile environment to inhibit their growth, shorten their survival or kill them. The use of a combination of preservation techniques to ensure food safety and stability is, of course, not new. In smoked products, for example, this combination includes heat, reduced moisture content and antimicrobial chemicals deposited from the smoke onto the surface of the food. Some smoked foods may also be dipped or soaked in brine or rubbed with salt before smoking, to impregnate the flesh with salt and thus add a further preservative mechanism. In jams and other fruit preserves, the combined factors are heat, a high solids content (reduced water activity) and high acidity. In vegetable fermentation, the desired product quality and microbial stability are achieved by a sequence of factors that arise at different stages in the fermentation process: the addition of salt selects the initial microbial population which uses up the available oxygen in the brine. This reduces the redox potential and inhibits the growth of aerobic spoilage microorganisms and favours the selection of lactic acid bacteria. These then acidify the product and stabilise it.

What is new is that consumers are less willing to accept some preservation techniques, for example the use of salt as a preservative, or the excessive use of others, such as the application of severe heat treatment, where these are seen to reduce sensory and nutritional quality. These constraints make it less easy for manufacturers to use traditional combinations of techniques to preserve foods. Food processors have been obliged to:

- develop new, more gentle techniques, such as 'minimal' processing technologies
- combine these 'gentler' techniques in more complex ways that ensure product safety and stability while not damaging a product's sensory and nutritional properties.

Hurdle technology provides a framework for combining a number of milder preservation techniques to achieve an enhanced level of product safety and stability.

7.2 The behaviour of microorganisms

Microorganisms react homeostatically to stress factors. When their environment is disturbed by a stress factor, they usually react in ways that maintain some key element of their physiology constant. Microorganisms undergo many important homeostatic reactions (see Table 7.1). Most stress reactions are active processes. In addition to genetic changes, they include some level of altered metabolism, and this often involves the expenditure of energy, e.g. to transport protons across the cell membrane, to maintain high cytoplasmic concentrations of 'osmoregulatory' or 'compatible' solutes.

Stress can produce varying results. In some cases, microorganisms strain every possible repair mechanism to overcome a hostile environment, become metabolically 'exhausted' and die. In these cases, microorganisms may die more quickly in apparently favourable conditions. As an example, salmonellae that survive the ripening process in fermented sausages will vanish more quickly if the products are stored at ambient temperature, and they will survive longer and

Table 7.1 Homeostatic responses to stress by microorganisms

Stress factor	Homeostatic response
Low levels of nutrients	Nutrient scavenging; oligotrophy; 'stationary-phase response'; generation of 'viable non-culturable' forms
Lowered pH	Extrusion of protons across the cell membrane; maintenance of cytoplasmic pH; maintenance of transmembrane pH gradient
Lowered water activity	Osmoregulation; accumulation of 'compatible solutes'; avoidance of water loss; maintenance of membrane turgor
Lowered temperature for growth	'Cold shock' response; changes in membrane lipids to maintain satisfactory fluidity
Raised temperature for growth	'Heat shock' response; membrane lipid changes
Raised levels of oxygen	Enzymic protection (catalase, peroxidase, superoxide dismutase) from H_2O_2 and oxygen-derived free radicals
Presence of weak organic acid preservatives	As lowered pH, and sometimes extrusion of the organic acid
Presence of biocides	Phenotypic adaptation; reduction in cell wall/membrane permeability
Ionising radiation (β, γ, X)	Repair of single-strand breaks in DNA
High hydrostatic pressure	Uncertain; possibly low spore water content
High voltage electric discharge	Low electrical conductivity of the spore protoplast
Competition from other microorganisms	Formation of interacting communities; aggregates of cells showing some degree of symbiosis; biofilms

possibly cause foodborne illness if the products are stored under refrigeration (Leistner, 1995). Salmonellae survive in mayonnaise at chill temperatures much better than at ambient temperature.

In contrast, some bacteria become more resistant or even more virulent under stress, because they generate stress shock proteins. The synthesis of protective stress shock proteins is induced by heat, pH, a_w, ethanol, oxidative compounds, etc., as well as by starvation. Stress reactions might have a non-specific effect because, due to particular stress, microorganisms also become more tolerant to other stresses, i.e. they acquire a 'cross-tolerance'. An example is reaction to changes in pH. In many foods the cytoplasmic pH of microorganisms will normally be one or two units higher than that of their environment (Booth, 1995; Booth and Kroll, 1989). The ability of microorganisms to grow at low pH values depends on their ability to prevent protons from crossing the cell membrane and entering the cytoplasm and, to a large degree, on their ability to export any protons that do gain access, thus maintaining a satisfactorily high intracellular pH. If net proton influx cannot be prevented, cytoplasmic pH will fall, leading to cessation of growth and death of the cell.

However, many bacteria, yeasts and moulds can react to acid conditions, so as to enhance their survival. The stress response following exposure to acid conditions results in adaptation such that the efficacy of proton export increases. This acid tolerance response then results in microorganisms that have been exposed to even mildly acidic conditions becoming able to grow or survive at low pH values that would otherwise have been lethal (Hill *et al.*, 1995). This adaptation is important with respect to the potential for growth of spoilage microorganisms in acid-preserved foods and is particularly important with respect to the survival of pathogenic bacteria, e.g. *E. coli* (Goodson and Rowbury, 1989; Miller and Kaspar, 1994), *Salmonella* (Foster, 1993), *L. monocytogenes* (Kroll and Patchett, 1992). Acid-adapted cells may also become tolerant to a range of other environmental stresses, e.g. in *S. typhimurium* (Leyer and Johnson, 1993). Acid-shocked *L. monocytogenes* became more acid tolerant, as expected, but also exhibited a rise in heat resistance (Farber and Pagotto, 1992). Conversely, the heat shock response that follows mild heating can result in cells becoming more acid tolerant (Rowbury, 1995).

The success of hurdle technology depends on ensuring metabolic exhaustion. Homeostatic stress responses often require the expenditure of energy by the stressed cells. Restriction of the availability of energy is then a sensible target to pursue. This probably forms the basis of many of the successful, empirically derived, mild combination preservation procedures exemplified by hurdle technology. As an example, if a food can be preserved by lowering the pH, then it is sensible also to include a weak acid preservative which will amplify the effect of the protons or to allow a milder, higher pH to be employed. It is sensible also to reduce the a_w as much as is organoleptically acceptable, so that the energy-demanding proton export is made more difficult by the additional requirement of the cells to osmoregulate. Then, if the food can be enclosed in oxygen-free vacuum or modified atmosphere packaging, facultative anaerobes will be further energy-restricted at a time when the various stress and homeostatic reactions are demanding more energy if growth is to proceed (Gould, 1995).

7.3 The range and application of hurdles

Potential hurdles for use in the preservation of foods can be divided into physical, physicochemical, microbially derived and miscellaneous hurdles (Leistner and Gorris, 1995). The most important hurdles, either applied as 'process' or 'additive' hurdles, are high temperature (F value), low temperature (t value), water activity (a_w), acidity (pH), redox potential (E_h), competitive microorganisms (e.g. lactic acid bacteria) and preservatives (e.g. nitrite, sorbate, sulphite). However, in addition, more than 50 hurdles of potential use for foods of animal or plant origin, which improve the stability and/or the quality of these products, have been identified. Some of these are shown in Table 7.2. An example of the issues involved in the use of hurdle technology can be seen in the case of chilled foods. In order to prevent the risks caused by *Clostridium*

Table 7.2 Examples of hurdles used to preserve foods

Type of hurdle	Examples
Physical hurdles	Aseptic packaging Electromagnetic energy (microwave, radio frequency, pulsed magnetic fields, high electric fields) High temperatures (blanching, pasteurisation, sterilisation, evaporation, extrusion, baking, frying) Ionising radiation Low temperatures (chilling, freezing) Modified atmospheres Packaging films (including active packaging, edible coatings) Photodynamic inactivation Ultra-high pressures Ultrasonication Ultraviolet radiation
Physico-chemical hurdles	Carbon dioxide Ethanol Lactic acid Lactoperoxidase Low pH Low redox potential Low water activity Maillard reaction products Organic acids Oxygen Ozone Phenols Phosphates Salt Smoking Sodium nitrite/nitrate Sodium or potassium sulphite Spices and herbs Surface treatment agents
Microbially derived hurdles	Antibiotics Bacteriocins Competitive flora Protective cultures

botulinum in chilled foods to be stored over 10 days, current recommendations include chilled storage combined with one of the following (Betts, 1996):

- minimum heat treatment of 90°C for 10 minutes or equivalent
- pH of 5 or less throughout the food
- salt level of 3.5% (aqueous) throughout the food
- a_w of 0.97 or less throughout a food; or
- any combination of heat and preservative factors that has been shown to prevent growth of toxin production by *C. botulinum*.

However, the use of any of these single hurdles is circumscribed by the potential to cause off-flavours in some chilled food products. If manufacturers are to achieve a product that accommodates consumer preferences while remaining safe and stable, they must achieve synergies through the combined use of these or other hurdles which, while they may not prevent the growth of pathogenic or spoilage microorganisms on their own, will do so collectively. Each product may require a different combination of hurdles, depending on a range of factors including the following:

- the initial microbial load of the product requiring preservation
- how favourable conditions are within the product for microbial growth
- target shelf-life.

Some hurdles (e.g. Maillard reaction products) influence the quality as well as the safety of foods, because they have antimicrobial properties and at the same time improve the flavour of the products; this also applies to nitrite used in the curing of meat. The same hurdle could have a positive or a negative effect on foods, depending on its intensity. For instance, chilling to an unsuitably low temperature will be detrimental to some foods of plant origin ('chilling injury'), whereas moderate chilling is beneficial for the shelf-life. Another example is the pH of fermented sausages, which should be low enough to inhibit pathogenic bacteria but not so low as to impair taste. If the intensity of a particular hurdle in a food is too small, it should be strengthened; on the other hand, if it is detrimental to the food quality, it should be lowered and an additional hurdle considered.

For refrigerated foods, chill temperatures are the major and sometimes the only hurdle. However, if exposed to temperature abuse during distribution, this hurdle breaks down, and spoilage or even food poisoning could occur. Additional hurdles can be incorporated as safeguards for chilled foods (Leistner, 1999). In developing countries, the application of hurdle technology for foods that remain stable, safe and good to eat even if stored without refrigeration is of particular importance and has made impressive strides. Examples include the development of novel high-moisture fruit products in Latin America, meat products in China and dairy products in India (Leistner, 2000). In developing countries, the use of hurdle technology can offset limitations in infrastructure, for example in the maintenance of a reliable cold chain.

Hurdles may be applied sequentially, for example as a series of processing steps, or in synchrony, for example in the way a product is formulated to achieve a certain level of water activity or acidity during processing. Some of the complexities of hurdle technology are shown in Fig. 7.1. This shows the use of six hurdles to preserve a food:

- heating (F)
- chilling (t)
- water activity (a_w)
- acidity (pH)

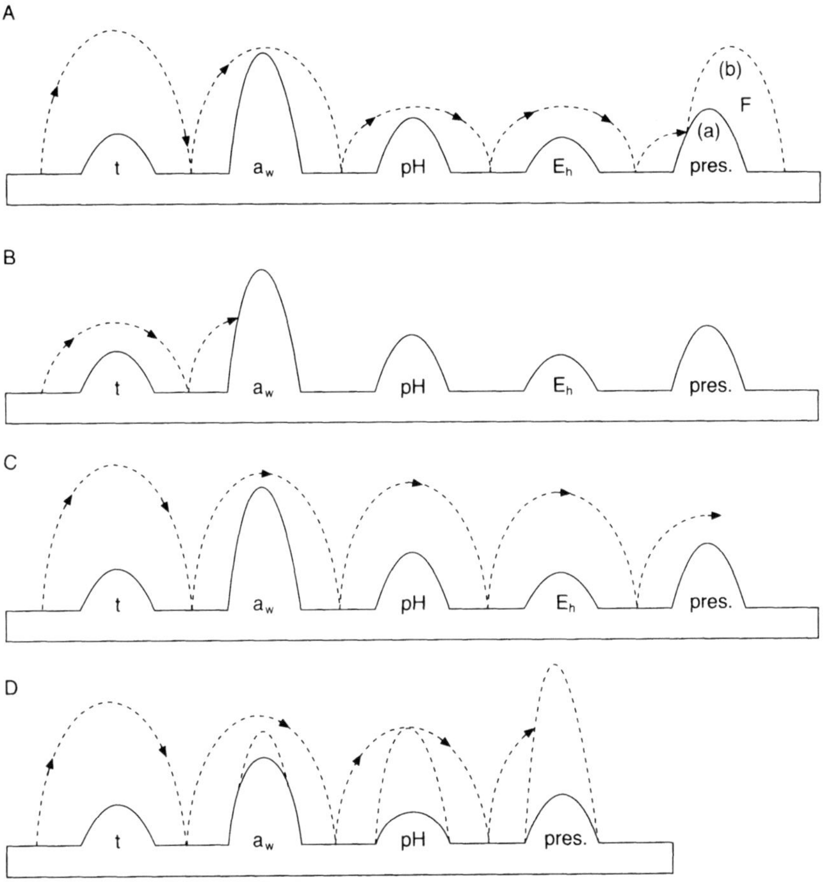

Fig. 7.1 The hurdle effect (adapted from Leistner, 1995).

- redox potential (E_h)
- preservatives (pres.)

Example A shows the use of hurdles of differing levels of intensity to bring microbiological growth under control. While gentle heating and chilled storage are not sufficient in themselves to inactivate spoilage or pathogenic microrganisms, combined with appropriate levels of water activity, acidity and redox potential, they weaken microbiological activity sufficiently to allow the relatively mild use of preservatives in giving the product its required shelf-life before consumption. In this case, particular emphasis is placed on the control of water activity and the use of preservatives. This set of hurdles means that a lower level of milder preservatives (a) can be used compared to the level required without the other hurdles (b).

A key factor is the initial microbial load in a product. If there are only a few microorganisms present at the start as in example B, a few or low hurdles are sufficient for the stability of the product. The aseptic processing of perishable foods is based on this principle. The same is true if the initial microbial load of a food (for example, high-moisture fruits) is substantially reduced (for example, by blanching with steam), because after such a reduction only a few microorganisms are present at the start, which are easily inhibited. On the other hand, as in example C, if due to bad hygienic conditions too many undesirable microorganisms are initially present, even the usual hurdles inherent in a product may be unable to prevent spoilage or food poisoning. Similarly a food rich in nutrients and vitamins will foster the growth of micro-organisms (so-called 'trampoline effect'). The hurdles in such a product must be enhanced otherwise they will be overcome. Example D illustrates the potential synergistic effect of hurdles, which create a multi-target disturbance of the homeostasis of the target microorganisms. A synergistic effect could occur if the hurdles in a food hit, at the same time, different targets (e.g. cell membrane, DNA, enzyme systems, pH, a_w, E_h) within the microbial cell, and thus disturb the homeostasis of the microorganisms present in several respects.

7.4 The use of hurdle technology in food processing

Hurdle technology as a concept has proved useful in the optimisation of traditional foods as well as in the development of novel products. However, it should be combined, if possible, with the HACCP concept and predictive microbiology. These three concepts are related. Hurdle technology is primarily used for food design, the HACCP concept for process control, and predictive microbiology for both process design and refinement. By considering these different approaches, an overall strategy for securing stable, safe and high quality foods should now be accomplished. This strategy could be applied in effective food design (Leistner, 1994a). Steps for effective food design are shown in Table 7.3.

Predictive microbiology is a promising concept which allows computer-based and quantitative predictions of microbial growth, survival and death in foods. However, the predictive models constructed so far handle a limited number of factors (hurdles) simultaneously. Factors considered to date include temperature, pH, a_w (due to salt or humectants), preservatives (e.g. nitrite, lactic acid) and CO_2). There are numerous other relevant hurdles to be considered, which are important for the stability, safety and quality of particular foods. Within these limits predictive microbiology does allow quite reliable predictions of the fate of microorganisms in food systems. Because not all factors may be taken into account, the limits indicated for growth of pathogens in foods are in general more prudent ('fail-safe') than the limits in real foods (Leistner, 1994b). Predictive microbiology will be an important tool in future food design, because it can narrow down considerably the range over which challenge tests with relevant microorganisms need to be performed. Although it will never render challenge testing obsolete, it may greatly reduce both time and costs spent in product development. Thus predictive microbiology should be an integral part of advanced food design (Leistner, 1994b).

Table 7.3 Steps for food design using an integrated concept, comprising hurdle technology as well as predictive microbiology and HACCP (Leistner, 1994b)

1. For the modified or novel food product the desired sensory properties and the desired shelf-life must be defined.
2. A tentative technology for the production of the food should be suggested.
3. The food is now manufactured according to this technology, and the resulting product analysed for pH, a_w, preservatives or other inhibitory factors and the temperatures for heating (if intended) and storage as well as the expected shelf-life are defined.
4. For preliminary stability testing of the suggested food product, predictive microbiology could be employed.
5. The product is now challenged with food-poisoning and spoilage microorganisms using somewhat higher inocula and storage temperatures than 'normal'.
6. If necessary, the hurdles in the product are modified, taking the homeostasis of the microorganisms and the sensory quality of the food (i.e. 'total quality') into consideration.
7. The modified product is again challenged with relevant microorganisms, and if necessary the hurdles are modified once more. Predictive microbiology for assessing the safety of the food might be helpful at this stage, too.
8. Now that the established hurdles of the modified or novel food are exactly defined, including tolerances, the methods for monitoring the process are defined (preferably physical methods should be used).
9. The designed food should be produced under industrial conditions because the possibilities for a scale-up of the proposed process must be validated.
10. For the industrial process the critical control points (CCPs) and their monitoring has to be established, and thus the manufacturing process should be controlled by HACCP.

Hurdle technology may be more broadly applicable to quality aspects of foods (Leistner, 1994a). However, McKenna (1994) emphasised that while hurdle technology is appropriate for securing the microbial stability and safety of foods, the total quality of foods is a much broader field and encompasses a wide range of physical, biological and chemical attributes. The concept of combined processes should work towards the total quality of foods rather than the narrow but important aspects of microbial stability and safety. But at present the tools for applying hurdle technology to total food quality are still not adequate, and this is equally true for predicting food quality by modelling.

7.5 Hurdle technology in practice: some examples

During the last decade, minimally processed, high-moisture, ambient stable fruit products (HMFP) have been developed in a number of Latin American countries. This novel process has already been applied to peach halves, pineapple slices, mango slices and purée, and papaya slices, as well as to whole figs, strawberries and pomalaca (Alzamora *et al.*, 1993; Alzamora *et al.*, 1995).

Preservation was based on the combination of a mild heat treatment (blanching for 1–3 minutes with saturated steam), slight reduction in a_w (to 0.98–0.93 by addition of glucose or sucrose), lowering of pH (to 4.1–3.0 by addition of citric or phosphoric acid), and the addition of antimicrobials (1,000 ppm potassium sorbate or sodium benzoate plus 150 ppm sodium sulfite or sodium bisulfite) to the syrup of the products. During storage of HMFP, the sorbate and, in particular, the sulfite levels decreased, and the a_w fell (i.e. the a_w hurdle increased), due to the hydrolysis of sucrose (Alzamora *et al.*, 1995).

The resulting products were still rated well by a consumer panel after three months of storage at 35°C for taste, flavour, colour and texture. They also proved shelf-stable and safe for at least three to eight months of storage at 25–35°C. Due to the blanching process, the initial microbial counts were substantially reduced and during the storage of the stabilised HMFP the number of surviving bacteria, yeasts and moulds decreased further, often below the detection limit (Sajur, 1985; Alzamora *et al.*, 1993; Alzamora *et al.*, 1995; Tapia de Daza *et al.*, 1995; Lopez-Malo *et al.*, 1994; Guerrero *et al.*, 1994; Argaiz *et al.*, 1995; Tapia de Daza *et al.*, 1996).

Raw vegetables contain a variety of bacteria, yeast and mould species, which can cause spoilage. Psychrotrophic pathogens (such as *L. monocytogenes, Aeromonas hydrophila* and *Yersinia enterocolitica*) might also develop if refrigeration is not adequate. In cooked sous-vide preparations, mild heat treatment eliminates vegetative microorganisms, but spore-formers may survive. For the safety of this group of minimally processed vegetables, non-proteolytic *C. botulinum* is of particular concern because it might grow well in refrigerated products in which the competitive flora has been inactivated by heat.

Enterobacteriaceae and *Pseudomonas* species constitute the major populations in raw vegetables before and after controlled-atmosphere (MAP) storage at 8°C. Growth of *L. monocytogenes* is inhibited, depending on the initial numbers of the pathogen, the type of produce and the size of the competitive spoilage flora. Reducing the initial microbial load by decontamination of the raw materials helps to minimise microbial spoilage. However, *L. monocytogenes* grows better on disinfected produce than on non-disinfected or water-rinsed produce, demonstrating the importance of avoiding recontamination with pathogens after disinfection. If refrigeration is not optimal, additional hurdles may be necessary. Lactic acid bacteria can be added or natural preservatives, derived from herbs and spices, which inhibit food poisoning and spoilage bacteria, yeasts and moulds. Edible coatings prepared from carbohydrates, proteins, or fats, applied directly to the surface of the product, act as a physical barrier to food contamination and inhibit microbial growth if antimicrobial compounds are added (Gorris, 2000).

Minimally processed foods with a longer shelf-life (up to 42 days at 1–8°C) are the so-called refrigerated processed foods of extended durability (REPFEDs) (Gorris and Peck, 1998). REPFEDs are heated to 65–95°C, which should eliminate cells of vegetative bacteria but not bacterial spores. Surviving spore-formers that germinate during storage of the products will be able to proliferate without competition from the bacteria previously present. REPFEDs are mostly packed under vacuum or in anaerobic atmosphere. The main threat for REPFEDs are microorganisms that produce heat-resistant spores and grow in the absence of oxygen at refrigeration temperatures (in particular non-proteolytic *C. botulinum*), which can multiply and form toxin at temperatures as low as 3.0°C (Carlin and Peck, 1996; Peck, 1997). Because optimal refrigeration may not be achieved throughout the supply chain, additional hurdles, such as organic acids, bacteriocins, natural preservatives derived from herbs and spices, or ultra-high pressure treatment may be required (Gorris and Peck, 1998).

7.6 The development of new hurdles: some examples

The development of new hurdles and the more effective use of existing hurdles are active areas of research. The following three examples illustrate the kinds of research being undertaken.

7.6.1 Breaking the permeability barrier of the Gram-negative bacterial outer membrane

Gram-negative bacteria can pose special problems in food safety and hygiene due to the relative resistance of these bacteria to various antimicrobials, including disinfectants, enzymes, hydrophobic antibiotics and bacteriocins such as nisin. This resistance is largely due to the presence of a unique membrane as the outermost layer of the cell envelope. The outer membrane (OM) functions as a potent permeability barrier to a number of external agents. The OM is specifically able to exclude hydrophobic molecules and macromolecules. The former are not able to enter the OM, since the outer surface of this membrane is exclusively built of lipopolysaccharide (LPS) molecules that have hydrophilic carbohydrate chains protruding from the membrane surface. Access for small nutrient molecules into the cells is provided by pore proteins (porins) having various degrees of selectivity for molecules to be transported; these pores are generally too small to allow entry of larger molecules such as bacteriocins or enzymes. The OM is present in all Gram-negative bacteria, whereas in Gram-positive species a complex cell wall surrounds the bacterial cell without provision of a comparable permeability barrier function. Once broken, the disrupted OM could be expected to let through a number of substances which could then act antimicrobially after reaching the deeper parts of the cell. Agents that cause OM disruption without being directly bactericidal are termed 'permeabilisers'.

EDTA as a permeabiliser

The inability of the bacteriocin nisin to affect Gram-negative species and, consequently, their sensitisation to nisin has been addressed in several papers (Helander *et al.*, 1997a; Cutter and Siragusa, 1995; Delves-Broughton, 1993; Stevens *et al.*, 1991; Schved *et al.*, 1994). Nisin affects Gram-positive bacteria, but is unable to penetrate the Gram-negative OM. However, it has long been known that chelating agents, notably ethylenediaminetetraacetic acid (EDTA) can disrupt the OM and render Gram-negative bacteria sensitive to agents that normally are ineffective against them. The action of EDTA results from its ability to sequester divalent cations from the OM, such as Mg^{2+} ions that bridge negatively charged constituents (LPS, protein) together in the outer leaflet of the OM. The loss of these cations' stabilising effect results in considerable release of LPS and other OM constituents from the cell. Phospholipids are exposed and a 'hydrophobic pathway' attracting hydrophobic substances is created. EDTA is known in some cases to release up to 40% of the LPS molecules, presumably creating large patches of disrupted surface allowing the passage of macromolecules through the exposed phospholipid (bi)layer. Accordingly Gram-negative bacteria can become sensitive to lysozyme upon treatment with EDTA. A schematic representation of the action of chelators on the OM is shown in Fig. 7.2.

Other permeabilisers

Recent research has focused on a number of permeabilisers and their mode of action, looking for food-grade substances to control harmful Gram-negative bacteria such as *Escherichia coli* 0157:H7, *Pseudomonas aeruginosa* or

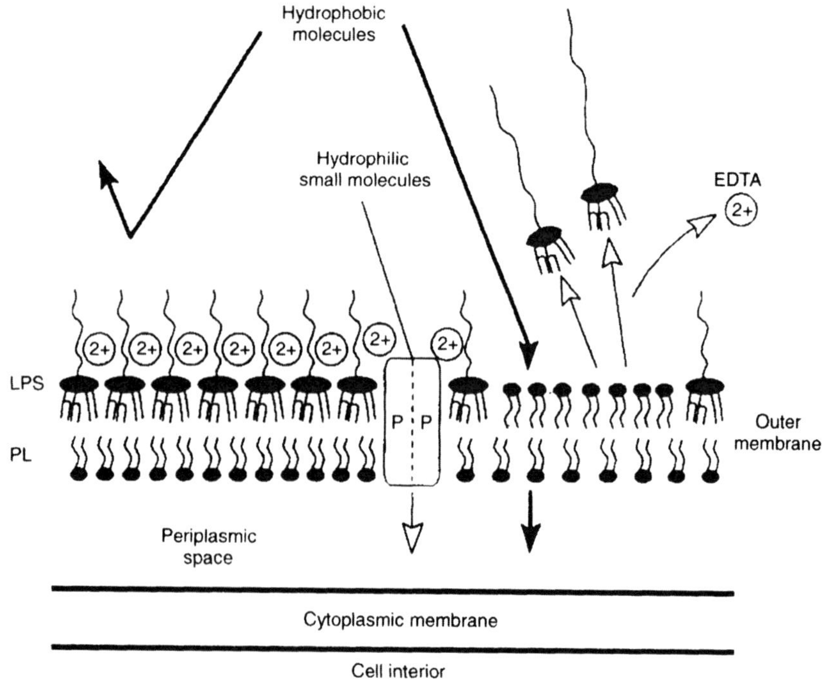

Fig. 7.2 Gram-negative bacterial outer membrane showing the intact structure on the left of the porin protein (PP) and a structure destabilised by the action of EDTA on the right.

Salmonella species (Helander *et al.*, 1997b; Helander *et al.*, 1998a; Helander *et al.*, 1998b). Organic acids (lactic, propionic, acetic acid) are naturally present in fermented food products, and are acceptable as food ingredients (citric acid). Lactic acid, for instance, functions as a strong permeabiliser for Gram-negative bacteria, including *E. coli* 0157:H7, which is known for its acid tolerance. Acidic pH as such (as adjusted by hydrochloric acid) does weaken the OM within a pH range of 3–5, depending on the bacterial species. Lactic acid, however, at 5–10 mM concentrations yielding pH 4 or 3.6, respectively, always brings about significantly stronger permeabilisation than that obtained by HCI at the same pH value. This behaviour can be attributed to the weak acid nature of lactic acid, which remains largely un-ionised at the above pH, but it is also possible that lactic acid functions as a chelator to some extent, since the 'additional' permeabilising effect of lactic acid is typically abolished by adding Mg^{2+} ions in the system. It can therefore be concluded that lactic acid can function as an agent that sensitises Gram-negative bacteria to other antimicrobials. Such additional antimicrobial effect can be expected to be naturally present in food products having undergone fermentation by lactic acid bacteria, explaining in part the great preservative potential of naturally fermented products.

7.6.2 Natural enzyme-based antimicrobial systems

According to Holzapfel *et al.* (1995), the excretion of an antimicrobial enzyme, such as lysozyme or lysostaphin, may improve the activity spectrum of protective cultures, and may render such a strain more suitable for application under versatile conditions. As antimicrobial metabolites, many enzymes serve in nature to protect a biological system against invasion of certain microorganisms. Typical examples are lysozyme in egg and lactoperoxidase in milk. Lysozyme is a bacteriolytic enzyme found in natural systems, including tears, saliva, plant tissues, milk and eggs. Food systems that contain such antagonistic enzymes possess an intrinsic stability against microorganisms (Table 7.4). Such purified antagonistic enzymes can be used as biopreservatives. Although antimicrobial enzymes represent a natural alternative to traditional preservation agents, so far their use has been limited because of their low cost efficiency. Two classes of

Table 7.4 Naturally occurring antagonistic enzyme systems in foods (Holzapfel *et al.*, 1995)

Enzyme	Inhibited microorganisms	Mechanisms of action
Lysozyme	Mainly Gram-positive	Lysis
Glucose oxidase	Gram-positive	H_2O_2, pH
Lactoperoxidase	Gram-positive Gram-negative	Oxidation
Lactoferrin	Gram-positive Gram-negative	Iron binding

enzymes are relevant, namely hydrolases and the oxidoreductases (Fuglsang *et al.*, 1995). The substrate for the hydrolases are the structural components of the cell walls of microorganisms; degradation of these components inactivates the cells. The oxidoreductases exert their effect by the generation *in situ* of reactive molecules that can destroy vital proteins in cell.

Applications of lysozyme

One or more lantibiotics and lysozyme, in combination, can be used at significantly lower concentrations than either component alone to yield a similar degree of control of prokaryotic microorganisms. The synergism of lantibiotics and lysozyme may be a result of each component making the other's target of action more accessible. By degrading the structural component of microbial cell walls lysozyme may increase the accessibility of the lantibiotics to the sensitive cell membranes or particular receptor molecules which are otherwise buried beneath the cell surface (Monticello, 1990).

Both class I and II bacteriocins act on the cytoplasmic membrane of target organisms by forming pores in the membrane, resulting in growth inhibition (Abee *et al.*, 1995; Bruno *et al.*, 1993; Chikindas *et al.*, 1993). The mechanisms of cell lysis by *N*-acetylhexosaminidases such as egg-white lysozyme involve the hydrolytic cleavage of the structural component of the cell wall. In the case of bacteria, the enzyme cleaves (1–4) linkages between *N*-acetylmuramic acid and *N*-acetyl glucosamine polymer of the peptidoglycan (Conner, 1993). Lysozyme is especially active against the outgrowth of clostridia spores but it also shows activity against other pathogenic or toxigenic bacteria, such as *Bacillus* and *Listeria* (Holzapfel *et al.*, 1995).

Gram-positive bacteria are more susceptible to action of lysozyme than Gram-negative bacteria, which contain a lipopolysaccharide layer in their outer membrane which prevents the entry of lysozyme to polysaccharide layer of the cell wall. However, susceptibility of Gram-negative bacteria to lysozyme or to bacteriocins can be increased by exposing the cell to stress. This can be brought about by starving cells at abnormal pH, by treatment with polymyxin B, EDTA or by osmotic injury (Conner, 1993). Hughey *et al.* (1989) investigated the activity of lysozyme against *Listeria monocytogenes* in vegetables, meat and dairy foods. They concluded that lysozyme (particularly with EDTA) could be a useful preservative for some foods, such as those prepared from fresh vegetables. Samuelson *et al.* (1985) developed a dip system with lysozyme and EDTA for eliminating salmonellae from broiler parts.

Applications of lactoperoxidase

Lactoperoxidase is one of several non-specific antimicrobial systems present in bovine milk. The active lactoperodidase system includes three primary components: the lactoperoxidase enzyme, thiocyanate (SCN-) and hydrogen peroxide (H_2O_2) (Daeschel and Penner, 1992). The lactoperoxidase system is antimicrobial due to the production of toxic, short-lived, oxidation products when lactoperoxidase catalyses the oxidation of thiocyanate by hydrogen peroxide. In the presence of both substrates the lactoperoxidase system oxidises thiocyanate to hypothiocyanate (OSCN-). This is a strong oxidising agent which in turn can oxidise essential sulfhydryl groups in bacterial proteins, but is harmless to host cells (Kamau *et al.*, 1990; Holzapfel *et al.*, 1995). Thiocyanate is present in milk and small amounts of hydrogen peroxide must be added to milk or produced by accompanying lactic acid bacteria.

Kamau *et al.* (1990) have shown that the lactoperoxidase system enhances the thermal destruction of *Listeria monocytogenes* and *Staphylococcus aureus* suspended in milk, and they have proposed activation of the lactoperoxidase system, followed by heating, as a method to control milkborne pathogens. According to Daeschel and Penner (1992), one approach to optimise the application of the lactoperoxidase system is to identify/develop compatible bacteria capable of generating adequate amounts of hydrogen peroxide at target temperatures.

7.6.3 Antimicrobial packaging materials

Surface treatment by antimicrobial agents for products such as cheeses, fruits and vegetables is widely practised. More recently the idea of incorporating antimicrobial agents directly into packaging films intended to come into contact with foods has been developed. Thus, the packaging material can serve as a source of releasing preservatives or antimicrobial agents, or even prevent the growth of microorganisms by presenting antimicrobial properties by itself. The materials can be divided into two types: those from which an active substance migrates to the surface of food, and those that are effective against microbial growth without migration of the active agents to the food. The technology development work of various functional materials has been active since the latter half of the 1980s, mainly in Japan.

Application techniques of antimicrobial agent in packaging materials

The release of antimicrobial substances can be achieved conventionally by adding a permeable or porous sachet containing the substances into the package. Some preservatives can also be incorporated into or onto polymeric packaging materials to provide antimicrobial activity. These agents can be applied by impregnation, mixing or using various coating techniques. The active agents can be placed in intermediate layers or encapsulated to achieve slow release into food. A more sophisticated concept is the use of immobilised enzymes, and materials having chemically bound antimicrobial functional groups on the material surface.

Some functional groups that have antimicrobial activity have been introduced and immobilised on the surface of polymer films by modified chemical methods. Haynie *et al.* (1995) prepared a series of antimicrobial peptides covalently bonded to a water-insoluble resin which proved to have antimicrobial activity. Laser-induced surface functionalisation of polymers has recently been found to be an effective way of imparting functional groups of food packaging polymer

Table 7.5 Potential substances that may be used as antimicrobial agents in food packages (Hotchkiss, 1995)

Type of substance	Example of substance
Organic acids	Propionic, sorbic, benzoic acid
Bacteriocins	Nisin
Spice extracts	Thymol, p-cymene
Thiosulfinates	Allicin
Enzymes	Peroxidase, lysozyme
Proteins	Conalbumine
Isothiocyanates	Allylisothiocyanate
Antibiotics	Imazalil
Fungicides	Benomyl
Chelating agents	EDTA
Metals	Silver
Parabenes	Heptylparaben

surfaces to improve adhesion, to modify barrier properties and to give the polymers antimicrobial activity. Paik *et al.* (1998) recently presented the use of 193 nm UV irradiation using an UV excimer laser to convert amide groups on the surface of polyamide plastic to amines which have antimicrobial effect.

Potential substances for antimicrobial packaging materials

Substances that can be used as antimicrobial include, for instance, ethanol and other alcohols, organic acids and salts (sorbates, benzoates, propionates), bacteriocins, etc. Hotchkiss (1995) has prepared an inventory of antimicrobial agents of potential use in food packages (Table 7.5). In food contact applications the substances incorporated should be safe and slow to migrate. Generally, it is recognised that metallic ions of silver, copper, zinc and others are safe antimicrobial agents. Ag-substituted zeolite is the most frequently used antimicrobial agent in plastic materials in Japan. The use of silver and Ag-zeolite is described by Ishitani (1995). The purpose of the zeolite is to allow for slow release of antimicrobial metal ions into the surface of the food products. The antimicrobial activity of metals is due to minute quantities of ions formed. Silver nitrate forms silver ions in water solutions which have a strong antimicrobial activity. These ions are considered to have inhibitory activities on metabolic functions of respiratory and electronic transport systems of microbes, and mass transfer across cell membranes. Experiments with yeast revealed that Ag-zeolite shows almost the same degree of activity regardless of the oxygen present and the existence of light.

7.7 The future of hurdle technology

It is clear that the hurdle concept will be utilised more and more in food preservation as minimal processing becomes more popular in food manufacturing. New hurdles will certainly be found, for example antimicrobial agents from plants and mushrooms. So-called 'precision weapons' will be developed, for example active packaging materials that target the surface of the foodstuffs where spoilage occurs. There is also a range of new processing technologies, such as ultra-high pressure, high electric field pulses and light pulses, being researched at present. Many of these offer benefits to the food industry with respect to increased product quality and assured safety. However, there is a need for more information concerning their modes of action, effects on foods and food constituents, and effects on microbiological activity.

Hurdle technology is currently used as qualitative concept, and quantitative data on the impact of combined hurdles are still lacking. Further studies should focus on the efficiency of combinations of inhibiting factors in realistic food product formulations. It is important to remember that, among traditional hurdles, temperature is still the most important. More quantitative data are still needed on microbial inhibition as a function of heat treatment combined with other hurdles such as a_w and pH. Further research is needed on the influence of various hurdles (for example, reduced a_w, low pH) on the heat resistance of pathogens, since current pasteurisation standards are valid for products with high water activity and near neutral pH. In processes based on low heat (pasteurisation, sous-vide cooking) much work is still to be done for the optimisation of these processes for each food product. Similarly, there is still much to be done with regard to distribution and retail temperatures. Temperatures are typically too high during distribution and retailing and vary too much. An unbroken chill chain and consistent low temperatures are the easiest and mildest way to ensure good quality and safety. A relatively small change in temperature (for example, 8°C to 4°C) can radically improve the shelf-life and safety of many food products, such as red meat, poultry meat, fish and fish products.

7.8 References

ABEE, T., KROCKEL, L. and HILL, C. 1995. Bacteriocins: modes of action and potentials in food preservation and control of food poisoning. *Int. J. Food Micro.*, **28**(2): 169–85.

ALZAMORA, S.M., CERRUTTI, P., GUERRERO, S. and LOPEZ-MALO, A. 1995. Minimally processed fruits by combined methods. In: Barbosa-Canovas, G.V. and Welti-Chanes, J. (eds) *Food Preservation by Moisture Control: Fundamentals and Applications*. Lancaster, PA: Technomic, 463–92.

ALZAMORA, S.M., TAPIA, M.S., ARGAIZ, A. and WELTI, J. 1993. Application of combined methods technology in minimally processed fruits. *Food Res Internat.*, **26**: 125–30.

ARGAIZ, A., LOPEZ-MALO, A. and WELTI-CHANES, J. 1995. Considerations for the development and the stability of high moisture fruit products during storage. In: Barbosa-Canovas, G.V. and Welti-Chanes, J. (eds) *Food*

Preservation by Moisture Control: Fundamentals and Applications. Lancaster, PA: Technomic, 729–60.

BETTS, G. (ed.). 1996. Code of practice for the manufacture of vacuum and modified atmosphere packaged chilled foods with particular regard to the risks of botulism. Guideline No. 11, Campden & Chorleywood Food Research Association, Chipping Campden, 1996, p. 114.

BOOTH, I.R. 1995. Regulation of cytoplasmic pH in bacteria. *Microbial Rev.*, **49**: 359–78.

BOOTH, I.R. and KROLL, R.G. 1989. The preservation of food by low pH. In: Gould, G.W. (ed.) *Mechanisms of Action of Food Preservation Procedures.* Amsterdam: Elsevier Science, 119–60.

BRUNO, M., KAISER, A. and MONTVILLE, T.J. 1993. Depletion of proton motive force by Nisin in *Listeria monocytogenes* cells. *Appl. Env. Micr.*, **58**: 2255–9.

CARLIN, F. and PECK, M.W. 1996. Growth of, and toxin production by, non-proteolytic *Clostridium botulinum* in cooked vegetables at refrigeration temperatures. *Appl Environ Microbiol.*, **62**: 3069–72.

CHIKINDAS, M.E., GARCIA-GARCERA, M.J., DRIESSEN, A.J.M., LEDEBOER, A.M., NISSEN-MEYER, J., NESS, I.F., ABEE, T., KONINGS, W.N. and VENEMA, G. 1993. Pediocin PA-1, a bacteriocin from *Pediococcus acidilacitici* PAC1.0, forms hydrophilic pores in the cytoplasmic membrane if target cells. *Appl. Env. Micr.*, **59**: 3577–84.

CONNER, D.E. 1993. Naturally occurring compounds. In: Davidson, P.M. and Branen, A.L. (eds). *Antimicrobials in Foods.* 2nd edn. New York: Marcel Dekker.

CUTTER, C.N. and SIRAGUSA, G.R. 1995. Population reductions of Gram-negative pathogens following treatments with nisin and chelators under various conditions. *J Food Prot.*, **58**: 977–83.

DAESCHEL, M.A. and PENNER, M.H. 1992. Hydrogen peroxide, lactoperoxidase systems, and reuterin. In: Bibek R. and Daeschel, M. (eds) *Food Biopreservatives of Microbial Origin*, pp. 207–57. London: CRC Press.

DELVES-BROUGHTON, J. 1993. The use of EDTA to enhance efficacy of nisin towards Gram-negative bacteria. *Int Biodeterior Biodegr*, **32**: 87–97.

FARBER, J.M. and PAGOTTO, F. 1992. The effect of acid shock on the heat resistance of *Listeria monocytogenes. Lett Appl Microbiol.*, **15**: 197–201.

FOSTER, J.W. 1993. Beyond pH homeostasis: the acid tolerance response of salmonellae. *ASM News*, **58**: 266–70.

FUGLSANG, C.C., JOHANSEN, C., CHRISTGAU, S. and ADLER-NIESSEN, J. 1995. Antimicrobial enzymes: application and future in the food industry – review. *Trends Food Sci. Tech.*, **6**(12): 390–6.

GOODSON, M. and ROWBURY, R.J. 1989. Habituation to normally lethal acidity by prior growth of *Escherichia coli* at a sub-lethal pH value. *Lett Appl Microbiol*, **8**: 77–9.

GORRIS, L.G.M. 2000. Hurdle technology, a concept for safe, minimal processing of foods. In: Robinson, R., Batt, C. and Patel, P. (eds) *Encyclopedia of*

Food Microbiology. London: Academic Press.

GORRIS, L.G.M. and PECK, M.W. 1998. Microbiological safety considerations when using hurdle technology with refrigerated processed foods of extended durability. In: Ghazala, S. (ed.) *Sous Vide and Cook-Chill Processing for the Food Industry*. Gaithersburg, MD: Aspen Publishers, pp. 206–33.

GOULD, G.W. 1995. Homeostatic mechanisms during food preservation by combined methods. In: Barbosa-Canovas, G.V. and Welti-Chanes, J. (eds) *Food Preservation by Moisture Control: Fundamentals and Applications*. Lancaster, PA: Technomic, pp. 397–410.

GUERRERO, S., ALZAMORA, S.M. and GERSCHENSON, L.N. 1994. Development of a shelf-stable banana purée by combined factors: microbial stability. *J Food Prot.*, **57**: 902–7.

HAYNIE, S.L., CRUM, G.A. and DOELE, B.A. 1995. Antimicrobial activities of amphiphilic peptides covalently bonded to a water-insoluble resin. *Antimicrobial Agents and Chemotherapy*, **39**: 301–7.

HELANDER, I.M., VON WRIGHT, A. and MATTILA-SANDHOLM, T. 1997a. Potential of lactic acid bacteria and novel antimicrobials against Gram-negative bacteria. *Trends Food Sci & Technol.*, **8**: 146–50.

HELANDER, I.M., ALAKOMI, H., LATVA-KALA, K. and KOSKI, P. 1997b. Polyethyleneimine is an effective permeabilizer of Gram-negative bacteria. *Microbiology*, **143**: 3193–9.

HELANDER, I.M., LATVA-KALA, K. and LOUNATMAA, K. 1998a. Permeabilizing action of polyethyleneimine on *Salmonella typhimurium* involves disruption of the outer membrane and interactions with lipopoly-saccharide. *Microbiology*, **144**: 358–90.

HELANDER, I.M., ALAKOMI, H., LATVA-KALA, K., MATTILA-SANDHOLM, T., POL, I., SMID, E.J., GORRIS, L.G.M. and VON WRIGHT, A. 1998b. Characterization of the action of selected essential oil components on Gram-negative bacteria. *J Agric. Food Chem*, **46**: 3590–5.

HILL, C., O'DRISCOLL, B. and BOOTH, I.R. 1995. Acid adaptation and food poisoning micro-organisms. *Int J Food Microbiol.*, **28**: 245–54.

HOLZAPFEL, W.H., GEISEN, R. and SCHILLINGER, U. 1995. Biological preservation of foods with reference to protective cultures, bacteriocins and food-grade enzymes. *Int J Food Micr*, **24**: 343–62.

HOTCHKISS, J.H. 1995. Safety considerations in active packaging. In: Rooney, M.L. (ed.) *Active Food Packaging*. London: Blackie Academic & Professional, pp. 238–55.

HUGHEY, V.L., WILGER, P.A. and JOHNSON, E.A. 1989. Antimicrobial activity of hen white lysozyme against *Listeria monocytogenes* Scott A in foods. *Appl. Env. Micr.*, **55**: 631–8.

ISHITANI, T. 1995. Active packaging for food quality preservation in Japan. In: Ackermann, P., Jägerstad, M. and Ohlsson, T. (eds) *Foods and Food Packaging Materials: Chemical Interactions*. Cambridge, Royal Society of Chemistry, pp. 177–88.

KAMAU, D.N., DOORES, S. and PRUITT, K.M. 1990. Enhanced thermal destruction of

Listeria monocytogenes and *Staphylococcus aureus* by the lactoperoxidase system. *Appl. Env. Micr.*, **56**: 2711.

KROLL, R.G. and PATCHETT, R.A. 1992. Induced acid tolerance in *Listeria monocytogenes. Lett Appl Microbiol.*, **14**: 224–7.

LEISTNER, L. 1994a. Introduction to hurdle technology. In: Leistner, L. and Gorris, L.G.M. (eds) *Food Preservation by Combined Processes.* Final Report of FLAIR Concerted Action No. 7, Sub-group B, EUR 15776 EN, pp. 1–6.

LEISTNER, L. 1994b. User guide to food design. In: Leistner, L. and Gorris, L.G.M. (eds) *Food Preservation by Combined Processes.* Final Report of FLAIR Concerted Action No. 7, Sub-group B, EUR 15776 EN, pp. 25–8.

LEISTNER, L. 1995. Principles and applications of hurdle technology. In: Gould, G.W. (ed.) *New Methods for Food Preservation.* London: Blackie Academic & Professional, pp. 1–21.

LEISTNER, L. 1999. Combined methods for food preservation. In: Shafiur Rahman, M. (ed.) *Food Preservation Handbook.* New York: Marcel Dekker, pp. 457–85.

LEISTNER, L. 2000. Use of combined preservative factors in foods of developing countries. In: Lund, B.M., Baird-Parker, A.C. and Gould, G.W. (eds) *The Microbiological Safety and Quality of Food.* Vol. 1. Gaithersburg, MD: Aspen Publishers, pp. 294–314.

LEISTNER, L. and GORRIS, L.G.M. 1995. Food preservation by hurdle technology. *Trends Food Sci Technol.*, **6**: 41–6.

LEYER, G.J. and JOHNSON, E.A. 1993. Acid adaptation induces cross-protection against environmental stresses in *Salmonella typhimurium. Appl Environ Microbiol.*, **59**: 1842–7.

LOPEZ-MALO, A., PALOU, E., WELTI, J., CORTE, P. and ARGAIZ, A. 1994. Shelf-stable high moisture papaya minimally processed by combined methods. *Food Res Int.*, **27**: 545–53.

MCKENNA, B.M. 1994. Combined processes and total quality management. In: Leistner, L. and Gorris, L.G.M. (eds) *Food Preservation by Combined Processes.* Final Report of FLAIR Concerted Action No. 7, Sub-group B, EUR 15776 EN, pp. 99–100.

MILLER, L.G. and KASPAR, W. 1994. *Escherichia coli* O157:H7. Acid tolerance and survival in apple cider. *J Food Prot.*, **57**: 460–4.

MONTICELLO, D.J. 1990. Control of microbial growth with lantibiotic/lysozyme formulations. EP O 427 912 A1. Haarman & Reimer Corp., New Jersey.

PAIK, J.S., DHANASEKHARAN, M. and KELLEY, M.J. 1998. Antimicrobial activity of UV-irradiated nylon film for packaging applications. *Pack. Techn. Sci.*, **11**: 179–87.

PECK, M.W. 1997. *Clostridium botulinum* and safety of refrigerated processed foods of extended durability. *Trends Food Sci Technol.*, **8**: 186–92.

ROWBURY, R.J. 1995. An assessment of the environmental factors influencing acid tolerance and sensitivity in *Escherichia coli, Salmonella* spp. and other enterobacteria. *Lett Appl Microbiol.*, **20**: 333–7.

SAJUR, S. 1985. *Preconservacion de Duraznos por Metodos Combinados.* Argentina: Universidad Nacional de Mar del Plata; M.S. thesis.

SAMUELSON, K.J., RUPNOW, J.H. and FRONING, G.W. 1985. The effect of lysozyme and ethylenediaminetetraacetic acid on *Salmonella* on broiler parts. *Poultry Sci.*, **64**: 1488.

SCHVED, F., HENIS, Y. and JUVEN, B.J. 1994. Response of spheroplasts and chelator-permeabilized cells of Gram-negative bacteria to the action of the bacteriocins pediocin SJ-1 and nisin. *Int. J. Food Micr.*, **21**: 305–14.

STEVENS, K.A., SHELDON, B.W., KLAPES, N.A. and KLAENHAMMER, T.R. 1991. Nisin treatment for inactivation of *Salmonella* species and other Gram-negative bacteria. *Appl. Env. Micr.*, **57**: 3613–15.

TAPIA DE DAZA, M.S., ALZAMORA, S.M. and WELTI-CHANES, J. 1996. Combination of preservation factors applied to minimal processing of foods. *Crit Rev Food Sci Nutr.*, **36**: 629–59.

TAPIA DE DAZA, M.S., ARGAIZ, A., LOPEZ-MALO, A. and DIAZ, R.V. 1995. Microbial stability assessment in high and intermediate moisture foods: Special emphasis on fruit preservation. In: Barbosa-Canovas, G.V. and Welti-Chanes, J. (eds) *Food Preservation by Moisture Control: Fundamentals and Applications.* Lancaster, PA: Technomic, pp. 575–601.

8

Safety criteria for minimally processed foods

P. Zeuthen, Technical University of Denmark, Lyngby

8.1 Introduction

The growing demand for convenience and fresh-like, healthy foods is driving the industrialised markets for chilled prepared foods. Traditional means to control food spoilage and microbiological safety hazards, such as sterilisation, curing or freezing, are not compatible with market demands for fresh-like convenience food. Therefore, food manufacturing industries seek compliance with these consumer demands through application of new and mild preservation techniques such as refrigeration, mild heating, modified atmosphere packaging and the use of natural antimicrobial systems. Food types that fall into the category of minimally processed foods are, for example, fresh-cut vegetables, prepared sandwiches, ready-to-eat meals and chilled prepared foods.

Minimally processed foods are produced using techniques that are not intended to completely inactivate all microorganisms present in the raw material. Rather, no or only mild inactivation is achieved during production and their safety during shelf-life depends largely on the following:

- an appropriate refrigerated storage which should prevent the growth of any hazardous microorganisms
- restriction of shelf-life.

In addition to these factors, intrinsic hurdles to microbial growth present in the food product, either originating from the raw material or introduced during processing (e.g. pH, water activity, preservatives, modified atmosphere), may contribute to their safety. The diversity of the products and the combination of different technologies make traditional safety concepts not always applicable for minimally processed foods.

This chapter will use the introduction and the definition given by the EEC FAIR Concerted Project CT 96-1020. Minimally processed foods are food products:

- that undergo a mild preservation process
- where product temperatures during processing fall between 0°C and 100°C
- that rely on refrigerated storage and distribution
- that have a water activity higher than 0.85
- that have a pH higher than 4.5.

The limiting values of pH and water activity are based on the growth limits of *Cl. botulinum* (pH>4.5) and *S.aureus* (a_w 0.85). This definition is consistent with the definition of the European Chilled Food Federation but slightly different from the definition of the Codex Alimentarius Commission. The Codex Committee on Food Hygiene describes the scope of its *Code of Hygienic Practice for Refrigerated Packaged Foods with Extended Shelf Life* (ALINORM 99/13, Appendix III) as follows:

> Products that
> - Are intended to be refrigerated during their shelf life to retard or prevent the proliferation of undesirable microorganisms
> - Have an extended shelf life of more than 5 days (*for products having a shelf life of 5 days or less, the Codex Code of Hygienic Practice for Pre-cooked Foods in Mass Catering should be consulted* (CAC/RCP 39-1993))
> - Are heat treated or processed using other treatments to reduce their original microbial population. (New technologies such as microwave heating, ohmic heating, oscillating magnetic field, high hydrostatic pressure, irradiation, etc. may provide equivalent treatment).
> - Are low acid, that is, with pH>4.6 and have high water activity a_w>0.92
> - May use hurdles in addition to heat or other treatments and refrigeration, in order to retard or prevent the proliferation of undesirable microorganisms
> - Are packaged, not necessarily hermetically, before or after processing (heat or other preservation treatments)
> - May or may not require heating prior to consumption.

The term 'chilled pasteurised foods' is used in this chapter to refer to minimally processed foods that have received a heat treatment or have been processed using other methods to reduce their original microbial population.

When considering the built-in safety in traditionally preserved food products, the safety relies on factors such as shelf-stable canned foods, which have undergone a heat-sterilising process, or deep-frozen foods, which are constantly maintained at freezing temperatures. It is therefore self-evident that the community at large is concerned about the safety and the wholesomeness of minimally processed foods.

When discussing safety of food products it is also important to comprehend the differences between 'safe' and 'wholesome'. The former relates exclusively to the impact that the intake of a food may have on the health of a person, whereas the latter also comprises the eating qualities of food products. Thus a food product may be discoloured or slightly rancid and yet edible although not 'wholesome' as such. The discussion on safety criteria will therefore be related to this only.

8.2 Safety problems with minimally processed foods

When reading through the definitions above it is readily understood that minimally processed foods have built-in weaknesses. Proper measurements of physical, physiochemical and even microbial parameters sometimes present difficulties. Thus it may be difficult, if not impossible, to measure the true pH or the water activity on a borderline between two adjacent layers of a food product although such measurements may be critical for the assessment of how 'safe' the food product is. To know exactly the pH and the water activity is very important in view of the interrelationship between these parameters and the storage temperature.

Many surveys have shown that it can by no means be taken for granted that foods are maintained under properly refrigerated conditions. However many efforts are made to refrigerate food products correctly, we still have a long way to go. Bøgh-Sørensen and Olsson (1990) give a good illustration of this in their overview. It is generally agreed that the storage temperatures at the primary producer are not generally poor. However, during transit of chilled foods through the chill chain, there are several reports on temperature abuses, in particular when raw materials have to be chilled from body heat temperature (red meat) or ambient temperature (vegetables and fruit). Sørensen *et al.* (1985) measured product temperatures of vacuum-packed sliced meat products on arrival at supermarkets. Very often product temperatures above 5°C were measured. The highest average temperature was 14.5°C. The survey, among many others, shows that product temperatures are often correlated with ambient temperatures, indicating that there are several possibilities for improving the chill chain, for example by improving the procedures used in loading and unloading. James and Evans (1990) made a survey of temperatures in the retail and domestic chilled chain. In seven chilled display cabinets for delicatessen, they found average product temperatures ranging from 0.4°C to 9.7°C with a maximum temperature of 13.0°C. Further, the fluctuations in temperature were as large as 11.0°C. In domestic refrigerators they found 3–10.4°C as an overall mean in 21 units. Over 85% of the appliances operated with a mean temperature above 5°C. Similar results regarding chilling of foods at retail level are given by Olsson (1990).

Greer *et al.* (1994) studied temperature distributions in retail cabinets. The coldest average temperatures, 1.7°–4.5°C, were recorded at the rear of the cabinets, and the warmest, 6.1°–10°C, at the centre of the case. The average temperatures increased with upward location of the foods, from 1.7°–6.1°C at 17 cm below to 4.5°–10°C at 4 cm above the load line. Further, a survey of air temperatures in four commercial retail cases indicated that temperatures of ≤ 4°C cannot be maintained throughout existing retail cabinets.

Willcox *et al.* (1994) also point out that retail display cabinets are critical points in the cold chain of minimally processed, modified atmosphere packaged vegetables. Apart from the fact that they found packages where the 'sell by' date was exceeded, the temperature performance of the display cabinets was fairly strongly influenced by both ambient temperature and day/night regime. Temperature differences of greater than 5°C were measured on the decks, and the temperature in one part of the cabinet increased towards the end of the day by 4°C and towards the end of the week by almost 7°C. Klepzig *et al.* (1999) investigated the influence of interruption in chilling on growth of spoilage and pathogenic bacteria on iceberg lettuce. In some experiments with artificial contamination where chilling was interrupted, they found growth of *Salmonella lexington, Escherichia coli* and *Listeria monocytogenes*, provided the interruption at 25°C lasted longer than six hours. On the other hand, interruption of chilling with 15°C initiated propagation of *Escherichia coli* and *Salmonella lexington* only if the interruption lasted 12 hours or more. However, they did not find any increase in numbers of *Listeria monocytogenes* during a temperature abuse at 15°C for 24 hours.

The importance of the limits for the parameters temperature, pH, and a_w are shown in Table 8.1. It will be seen that several food-poisoning organisms will grow if, for example, chilled foods are left at temperatures just slightly above the 'legal' limits, which they often are, according to Bøgh-Sørensen and Olsson (1990). The pathogens of most concern for chilled pasteurised foods are sporogenic bacteria that are able to grow at refrigeration temperatures, such as non-proteolytic *C.botulinum* and *B. cereus.* The microbial safety of these foods is likely to rely on a combination of hurdles (see Chapter 7).

Providing that post-process contamination can be prevented, some non-sporogenic pathogens that are quite resistant to mild heat treatments may survive when pasteurisation is not completely effective or not executed properly. An example of this is *L. monocytogenes.* Although growth of this bacterium is often retarded in the presence of intrinsic hurdles in the foods, an extended shelf-life allows more time for growth to reach hazardous levels. Mossel and Struijk (1991) thus found growth to 10^{-6} CFU/g in two weeks at 3°C. Similarly, Graham *et al.* (1996) found, in a major study on growth from spores of non-proteolytic *Cl. botulinum*, that at a temperature of 5°C, a pH of 6.12 and a concentration of NaCl of 2%, the experimental doubling time was 32 hours after a lag time of 290 hours. At a temperature of 7°C, a pH of 5.80 and at the same NaCl concentration, the experimental doubling time was 14 hours and the lag time 180 hours.

Modern processing of food products often involves a minimal heat treatment at 50–95°C under 'sous-vide' and refrigeration. Such a treatment will make an extended storage possible. Although these foods are not sterile, the storage life is

Table 8.1 Growth limits for some pathogenic bacteria

Microorganism	*Min. temp. (°C)*	*Min. pH*	*Min. a_W*	*Aerobic/anaerobic*
Mesophilic *B.cereus*	15	4.3	0.95	Facultative
Psychrotrophic *B. cereus*	4	4.3	0.95	Facultative
Campylobacter jejuni	32	4.9	0.99	Microaerophilic
Mesophilic, *proteolyctic Cl.botulinum*	10	4.6	0.93	Anaerobic
Non-proteolytic (psychrotrophic) *Cl.botulinum*	3	5.0	0.97	Anaerobic
Cl. perfringens	12	5.0	0.95	Anaerobic
Escherichia coli	7	4.4	0.95	Facultative
Escherichia coli 0157	6.5	4.5	0.95	Facultative
Listeria monocytogenes	0	4.3	0.92	Facultative
Salmonella	7 (some strains 5.2)	4.0	0.94	Facultative
S. aureus	6 (10 for toxin production)	4.0 (4.5 for toxin production)	0.83 (0.9 for toxin production)	Facultative
Vibrio cholerae	10	5.0	0.97	Facultative
Vibrio parahaemolyticus	5	4.8	0.94 (halophile)	Facultative
Yersinia enterocolitica	−1	4.2	0.96	Facultative

Source: Adapted from CA FAIR CT96-1020.

often found to be up to six to seven weeks. The principal microbial hazard in foods of this type is growth of and toxin production by non-proteolytic *Cl. botulinum*. Lysozyme has been shown to increase the measured heat resistance of non-proteolytic *Cl. botulinum* spores. However, the heat treatment guidelines for prevention of risk of botulism in these products have not taken into consideration the effect of lysozyme, an enzyme that can be present in many food products. Fernández and Peck (1999) made a study to assess the botulism hazard, the effect of heat treatments at 70–90°C combined with refrigerated storage for up to 90 days on growth from 10^{-6} spores of non-proteolytic *Cl. botulinum* in an anaerobic meat medium containing 2400 U of lysozyme/ml. At a storage temperature no higher than 8°C, they found the following heat treatments each prevented growth and toxin production during 90 days: 70°C for $\geq$ 2,545 min., 75°C for $\geq$ 463 min., 80°C for $\geq$ 230 min., 85°C for $\geq$ 84 min., and 90°C for $\geq$ 33.5 min.

As seen above, it is sometimes possible to indicate through a model exactly how long a heating time at a specified temperature is required to ensure the safety of a food product, but in other cases this is not possible. Thus the development of effective quantitative microbiological models have to be made use of in other cases. This can be done on the availability of data on consumers' exposure to a biological agent and the dose–response relationship that relates levels of the biological agent ingested with frequency of infection or disease. Previously such information has been acquired from human volunteer feeding studies. An example of this is quantification of staphylococcal enterotoxin necessary to provoke staphylococcal food poisoning. However, this method is not feasible for pathogens that have a significant risk of being life threatening or for which morbidity is primarily associated with high-risk populations (i.e. immuno-compromised persons). Thus Buchanan *et al.* (1997) proposed that purposefully conservative dose–response relationships can be estimated on the basis of combining available epidemiological data with food survey data for a ready-to-eat product, which would often also cover minimally processed foods. In their studies they exemplified this by combining data on the incidence of listeriosis in Germany with data on the levels of *Listeria monocytogenes* in smoked fish to generate a dose–response curve for this foodborne pathogen. Such models are extremely useful but it will easily be understood that to cover any general standard for quantification of *Listeria monocytogenes,* several similar models have to be added to these results. Further, this study relates to one particular bacterium only. Therefore there are plenty more studies to perform in this field to complete a safety criterion. In conclusion, information about proper handling and processing procedures is available to a very large extent, but in practice abuses occur very often.

8.3 Fresh fruit and vegetables

According to the final report of CA FAIR CT96-1020, one category of minimally processed foods is products without a pasteurisation or equivalent inactivation procedure during preparation (e.g. fresh-cut vegetables). In these foods, non-spore-forming as well as spore-forming pathogens should be considered as potential hazards. Here, the presence of pathogenic bacteria, viruses and parasites in the product can be prevented only by good agricultural practices (GAP) and good manufacturing practices (GMP). In order to prevent or control growth of foodborne pathogens in these foods, storage temperature and shelf-life are critical factors.

Essentially, fresh-cut vegetables consist of raw fruits or vegetables that have undergone a minimal processing such as peeling, slicing or shredding before being ready for use. These products are usually packed in sealed plastic pouches or trays. The major features regarding quality and safety of these food products are: (a) they are raw and the plant tissue remains alive throughout the entire shelf-life; (b) the plant tissue may be damaged by the minimal process to various extents; (c) the packaging protects the product from microbial contamination and permits an extension of shelf-life; (d) fresh-cut vegetables are increasingly processed under quality assurance systems; (e) it is mandatory to make use of HACCP procedures and Good Hygienic Practices in the production of fresh-cut vegetables.

In most cases, retention of sensory quality of the product depends on normal metabolism of the plant tissue. Therefore, the minimal processing of raw vegetables dramatically increases the physiological activity and leads to senescence or deviation in the metabolism, such as fermentation of the products. Refrigeration of fresh-cut fruits and vegetables is essential to delay biochemical and physiological deterioration triggered by the stress of processing. In addition, fresh-cut fruits and vegetables are often packaged to retain their quality by limiting water loss and by creating a proper modified atmosphere. However, this may generate atmospheric modifications by respiration of the plant tissue, which in turn may cause fast fermentative changes and spoilage. Since respiration is strongly affected by any shift in temperature, modified atmosphere packaging is not applicable in this case if the storage temperature is fluctuating.

Since most fresh-cut fruits and vegetables have a pH that is favourable for growth and survival of pathogenic bacteria, and in addition maintain a high a_w when packaged, refrigeration and restriction of shelf-life are the only measures that can prevent or reduce growth of foodborne bacterial pathogens. At ambient temperatures, these products in most cases support growth of many foodborne bacterial pathogens, as pointed out by Nguyen-the and Carlin (1994). However, even when kept under refrigeration, consideration should be given to pathogens with low infectious dose such as *Escherichia coli 0157:H7* which may not grow but survive at refrigeration temperatures, or to pathogenic viruses and parasites that survive in foods. In the case of viruses, refrigeration may extend their survival on the surface of fresh-cut vegetables (see Konowalchuk and Speirs, 1975). Several researchers have proved the presence of *Listeria monocytogenes* on raw vegetables, e.g. Carlin *et al.* (1995). In the case of this bacterium, growth would be possible even if the storage temperature were maintained at 3–4°C.

8.4 Shelf-life evaluation

It may seem odd to include anything about shelf-life in this chapter as it was specifically stated that this chapter deals with safety of minimally processed foods only. Yet it seems in place to mention the subject, because shelf-life is invariably mentioned together with safety of foods. Of course, even if a food is tainted or spoiled, this is no proof that it is unsafe, although this is often maintained in practical life.

Shelf-life is defined as the period during which the product maintains its microbiological safety and sensory qualities, including visual appearance, under expected storage conditions (CT96-1020). It should be noted that Codex Alimentarius refers to *a specific storage temperature*. It is based in particular on identified hazards for the product, heat or other preservation treatments, packaging method, storage conditions and other hurdles or inhibiting factors that may be used. The shelf-life is determined under the responsibility of the manufacturer by performing shelf-life assessment and validation studies.

Minimally processed foods are produced using techniques that are insufficient to ensure their commercial stability for extended periods. Refrigeration and restriction of shelf-life are principal requirements for the control of food spoilage and retardation of pathogen growth in these foods. The legislative requirements and recommendations for temperature control during manufacturing, heating, cooling and chilled storage are abundant, but there are no rules in food legislation on how long food should last. It is, however, the responsibility of the manufacturer to determine the shelf-life of the product manufactured and to ensure that the product is safe throughout its assigned shelf-life. Very little information is available for producers of minimally processed foods on the concept of shelf-life, and the knowledge of product storage conditions and consumer handling is currently limited. In particular these factors are not sufficiently considered in shelf-life protocols. Since shelf-life evaluation requires extensive knowledge in the field of food microbiology and technology, it is advisable that guidelines on shelf-life evaluation of minimally processed foods will be much more readily available than is the case today, since it also pertains to food safety. When reading the EEC legislation, this defines both shelf-life as 'the date until which the foodstuff retains its specific properties when properly stored' (79/112/EEC) and product safety as follows:

> Safe product shall mean any product which, under normal or reasonably foreseeable conditions of use, *including duration*, does not present any risk or only the minimum risks compatible with the product's use, considered as acceptable and consistent with a high level of protection for safety and health ...

Since there seems to be some confusion even in the public legislation between safe and acceptable products from a quality point of view, as well as insufficient knowledge in practice about how to ensure safe food products, it is clear that there are limitations in how safe food products can be ensured.

A few guideline documents on shelf-life evaluation for minimally processed foods, which are intended for use by food manufacturers, are issued. The guidelines on shelf-life evaluation for chilled foods produced by the CFDRA (Campden and Chorleywood Food and Drink Research Association) Shelf Life Working Party (1991) have been prepared by food manufacturers and retailers in conjunction with members of CFDRA. The primary objective of the guidelines is to ensure that shelf-lives are set to ensure the safety of chilled food. It was anticipated that the protocol should be used throughout the UK chilled food industry unless an assessor is able to demonstrate the validity of alternative methods or testing.

In 1995 a protocol for the validation of shelf-life for chilled foods has been drawn up by the French trade organisation SYNAFAP (Syndicat National des Fabricants de Plats Préparés Frais). The objective of the protocol is to validate the evaluation of a product's shelf-life from the point of view of microbiological safety. The scope of the protocol is limited as it relates only to validation of shelf-life but not to assessment. The protocols from both CFDRA and SYNAFAP are implemented in three phases: a prototype phase, an industrial pilot phase and a full-size production phase. Both protocols strongly emphasise the use of hazard analysis throughout the shelf-life assessment and validation. Hazard analysis is performed to identify the type of microorganisms to take into account and the acceptability thresholds for each of these microorganisms. Both guidelines stress the importance of using statistics, not only for evaluation of the anticipated mean shelf-life but also in order to estimate the range of shelf-lives that would apply under a given set of storage conditions. To do this accurately would require a large number of trials. Over 200 trials would be required and only a few are recommended. For economic reasons this is one of the typical instances where an agreement between theory and practice is difficult to achieve, if not for any other reasons.

8.5 Current legislative requirements: the EU

For several years the community has realised that there are safety problems involved with the way we are offering several prepared foods for immediate consumption, including food types like minimally prepared food products. Therefore it has been natural to legislate in the field in order to minimise the risks. This legislation has of course made use of the information on growth and survival of pathogenic bacteria and the presence of other toxic substances such as bacterial spores and mycotoxins. However, to implement legislation takes time, and knowledge on various noxious substances is increasing. Therefore the present legislation is lagging behind and is uneven. In the so-called Inventory Report of FAIR CT96-1020 an attempt is made to give an account of the present legislation and important guidelines in the field. Within Europe, the build-up of food legislation is both vertical and horizontal.

The horizontal legislation covers all food products, either to protect public health or for the protection of other consumer interests, such as the provision of information or the prevention of misleading trade practices (e.g. with regard to labelling). The vertical legislation is also called 'recipe' legislation, laying down detailed specifications for a specific type of food product, such as fish or meat products. With regard to legislation for processed foods the EEC has divided the legislation into five main sections: 'Food Control' (89/397/EEC) is the section of interest in this connection. Part of the vertical legislation on primary products is of course also of great interest, since parts of it deal with hygienic measures necessary for the production of safe food products. They are of special interest because the food products in question are minimally processed and thus often base their safety on chilling and raw materials containing only few or no pathogens. In essence, the Directive is based on the obligation of implementation of HACCP principles and some general hygiene requirements.

Table 8.2 A summary of the general hygiene requirements according to the EEC Directive 93/43/EEC

Topic	Legal requirements
Requirements for production rooms	• Adequate construction and facilities for cleaning and disinfection • Protection against cross-contamination • *Suitable* temperature conditions for hygienic processing • *Suitable* temperature conditions for storage
Transport	• Vehicles only for transport of foodstuffs or adequate separation between products • *Appropriate* temperature during transport, eventually monitoring of temperature during transport
Equipment requirements	• Adequate cleaning and disinfection
Food waste	• Adequate provisions for removal and storage of food waste
Water supply	• Use of potable water • Water unfit for drinking only if no direct contact with the food product (e.g. for steam generation)
Personal hygiene	• Protective clothing *if necessary* • Provision of hot and cold running water
Training	• Responsibility of food business operator • Obligation for adequate training of personnel

In European hygiene legislation the vertical/horizontal sectioning also exists. For food products of animal origin, vertical directives prescribe in some detail the hygiene requirements that must be observed. All other food products are covered by the general directive on food hygiene (Directive 93/43/EEC).

Minimally processed foods can contain ingredients of different origin and are usually combinations of ingredients from different product categories. Therefore, different legislation is in force, which complicates the application of European food law in practice. The general Food Hygiene Directive 93/43/EEC is summarised in Table 8.2.

The vertical directives are specific for a product group. They contain more detailed requirements such as approval of establishments, stricter temperature condition, and official controls for some products of animal origin. As an example of this, the Directive 92/5/EEC on meat products is summarised in Table 8.3. As of 1997 only half of the European states have implemented the legislation, but most of the states are on the verge of passing the laws now. An example of some of the outstanding differences are the legal time–temperature requirements for cooling and chilled storage of minimally processed foods (Table 8.4). With regard to microbial norms, these vary from one state to

Table 8.3 Summary of Directive 92/5/EEC

Topic	Legal requirements
Approved processing establishment	According to prescribed treatment and hygiene requirements of the Directive
Obligation for 'own checks' (only the fishery products Directive explains about HACCP principles). In 1997 HACCP requirements were generally adopted in the so-called Green Paper.	• Identification of critical points • Monitoring of critical control points • Taking samples for analysis in a laboratory in order to check cleaning and disinfection • Keeping written records of control for at least six months after the 'use before' date of the product.
Layout and equipment	• 12°C • Separate room (place) for storing cleaning and disinfection materials and products
Equipment requirements	• Adequate cleaning and disinfection • Hot water for cleaning tools at a temperature of not less than 82°C
Requirements for personal hygiene and for staff	• Suitable clean working clothes • Hair completely enclosed by headgear • Required medical certificate for all personnel • Separate hand washing facilities • Taps must not be hand-operable • Facilities with cleaning and disinfection products and hygienic means of drying hands
Special additional conditions for meat-based prepared meals:	
Cold storage of ingredients	• *Suitable* temperature conditions for storage
Temperature conditions of preparation	• Minimal time of keeping meat product between 10°C and 63°C, e.g. immediately mix meat product with other meal ingredients or first refrigerate the meat product to 10°C or less before mixing
Minimum heating	• At least equivalent to pasteurisation • Parameters being approved by authority
Cooling requirements	• Cooling to 10°C (internal temperature) in less than 2 hours (authorities may derogate from this in special cases) • Cooling to storage temperature as soon as possible
Chilled storage	Storage temperature not specified
Maximum shelf-life	• Clear indication of date of manufacture • Shelf-life not specified

Table 8.4 Legal time–temperature requirements for cooling and chilled storage of minimally processed foods in different European states

Country	Chilled storage	Cooling after heating
Belgium	Max. 7°C	Immediately
Denmark	5°C, MAP of minced meat: 2°C	• From 65° to 10°C in 180 min • If sold within 3 hours, no requirements
Finland	Meat-based products: 6°C; other chilled products: 8°C	To 10°C in 120 min, exceptions allowed
France	Depends on stage of production, e.g. in retail, storage at ≤ 4°C	To 10°C in 120 min (usually from 63°C to 10°C in less than 120 min)
Italy	Meat products: −1–7°C Fish products: 0–4°C	For meat products to 10°C in 120 min
Spain	0–3°C	Not specified, specific for each product
Sweden	< 8°C	Not specified
The Netherlands	Max. 7°C	As quick as possible
UK	8°C	As quick as possible
Norway (not EU member)	−1 to 4°C; retail: < 7°C	• From 60°C to 7°C in 4 hours (in general) • To 10°C in 120 min (for meat as ingredient in ready-to-eat meals

Note: This table is not exhaustive. Additional temperature requirements can be found in the different national GMP codes.

another. A complete listing of these would be too extensive here. In addition, these norms are under revision in several states, but differences still exist.

Many other more or less official authorities have presented good manufacturing practices or codes on the handling of perishable foods. Among these are, first of all, the Codex Committee on Food Hygiene, which has worked out a code of hygienic practice for refrigerated packaged foods with extended shelf-life. The code must be used in addition to General Principles of Food Hygiene (ALINORM 97/13A) by Codex Alimentarius. It is interesting to notice that definitions differ from Codex to other legislative bodies. Granted, the Codex *Code of Hygienic Practice for Refrigerated Packaged Foods with Extended Shelf Life*, as described in the introduction to this chapter, differs slightly in the definition of the types of foods dealt with here, but heating temperature requirements are different (and less precise). The pH limit is comparable, but Codex sets a limit in water activity at 0.92, whereas the FAIR CT96-1020 has found it necessary to set up a limit at 0.85 to cover absence of growth of *Staphylococcus aureus*.

At the European level, the European Chilled Food Federation (ECFF) has worked out extensive 'guidelines for the hygienic manufacture of chilled foods'. The guideline is very specific in its definitions regarding heating, packaging (or no packaging), possibility of after-contamination and specification of risk of survival or growth of pathogenic bacteria in various food categories. At national level the so-called Inventory Report of CT96-1020 has identified 31 existing GMP codes, guidelines, safety reports, etc., for minimally processed foods. Most of these deal with manufacturing practices, but a few comprise GMPs on retail and catering. Nearly all of them are directed towards specific sectors such as handling of products of vegetable or animal origin. It would be beyond the scope of this chapter to go into further details of these guidelines. The general hygiene requirements in the codes all deal with the following:

- design and construction of food handling and storage areas
- material and equipment
- cleaning and disinfection
- pest control
- personal hygiene and training
- correct working instruction, good manufacturing practice.

In addition the codes, as a rule, include the HACCP principles. Some codes, e.g. the European Chilled Food Federation, divide the manufacturing premises into different sections:

- good manufacturing practice area (GMPA)
- high care area (HCA)
- high risk area (HRA).

The purpose of this is to be able to separate specific operations to specific rooms, e.g. to minimise recontamination during processing. Examples of different hygiene standards according to the area are discussed below (see Table 8.5).

In the inventory report, 19 codes and draft codes have been compared with regard to process parameters for cook-chill products. The comparison deals with chilled storage of ingredients, preparation, pasteurisation, post-process chilling, storage, distribution/retail storage, shelf-life and reheating. It is interesting to note that the requirements under each item differ. Some examples are shown in Table 8.6.

The safety of minimally processed foods depends, of course, on whether pathogenic bacteria are present, or, alternatively, whether they are present in such small numbers that they do not constitute any health hazard, if the food is handled as prescribed. The problem here is that it implies that all presumptions regarding, for example, temperature conditions, shelf-life, hygienic measures, etc., are the same. As shown above, this is not so.

When going through the various GMPs and national and international legislation, large differences exist between numbers of permissible pathogens in foods, including various types of minimally processed foods. The FAIR CT96-1020 Inventory Report has compared norms from four GMPs for cook-chill products. The differences are shown in Table 8.7.

Table 8.5 Different hygiene standards according to area

GMPA	HCA	HRA
	Hand washing facilities	
Taps should not be hand-operable. Disposable towels are strongly recommended	Taps must not be hand-operable. Disposable towels must be used	
	Equipment (example)	
Wheeled trolleys etc. should be subject to scheduled cleaning and disinfection	Wheeled trolleys etc. should not cross from GMP to HCA unless adequate provision is made to clean and disinfect	Wheeled trolleys etc. must not enter the area unless adequate provision is made to clean and disinfect
	Temperature control	
Temperature should be monitored and recorded	Temperature must be monitored and recorded	
	The air handling system should be capable of maintaining the rooms at a temperature of 12°C	The air handling system must be capable of maintaining the rooms at a temperature of 12°C
	Assembly and packaging	
Methods of assembly of raw ingredients should be determined with due regard to GMP	Pre-treatment of raw materials intended to reduce the microbial load and decontamination should be completed in the HCA	All ingredients brought into this area must be pre-cooked. The room for assembling of final product must be maintained to the highest standards

8.6 Microbiological risk assessment

In a paper by Notermans and Mead (1996) they explain how the various surveillance systems were developed. Since the traditional surveillance of end-products was inadequate for detecting unsafe batches, the situation was improved by introducing GMPs, although GMPs merely reflect general guidelines instead of providing an objective approach to risk assessment. The concept was then extended by introduction of the hazard analysis critical control point (HACCP) system. With HACCP, control procedures are directed at specific operations that are crucial to food safety and the effects of which on product contamination can be quantified. In this kind of surveillance, the control is proactive since remedial action is taken in advance. However, the HACCP system is still largely applied on a qualitative basis. For proper risk management, a more quantitative determination of the hazards associated with food consumption is necessary.

Table 8.6 Differences in handling requirements of minimally processed foods as stated in 19 public codes

Chilled storage of ingredients	From 'not specified' to '0 to 3°C' or specific handling of different types of foods
Preparation	From a room temperature of 12°C requirements to food temperature, usually 5°C to 10°C
Pasteurisation	From 10 min at 90°C (with *Cl. bot.* as ref. microorganism) to cook to min 75°C
Post-process chilling	From 'as quick as possible' to 'to 10°C in 120 min'
Storage temperature	From '0–3°C' to '$\leq$ 7°/$\leq$ 8°C'
Distribution/retail storage	From '0–3°C' to '$\leq$ 10°C'
Shelf-life	From 'not stated' to max. 42 days; max. 3 weeks outside facility; max. 1 week in home refrigerator
Reheating	From 'to temperature suitable for organoleptic purposes' to 'min. core temperature of 75°C (start reheating within 30 min of removal from refrigerator)'

Table 8.7 Differences in microbiological norms for pathogens in cook-chill products

Listeria monocytogenes	From absent in 0.01g to < 100/g
Salmonella	From 'absent' to absent in 25 g
Staphylococcus (coagulase+)	From 'absent' to $\leq$100000/g*
Clostridium perfringens	From not mentioned to $\leq$100000/g*
Campylobacter	From not mentioned to absent in 25 g

Note: * refers to a national legislation but the GMP advises applying stricter norms than legally prescribed.

Risks are a normal part of life. Over time, some increase while others decrease with the change in social behaviour, environmental changes and modern technology (Kindred, 1993). The purpose of risk analysis is to create informed risk management in decision making. This is achieved through the use of three components:

1. Risk analysis.
2. Risk assessment.
3. Risk communication.

The risk assessment forms the base of risk analysis, since it is the scientific process through which is gathered and interpreted scientific information from many sources for the characterisation and estimation of the risk involved. The risk management then evaluates the information about hazards and risks through the risk assessment together with other important factors that may influence management decisions. It also selects any legal actions to be taken. The risk communication consists of communication about the hazards, risk and management actions with the various involved parties (Kindred, 1996).

Over time, risk analysis has been developed and used in various areas. Thus the National Research Council of the USA published guidelines for chemical risk assessment and management (Anon., 1983). More recently, guidelines for other areas have been made, including application of risk analysis techniques to biological, including microbial, hazards.

On an international basis, the same ideas have been developed, both nationally and by, for example, Codex Alimentarius. Microbiological risk assessment has been defined by the Codex Alimentarius Commission (1998). Within this concept 'risk' is defined as 'a health effect caused by a hazard in a food and the likelihood of its occurrence'. According to International Life Science Institute (1998), risk analysis is composed of three elements: microbiological risk assessment, microbiological risk management and microbiological risk communication. To assess a risk, these three elements will be in continuing contact. Based on that, the risk assessment will make hazard identification, hazard characterisation, exposure assessment and risk characterisation; it will assess policy alternatives and select and implement appropriate options, whereas the risk communication, based on the dialogue with the other partners, will make interactive exchange of information and opinions in a permanent dialogue, in particular with the consumers and industries.

Codex Alimentarius has taken a slightly different approach. Accordingly, microbiological risk assessment is a process composed of four steps:

1. Hazard identification.
2. Hazard characterisation.
3. Exposure assessment.
4. Risk characterisation.

This approach is designed for a public authority and is not useful for private industry. The input from the private sector is that it will be able to contribute data and practical experience, but first of all to manage the manufacture of safe foods by applying the various safety concepts, including HACCP principles and GMP.

With regard to application of microbiological risk assessment to minimally processed foods in practice, in **hazard identification** the major types of concern are: *Bacillus cereus, Campylobacter jejuni, Clostridium botulinum, Clostridium perfringens, Escherichia coli, Escherichia coli 0157, Listeria monocytogenes, Salmonella, Staphylococcus aureus, Vibrio cholerae, Vibrio parahaemolyticus* and *Yersinia enterocolitica.* It is not possible to make the list exhaustive. The **hazard characterisation** deals with determination of an infective dose, the impact of the composition of food, the virulence of strains and the consumers at risk, and is therefore closely related to the **exposure assessment**, which is the determination of the numbers and quantities of pathogens ingested by the consumer. After identification of the organisms, information is needed on their numbers and distribution in raw materials for food components. Such information provides the basis for determining the effects of processing and product composition on the level of contamination of the final product at the time of consumption. For this assessment, use can be made of several techniques, such as surveillance tests, storage tests and microbiological challenge testing, each with its own characteristics and field of application. Also, use can be made of mathematical models to predict growth or death of the organisms of interest (Notermans and Mead, 1996). When the frequency distributions of potentially hazardous organisms in the food are known, it is necessary to determine whether or not these levels are acceptable. Such information can be obtained by assessing the dose–response relationship. This is difficult because information is scant. However, some is available on the basis of analysing foodborne disease outbreaks and from human volunteer studies. Black *et al.* (1988) have demonstrated this with *Campylobacter jejuni*, where the infectivity and probability of disease was studied, using human volunteers. They concluded that even small numbers of *Campylobacter jejuni* may cause infection, but it was not possible to state any exact infectious dose.

Human volunteer feeding studies are not always feasible if the disease can cause a significant risk of being life threatening or if the morbidity is primarily associated with high-risk persons (i.e. immunocompromised persons). When this is the case, it is proposed that a more conservative dose–response relationship can be estimated on the basis of combining available epidemiologic data with food survey data and a ready-to-eat product. Buchanan *et al.* (1997) published a practical example of this. The authors combined data on the incidence of listeriosis in Germany with data on the levels of *Listeria monocytogenes* in smoked fish to generate a dose–response curve for this foodborne pathogen. They conclude that the use of epidemiologic and microbiological food survey data in conjunction with an exponential model appears to be an alternative means of estimating dose–response relationship for foodborne pathogenic bacteria that are not amenable to human volunteer feeding studies.

Based on the information obtained, it will be possible to obtain a **risk characterisation.** According to Notermans and Mead (1996), risk characterisation is defined as the ranking of disorders according to severity, perception and economic and social consequences, enabling a decision to be made on the acceptance of a particular risk. It should also include determination of causative factors contributing to the risk from a particular food product. But of course, the risk characterisation will also include some uncertainties, because it is based on a set of assumptions. However, these steps will be highly useful for the risk manager in the risk management, when it has to evaluate the efficiency of various control measures to be taken.

It must be emphasised that risk management is not of a scientific nature, but it deals with the necessary measures to be taken as a consequence of the hazard in question. It is compelled to select and implement risk control options, because so many assumptions and limitations are involved. These include limitations of the scientific information (uncertainty, etc.), limited analytical resources, complexity (such as multiple agents and interacting factors), and unknowns. At the legislative level, the challenge is to balance the scientific evidence in the risk assessments with public concerns, economic interests and even political concern. After risk control options have been selected and implemented, risk management must then evaluate the outcome of the action (Kindred, 1996).

8.7 Future developments

Consumer demands today call for microbiologically safe foods, offering convenience, yet fresh-like sensory and nutritional characteristics. In this respect, research naturally is directed into optimisation of conventional thermal technologies and of more innovative thermal technologies (e.g. microwave heating, ohmic heating) with regard to heat transfer profiles and efficiency. In addition, several other novel technologies have emerged, such as irradiation and high pressure processing, which have no, or only a minimal, thermal effect. In this context, only physical non-thermal processes are considered that stabilise food products by reducing the microbial load. The novel technologies comprise high pressure processing (HPP), pulsed electric fields (PEF), high intensity pulsed light, oscillating magnetic fields and ultrasonic (sonication) treatments. With regard to many of these technologies, there is a lack of detailed scientific knowledge on the actual impact on pathogenic microorganisms, as well as of practical scale applications to evaluate the real potential of these technologies in the area of minimal processing in much detail. In some quarters irradiation is also accounted for as belonging to this group of preservation methods. This is doubtful, since irradiation has been known for so many years and has been examined so intensively – because of public concern about irradiation in general – that it would be wrong to maintain that the method is not investigated properly.

The potential of non-thermal new technologies is to stabilise food products adequately while retaining product quality characteristics. In particular the improvement of quality characteristics or the improved retention of quality aspects of the raw materials in the end product has provoked significant industrial interest. This interest initially stimulated the idea that these innovative technologies could be a very valuable alternative to the more conventional thermal food processing technologies.

The techniques of novel technologies are dealt with in detail in other chapters in this book, but research into these technologies is gradually increasing. It has been recognised that none of them can be considered as perfect for all purposes and thus would completely be able to replace, for example, thermal processing. However, for specific purposes they have turned out to be excellent from an eating quality point of view. Considering how much research there is still to be done when minimally processed foods are manufactured using conventional methods, it would be premature to indicate any final safety criteria for these types of foods using novel technologies. This, however, means that there will be much research to carry out in this particular field in the near future. Industry is very eager to make use of many of the new technologies.

The information shown above has clearly shown that much has been achieved in the area of securing safety in production, storage and marketing of minimally processed foods, especially within the past decade. However, absolutely safe food does not exist. Therefore the safety of food will be based on controlling all steps in the food production process.

In this connection it is of importance that safety measures are moving in a quantitative direction, whereas quality control earlier was much more directed towards identification of hazards as such and no more. Besides, securing safety in foods in general, including in minimally processed foods, has become much more integrated in the manufacturing and marketing of foods. From identification of hazards, particularly in connection with the occurrence of pathogenic bacteria and their toxins, as well as toxogenic spores, a quantitative approach has been taking over, covering all aspects of microbial safety. This has been achieved first by introducing GMP, but it must be admitted that GMP merely reflected general guidelines instead of providing an objective approach towards securing safety of foods. Next, the HACCP system was introduced. This enabled the control procedures to be directed at specific operations, which are crucial to securing food safety and the effects of which on product contamination can be quantified. To reach an even better level of being able to measure food safety on a quantitative basis, it is necessary to incorporate HACCP into a proper system of quantitative risk analysis. Even then it must be admitted that even this system is not infallible (Notermans and Mead, 1996).

With regard to shelf-life evaluation, some of the concluding statements and recommendations of the CT96-1020 report are as follows:

- It is the responsibility of the manufacturer to ensure the safety of products throughout the assigned shelf-life. Manufacturers should be fully informed of their responsibility regarding shelf-life assessment and validation.

In most industrialised countries food legislation prescribes some kind of product shelf-life indication for the benefit of the consumer, but these demands are often founded in partly a requirement for safety and partly a pure attempt to increase the overall level of (eating) quality of foods. The food trade wishes of course that their products are safe to eat, but economic factors regarding overall acceptability are considered of equal importance. To learn all aspects of shelf-life evaluation is very cumbersome, if not impossible in a competitive market. Overcoming this problem alone is an overlooked area of research.

- The present guidelines for shelf-life assessment are of high quality and suitable for use, in particular for larger enterprises, but they are too little known by the industry at large and they should be uniform. This is another problem to overcome.
- Models to predict the growth of microorganisms during chilled storage, microbial challenge tests and other scientific validation studies should be further developed.

Predictive microbiology has been researched extensively and with great success during the past decades. Many of the studies have been carried out in model media, which of course was necessary and useful, but models cannot always cover the specific cases that exist in some foods, each with their particular micro-ecology and physiochemical composition. In spite of the fact that much research already exists, much work has still to be done in the area.

In principle, minimally processed food products can be divided in two categories. One category is products with some kind of mild treatment, physically, such as pasteurising or treatment by novel technology, or chemically, such as treatment with acidification or other similar treatment. The other category of foods receives literally no treatment, such as fresh-cut vegetables, which, besides hygienic handling, perhaps including chlorine washing, rely for safety on chilling, perhaps packaging in a modified atmosphere and a short shelf-life under control.

Within the first category many studies have been performed regarding the effects of hurdles on the safety and shelf-life of the products. Until recently, however, examination of the hurdle technology was a qualitative concept. Quantitative data on the impact of combined hurdles is now becoming available for chilled pasteurised foods. Examples of this are Fernandez and Peck (1997, 1999) and Graham *et al.* (1996).

Fernandez and Peck (1997, 1999) reported the effect of prolonged heating at 70–80°C and at 70–90°C and incubation at refrigeration temperatures on growth from spores and toxigenesis by non-proteolytic *Clostridium botulinum*. In the latter project the presence of lysozyme was also examined. Graham *et al.* (1996) also reported on the inhibitory effect of combinations of heat treatment, pH and sodium chloride on growth from spores of non-proteolytic *Clostridium botulinum*. Such studies are extremely useful for filling out the gaps existing in the knowledge on quantitative data, but many more studies are necessary in the future. Similarly, in the second category several reports are published, e.g. by Lakakul *et al.* (1999) or by Beuchat and Brackett (1990). Lakakul *et al.* carried out studies regarding modelling respiration of apple slices in modified atmosphere packages and thus added valuable exact data to the knowledge in the field, and so did Beuchat and Brackett in their studies. However, there is still a large gap in the availability of data before a full protocol can be presented, so here also studies are necessary.

Further research should also be conducted on other inhibitory factors in order to establish safe levels, which could be recommended for novel hurdles. These include preservatives like lactate, nitrite or bacteriocins which may offer interesting opportunities in view of their natural occurrence in some foods.

In handling and storage of minimally processed foods, proper refrigeration is indispensable. Reports on the influence on shelf-life caused by chilling failures are numerous, but although from a professional view it is recommended to store, for example, chilled pasteurised foods in the order of 0–1°C, thereby controlling product safety as well as quality, control of such low temperatures may not be achieved by today's distribution chain. Besides, such temperatures are also very seldom observed at retail level, as shown earlier. One way of better temperature control would be the introduction of temperature indicators on individual packages, but although such indicators exist, they are still too expensive to be used generally at retail level. Practical engineering development work is necessary, both in order to design better chilling cabinets to be used at retail level as well as the development of reliable temperature indicators at lower prices.

This chapter has dealt with safety criteria for minimally processed foods. Conventionally, only pathogenic microorganisms are considered. However, growth of toxigenic moulds or the presence of virus or protozoan parasites have not been taken into account. In one way this is wrong, since all categories are potentially hazardous for the consumer. In the case of toxigenic moulds, they would be potentially dangerous if their mycotoxins were present in the raw material. They would remain, regardless of whether or not they were exposed to mild or severe processing. Growth of toxigenic moulds during the relatively short shelf-lives of this kind of food would not be possible.

Viral pathogens would be another potential hazard. The significance of the group with respect to foodborne disease is clear with the inclusion of Norwalk virus, Hepatitis A virus and other viruses within the top ten causes of foodborne disease outbreaks in the USA (1983–87; Cliver, 1997). However, despite their significance, data regarding the effects of food production environments, food preparation and storage conditions on the survival and infectivity of viruses are extremely limited, partly through the complexity of viral detection assays. Here again is another under-researched area, which normally is not considered at all when setting up safety criteria for foods in general. The protozoan pathogens are another group of foodborne pathogens, which have received too little attention in developed countries. However, in the industrialised part of the world, disease outbreaks due to protozoa are very rare, although they are known (Beuchat, 1996). This problem is also not accounted for when safety criteria for, in particular, fresh-cut vegetables are set up.

8.8 References and further reading

ANON. 1983. Risk assessment in the federal government: managing the process. National Research Council. National Academy Press. Washington, DC.

BEUCHAT, L.R. 1996. Pathogenic microorganisms associated with fresh produce. *Journal of Food Protection*, **59**, 204–16.

BEUCHAT, L.R. and BRACKETT, R.E. 1990. Survival and growth of *Listeria monocytogenes* on lettuce as influenced by shredding, chlorine treatment, modified atmosphere packaging and temperature. *Journal of Food Science*, **55**, 755–8, 870.

BLACK, R.E., LEVINE, M.M., CLEMENTS, M.L., HIGHES, T.P. and BLASER, M.J. 1988. Experimental *Campylobacter jejuni* infections in humans. *Journal of Infectious Diseases*, **157**, 472–9.

BØGH-SØRENSEN, L. and OLSSON, P. (1990) The chill chain. In *Chilled Foods*, ed. T.R. Gormley. Elsevier Science Publishers, Barking, pp. 245–69.

BUCHANAN, R.B., DAMERT, W.G., WHITING, R.C. and VAN SCHOTHORST, M. 1997. Use of epidemiologic and food survey data to estimate a purposefully conservative dose–response relationship for *Listeria monocytogenes* levels and incidence of listeriosis, *Journal of Food Protection*, **60**, 918–22.

CARLIN, F., NGUYEN-THE, C. and ABREU DA SILVA, A. 1995. Factors affecting the growth of *Listeria monocytogenes* on minimally processed fresh endive. *Journal of Applied Bacteriology*, **78**, 636–46.

CFDRA Shelf Life Working Party. 1991. Evaluation of shelf life for chilled foods. Technical Manual No. 28.

CLIVER, D.O. 1997. Foodborne viruses. *In Food Microbiology: Fundamentals and Frontiers*, ed. M.P. Doyle, L.R. Beuchat and T.J. Montville. ASM Press, Washington DC, pp. 437–46.

CODEX ALIMENTARIUS COMMISSION. 1998. Draft principles and guidelines for the conduct of microbiological risk assessment. ALINOR 99/13A.

FAIR CONCERTED ACTION FAIR CT96-1020. 1999. Harmonization of Safety Criteria for Minimally Processed Foods, Rational and Harmonization Report.

FERNÁNDEZ, P.S. and PECK, M.W. 1997. Predictive model describing the effect of prolonged heating at 70°C to 80°C and incubation at refrigeration temperatures on growth and toxigenesis by non-proteolytic *Clostridium botulinum*. *Journal of Food Protection*, **60**, 1064–71.

FERNÁNDEZ, P.S. and PECK, M.W. 1999. A predictive model that describes the effect of prolonged heating at 70 to 90°C and subsequent incubation at refrigeration temperatures on growth from spores and toxigenesis by nonproteolytic *Clostridium botulinum* in the presence of lysozyme. *Applied and Environmental Microbiology*, **65**, 3449–57.

GRAHAM, A.F., MASON, D.R. and PECK, M.W. 1996. Predictive model of the effect of temperature, pH and sodium chloride on growth from spores of non-proteolytic *Clostridium botulinum*. *International Journal of Food Microbiology*, **31**, 69–85.

GREER, C.C., GILL, C.O. and DILTS, B.D. 1994. Evaluation of the bacteriological consequences of the temperature regimes experienced by fresh chilled meat during retail display. *Food Research International*, **27**, 371–7.

ILSI. 1998. Principles for the development of risk assessment of microbiological hazards under Directive 93/43/EEC concerning the hygiene of foodstuffs. European Commission. September 1997.

JAMES, S. and EVANS, J. 1990. Temperatures in the retail and domestic chilled chain. *In Processing and Quality of Foods*, vol. 3, 3.273–3.278, Elsevier Science Publishers, Barking.

KINDRED, T.P. 1993. An overview of risk assessment. Proceedings, 97th Annual Meeting, Las Vegas, Nevada, The United States Animal Health Association, Richmond, VA, 205–9.

KINDRED, T.P. 1996. Risk analysis and its application in FSIS. *Journal of Food Protection*, Supplement, 24–30.

KLEPZIG, I., TEUFEL, P., SCHOTT, W. and HILDEBRANDT, G. 1999. Auswirkungen einer Unterbrechung der Kühlkette auf die mikrobiologische Beschaffenheit von vorzerkleinerten Mischsalaten. *Archiv für Lebensmittelhygiene*, **50**, 73–120.

KONOWALCHUK, J. and SPEIRS, J.I. 1975. Survival of enteric viruses on fresh vegetables. *Journal of Milk and Food Technology*, **38**, 469–72.

LAKAKUL, R., BEAUDRY, R.M. and HERNANDEZ, R.J. 1999. Modelling respiration of apple slices in modified-atmosphere packages. *Journal of Food Science*, **64**, 105–10.

MOSSEL, E.A.A. and STRUIJK, C.B. 1991. Public health implications of refrigerated pasteurised ('sous vide') foods. *International Journal of Food Microbiology*, **13**, 187–206.

NGUYEN-THE, C. and CARLIN, F. 1994. The microbiology of minimally processed fresh fruits and vegetables. *Critical Revue of Food Science Nutrition*, **34**, 371–401.

NOTERMANS, S. and MEAD, G.C. 1996. Incorporation of elements of quantitative risk analysis in the HACCP system. *International Journal of Food Microbiology*, **30**, 157–73.

NOTERMANS, S., DUFRENNE, J., TEUNIS, P., BEUMER, R., TE GIFFEL, M. and PEETERS WEEM, P. 1997. A risk assessment study of *Bacillus cereus* present in pasteurized milk. *Food Microbiology*, **14**, 143–51.

OLSSON, P. 1990. Chill cabinet surveys. *In Processing and Quality of Foods*, vol. 3, 3.279–3.288, Elsevier Science Publishers, Barking.

SØRENSEN, I.-M.J., MOSBAK, S. and BØGH-SØRENSEN, L. 1985. Kølede kødprodukter (in Danish). Danish Meat Products Laboratory, Frederiksberg, Denmark.

SYNAFAP. 1995. Aide à la maîtrise de l'hygiène alimentaire des produits traiteurs frais et réfrigérés (containing the SYNAFAP protocol for the validation of shelf-life).

WILLCOX, F., HENDRICKX, M. and TOBBACK, P. 1994. A preliminary survey into the temperature conditions and residence time distribution of minimally processed MAP vegetables in Belgian retail display cabinets. *International Journal of Refrigeration*, **17**, 436–44.

8.9 Acknowledgement

Dr Toon Martens, Alma University Restaurants, project leader of the EEC FAIR Concerted Action CT96-1020 'Harmonization of safety criteria for minimally processed foods' is acknowledged for permission to make extensive use of the inventory report and the final report of the project.

9

Minimal processing in practice

Fresh fruits and vegetables

E. Laurila and R. Ahvenainen, VTT Biotechnology, Espoo

9.1 Introduction

Minimal processing of raw fruits and vegetables has two purposes (Huxsoll and Bolin, 1989):

1. Keeping the produce fresh, without losing its nutritional quality.
2. Ensuring a product shelf-life sufficient to make distribution feasible within a region of consumption.

The microbiological, sensory and nutritional shelf-life of minimally processed vegetables or fruits should be at least four to seven days, but preferably up to 21 days depending on the market (Ahvenainen, 2000; Wiley, 1994; Ahvenainen and Hurme, 1994). Commercial requirements for the manufacture of ready-to-use pre-peeled, sliced, grated or shredded fruit and vegetables are summarised in Table 9.1. The aim of this chapter is to:

- assess quality and safety aspects of minimally processed fruits and vegetables
- describe the key steps in the food chain, beginning with raw material and processing and ending with packaging, which affect the quality and shelf-life of minimally processed fruits and vegetables.

9.2 Quality changes in minimally processed fruits and vegetables

As a result of peeling, grating and shredding, produce will change from a relatively stable commodity with a shelf-life of several weeks or months to a perishable one that has only a very short shelf-life, as short as one to three days

Table 9.1 Requirements for the commercial manufacture of ready-to-use pre-peeled and/or sliced, grated or shredded fruit and vegetables

Working principle	Demands for processing	Customers	Shelf-life (days) at 5°C	Examples of suitable fruit and vegetables
Preparation today, consumption tomorrow	• Standard kitchen hygiene and tools • No heavy washings for peeled and shredded produce; potato is an exception • Packages can be returnable containers	Catering industry Restaurants Schools Industry	1–2	Most fruits and vegetables
Preparation today, the customer uses the products within 3–4 days	• Disinfection • Washing of peeled and shredded produce at least with water • Permeable packages; potato is an exception	Catering industry Restaurants Schools Industry	3–5	Carrot Cabbage Iceberg lettuce Potato Beetroot Acid fruits Berries
Products are also intended for retailing	• Good disinfection • Chlorine or acid washing for peeled and shredded produce • Permeable packages; potato is an exception • Additives	In addition to the customers listed above, retail shops can also be customers	5–7*	Carrot Chinese cabbage Red cabbage Potato Beetroot Acid fruits Berries

Note: * If longer shelf-life up to 14 days is needed, the storage temperature must be 1–2°C.

at chilled temperatures. During peeling and grating operations, many cells are broken, and intracellular products, such as oxidising enzymes, are released. Minimally processed produce deteriorates due to physiological ageing, biochemical changes and microbial spoilage, which may result in degradation of the colour, texture and flavour (Varoquaux and Wiley, 1994; Kabir, 1994).

9.2.1 Physiological and biochemical changes

The most important enzyme in minimally processed fruits and vegetables is polyphenol oxidase, which causes browning (Laurila, Kervinen and Ahvenainen, 1998; Varoquaux and Wiley, 1994; Wiley, 1994). Another important enzyme is lipooxidase which catalyses peroxidation, causing the formation of numerous bad-smelling aldehydes and ketones. Ethylene production can also increase and because ethylene contributes to the neosynthesis of enzymes involved in fruit maturation, it may play a part in physiological disorders of sliced fruits, such as softening (Varoquaux and Wiley, 1994).

With processing, the respiration activity of produce will increase by 20% to as much as 700% or more depending on the produce, cutting grade and temperature (Varoquaux and Wiley, 1994; Mattila *et al.*, 1995b). If packaging conditions are anaerobic, this leads to anaerobic respiration causing the formation of ethanol, ketones and aldehydes (Powrie and Skura, 1991).

9.2.2 Microbiological changes

During peeling, cutting and shredding, the surface of the produce is exposed to the air and to contamination with bacteria, yeasts and moulds. In minimally processed vegetables, most of which fall into the low acid range category (pH 5.8–6.0), high humidity and the large number of cut surfaces can provide ideal conditions for the growth of microorganisms (Willocx *et al.*, 1994).

The populations of bacteria found on fruits and vegetables vary widely. The predominant microflora of fresh leafy vegetables are *Pseudomonas* and *Erwinia* spp., with an initial count of approximately 10^5 cfu/g, although low numbers of moulds and yeasts are also present. During cold storage of minimally processed leafy vegetables, pectinolytic strains of *Pseudomonas* are responsible for bacterial soft rot (Varoquaux and Wiley, 1994; Willocx *et al.*, 1994). An increase in storage temperature and carbon dioxide concentration in the package will shift the microflora towards lactic acid bacteria (Garg *et al.*, 1990; Marchetti *et al.*, 1992; Brackett, 1994; Markholm, 1992; Hurme *et al.*, 1994; Ahvenainen *et al.*, 1994; Manzano *et al.*, 1995).

The high initial load of microbes makes it difficult to establish the cell number threshold beyond which the product can be considered spoiled. Many studies show that a simple correlation does not exist between spoilage chemical markers such as pH, lactic acid, acetic acid, carbon dioxide, sensory quality and total microbial cell load (Marchetti *et al.*, 1992; Hurme *et al.*, 1994; Ahvenainen *et al.*, 1994; Manzano *et al.*, 1995). In fact, different minimally processed fruit and vegetable products seem to possess different spoilage patterns in relation to the characteristics of the raw materials (Huxsoll and Bolin, 1989; Marchetti *et al.*, 1992).

Because minimally processed fresh fruits and vegetables are not heat treated, regardless of additives or packaging, they must be handled and stored at refrigerated temperatures, at 5°C or under in order to achieve a sufficient shelf-life and microbiological safety. Some pathogens such as *Listeria monocytogenes*, *Yersinia enterocolitica*, *Salmonella* spp. and *Aeromonas hydrophila* may still survive and even profilerate at low temperatures (Brackett, 1994; Riquelme *et al.*, 1994). On the other hand, minimally processed fruits are relatively safe when compared to other foods, as they are generally acidic enough to prevent growth of pathogens. The normal spoilage organisms in refrigerated produce are also usually psychrotrophic and therefore have a competitive advantage over most pathogens.

9.2.3 Nutritional changes

Little is known about nutritive value, i.e., vitamin, sugar, amino acid, fat and fiber content of minimally processed produce. Washing does not decrease the vitamin content (C vitamin and carotenes) of grated carrot, shredded Chinese cabbage or peeled potatoes significantly (Hägg *et al.*, 1996).

9.3 Improving quality

If products are prepared today and consumed tomorrow, very simple and inexpensive processing methods can be used. Most fruits and vegetables are suitable for this kind of preparation. Such products may also be suitable for catering, where they will undergo further processing. If, however, products need a shelf-life of several days, or up to one week and more, as is the case with the products intended for retailing, then more advanced processing methods and treatments are needed using the hurdle concept (Wiley, 1994; Ahvenainen and Hurme, 1994; Leistner and Gorris, 1995). The key steps are summarised in Table 9.2. Preservation is based on the synergies between individual steps such as these. These steps must also take place within a safe processing environment. Hygienic processing within a framework of good manufacturing practices and effective HACCP management is of utmost importance in preventing microbiological and other risks (Huxsoll and Bolin, 1989; Wiley, 1994; Ahvenainen and Hurme, 1994; Ahvenainen *et al.*, 1994; Zomorodi, 1990). Some of the key hazards and their methods of control within a HACCP framework are summarised in Table 9.3.

Table 9.2 The key requirements in the minimal processing of fruits and vegetables

- Raw material of good quality (correct cv. variety, correct cultivation, harvesting and storage conditions)
- Strict hygiene and good manufacturing practices, HACCP
- Low temperatures during working
- Careful cleaning and/or washing before and after peeling
- Water of good quality (sensory, microbiology, pH) used in washing
- Mild additives in washing for disinfection or browning prevention
- Gentle spin drying after washing
- Gentle cutting/slicing/shredding
- Correct packaging materials and packaging methods
- Correct temperature and humidity during distribution and retailing

9.4 Raw materials

It is self-evident that vegetables or fruits intended for pre-peeling and cutting must be easily washable and peelable, and their quality must be first class. The correct and proper storage of vegetables and careful trimming before processing are vital for the production of prepared vegetables of good quality (Wiley, 1994; Ahvenainen and Hurme, 1994; Kabir, 1994). The study of various cultivar varieties of eight different vegetables showed that not all varieties of the specified vegetable can be used for the manufacture of prepared vegetables. The correct choice of variety is particularly important for carrot, potato, swede and onion. For example, with carrot and swede, the variety that gives the most juicy grated product cannot be used in the production of grated products that should have a shelf-life of several days (Ahvenainen *et al.*, 1994). Another example is potato, with which poor colour and flavour become problems if the variety is wrong (Laurila, Hurme and Ahvenainen, 1998; Mattila *et al.*, 1995a). Furthermore, the results showed that climatic conditions, soil conditions and agricultural practices, e.g. fertilisation and harvesting conditions, can also significantly affect the behaviour of vegetables, particularly that of potatoes, in minimal processing (Ahvenainen *et al.*, 1998).

9.5 Peeling, cutting and shredding

Some vegetables or fruits, such as potatoes, carrots or apples, need peeling. There are several peeling methods available, but on an industrial scale the peeling is normally accomplished mechanically (e.g. rotating carborundum drums), chemically or in high-pressure steam peelers (Wiley, 1994). However, the results have shown that peeling should be as gentle as possible. The ideal method would be hand-peeling with a sharp knife. The relative effects of carborundum and knife peeling are shown in Fig. 9.1. Carborundum-peeled potatoes must be treated with a browning inhibitor, whereas water washing is enough for hand-peeled potatoes. If mechanical peeling is used, it should

Table 9.3 Hazards, critical control points, preventive and control procedures in processing and packaging of ready-to-use fruits and vegetables (Gorris, 1996)

Critical operational step	Hazards	Critical control point(s)	Preventive and control measures
Growing	Contamination with faecal pathogens Insects and fungal invasions	Cultivation techniques	• Use synthetic fertiliser* • Inspect the sources of irrigation water* • Use pesticides
Harvesting	Microbial spoilage and insect invasion Cross-contamination	Assessment of produce maturity Handling practices Temperature control Sanitation	• Harvest prior to peak maturity • Minimise mechanical injuries • Harvest in the morning or at night • Employ pickers trained in elementary hygiene
Transporting	Microbial growth Cross-contamination	Time/temperature Loading practices Produce Containers	• Keep the temperature low • Avoid long distance transport • Maintain uniform cooling in transport containers • Avoid damage, do not overload the containers • Separate sound and injured produce in the field • Use well-washed/disinfected metal or plastic containers
Washing	Contamination from water	Water Washing practices Dewatering	• Use potable water, test routinely for the presence of coliform bacteria • Control microbial contamination by chlorination and antimicrobial dipping • Do not overload the washing tanks/change the water periodically • Remove excess water
Sorting	Cross-contamination	Sorter Lightener Conveyor	• Employ sorter who has experience of the inspection of produce • Provide adequate lightning • Clean and disinfect periodically
Packaging	Microbial growth	Packaging film Relative humidity and temperature control	• Choose the permeability of film correctly • Analyse gas composition routinely by using simple techniques • Use fungicide impregnated film • Dewater the drenched produce carefully • Use films which have antifogging properties • Check product/storage temperature at regular intervals
Storage/distribution	Growth and spread of microorganisms	Temperature control Light Consumer practice	• Maintain the refrigeration of produce in the range of 0–5°C • Prevent moisture condensation by proper temperature control • Take the effect of light into consideration** • Provide labelling with instructions for storage conditions

Notes: * For the produce grown close to ground and consumed raw. ** Light may affect the gas composition in the packaging by inducing photosynthesis in green vegetables.

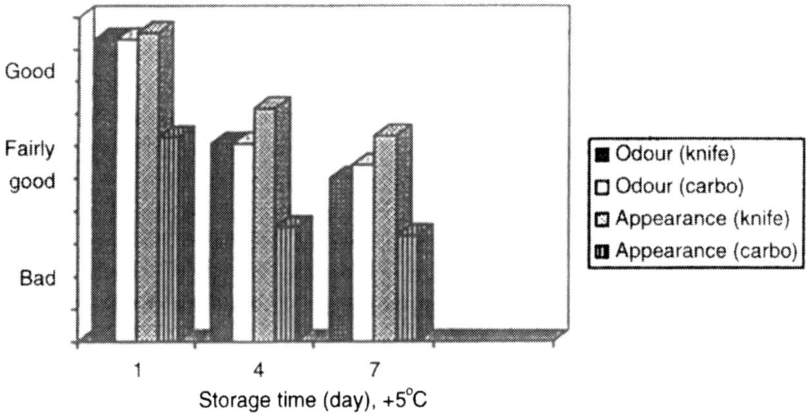

Fig. 9.1 The effect of peeling method and storage time on the odour and appearance of potato packed in a gas mixture of 20% $CO_2 + 80\%$ N_2 and stored at 5°C.

resemble knife-peeling. Carborundum, steam peeling or caustic acid disturb the cell walls of a vegetable which enhances the possibilities of microbial growth and enzymatic changes. Carborundum and knife peeling can be combined, with a first stage of rough peeling and then a second stage of finer knife peeling. Enzymatic peeling can be successful, for example in the case of oranges (Pretel *et al.*, 1998).

Many studies show that the cutting and shredding must be performed with knives or blades as sharp as possible and made from stainless steel. Carrots cut with a razor blade were more acceptable from a microbiological and sensory point of view than carrots cut with commercial slicing machines. It is clear that slicing with blunt knives impairs quality retention because of the increased breaking of cells and release of tissue fluid. A slicing machine must be installed solidly, because vibrating equipment may possibly impair the quality of sliced surfaces. Mats and blades used in slicing should also be disinfected, for example with a 1% hypochlorite solution.

9.6 Cleaning, washing and drying

Incoming vegetables or fruits, which are covered with soil, mud and sand, should be carefully cleaned before processing. A second wash must usually be done after peeling and/or cutting (Wiley, 1994; Ahvenainen and Hurme, 1994). For example, Chinese cabbage and white cabbage must be washed after shredding, whereas carrot must be washed before grating (Hurme *et al.*, 1994; Ahvenainen *et al.*, 1994). Washing after peeling and cutting removes microbes and tissue fluid, thus reducing microbial growth and enzymatic oxidation during storage. Washing in flowing or air-bubbling water is preferable to dipping into still water (Ohta and Sugawara, 1987). The microbiological quality of the

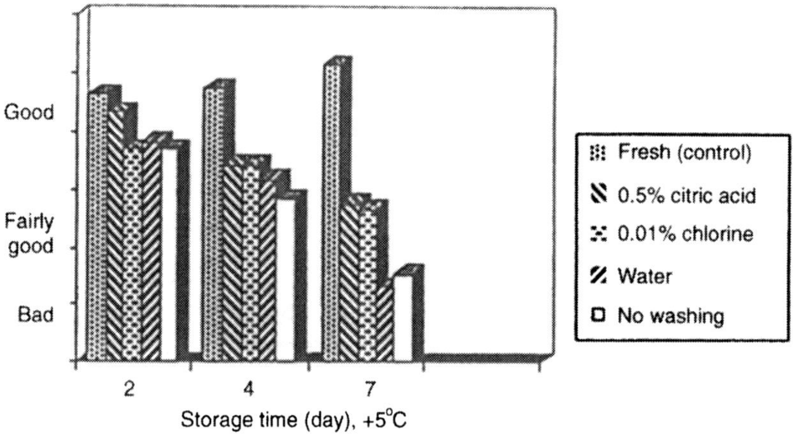

Fig. 9.2 The effect of washing solution and storage time on the odour of grated carrots packed in air and stored at 5°C.

washing water used must be good and its temperature low, preferably below 5°C. The recommendable quantity of water to be used is 5–10 l/kg of product before peeling/cutting (Huxsoll and Bolin, 1989) and 3 l/kg after peeling/cutting (Hurme *et al.*, 1994; Ahvenainen *et al.*, 1994).

Preservatives can be used in washing water for the reduction of microbial numbers and to retard enzymatic activity, thereby improving the shelf-life. 100–200 mg of chlorine or citric acid/l is effective in washing water before or after peeling and/or cutting to extend shelf-life (Wiley, 1994; Kabir, 1994; Hurme *et al.*, 1994; Ahvenainen *et al.*, 1994; O'Beirne, 1995). The relative effects of differing washing solutions are shown in Fig. 9.2. However, when chlorine is used, vegetable material should be rinsed. Rinsing reduces the chlorine concentration to the level of that in drinking water and means that sensory quality is not compromised (Hurme *et al.*, 1994). The effectiveness of chlorine can be enhanced by using a combination of low pH, high temperature, pure water and correct contact time (Wiley, 1994; Kabir, 1994). It seems that chlorine compounds reduce counts of aerobic microbes at least in some leafy vegetables such as lettuce (Wiley, 1994; Garg *et al.*, 1990), but not necessarily in root vegetables or cabbages (Garg *et al.*, 1990; Ahvenainen *et al.*, 1994). Chlorine compounds are of limited effectiveness in suppressing growth of *Listeria monocytogenes* in lettuce and cabbage (Skyttä *et al.*, 1996; Francis and O'Beirne, 1997). In addition, the use of some preservatives (e.g. chlorine compounds) is not necessarily allowed in all countries. Alternatives to chlorine include chlorine dioxide, per acetic acid ozone, trisodium phosphate and hydrogen peroxide (Sapers and Simmons, 1998). Hydrogen peroxide vapour treatment, for example, appears to reduce microbial counts on freshly cut cucumber, bell peppers and zucchini, extending shelf-life without leaving significant residues or compromising product quality. However, more research is still required to validate these treatments.

Washing water should be removed gently from the product (Wiley, 1994). A centrifuge seems to be the best method. The centrifugation time and rate should be chosen carefully (Zomorodi, 1990; Bolin and Huxsoll, 1991), so that the process removes loose water but does not damage vegetable cells.

9.7 Browning inhibition

A key quality problem for fruits and vegetables such as peeled and sliced apple and potato is enzymatic browning, and washing with water is not effective in preventing discoloration (Wiley, 1994; Mattila *et al.*, 1995a). Traditionally, sulphites have been used for prevention of browning. However, the use of sulphites has some disadvantages, in particular dangerous side-effects for asthmatics. For this reason, the FDA in the USA partly restricted the use of sulphites (Anon., 1991). At the same time, interest in substitutes for sulphites is increasing.

Enzymatic browning requires four different components: oxygen, an enzyme, copper and a substrate. In order to prevent browning, at least one component must be removed from the system. In theory, polyphenyloxidase (PPO)-catalysed browning of vegetables and fruits can be prevented by such factors as the following (Whitaker and Lee, 1995):

- heat or reaction inactivation of the enzyme
- exclusion or removal of one or both of the substrates (O_2 and phenols)
- lowering the pH to two or more units below the optimum
- adding compounds that inhibit PPO or prevent melanin formation.

Many inhibitors of PPO are known, but only a few have been considered as potential alternatives to sulphites (Vámos-Vigyázó, 1981). The most attractive way to inhibit browning would be 'natural' methods, such as the combination of certain salad ingredients with each other. Pineapple juice appears to be a good potential alternative to sulphites for the prevention of browning in fresh apple rings (Lozano-de-González *et al.* 1993; Meza *et al.* 1995). Washing in glycine betaine solution has been found to be effective in retaining sensory quality, particularly for pre-packed shredded lettuce (Hurme *et al.*, 1999).

Probably the most often studied alternative to sulphite is ascorbic acid. This compound is a highly effective inhibitor of enzymatic browning, primarily because of its ability to reduce quinones back to phenolic compounds before they can undergo further reaction to form pigments. However, ascorbic acid eventually oxidises to dehydro-ascorbic acid (DHAA), allowing quinones to accumulate and undergo browning, and is best used in combination with other substances such as citric acid. Dipping in hot ascorbic acid/citric acid solutions improved the shelf-life of pre-peeled potatoes to about two weeks. However, high concentrations of ascorbic acid (0.75%) have produced an unpleasant taste in fruits (Luo and Barbosa-Cánovas, 1995). Ascorbic acid derivatives, like AAP and AATP, have been used as browning inhibitors alone or in combination with other inhibitors for potatoes and apples (Sapers *et al.*, 1989; Sapers and Miller,

1992, 1993; Monsalve-Gonzalez *et al.*, 1993). Erythorbic acid, isomer of ascorbic acid, has been used as an inhibitor of enzymatic browning in combination with ascorbic acid or citric acid for potato slices (Dennis, 1993) and for whole abrasion peeled potatoes (Santerre *et al.*, 1991).

Citric acid acts as a chelating agent and acidulant, both of which characteristics inhibit PPO. Promising results have been obtained using citric acid and the combinations citric–ascorbic acid and benzoic–sorbic acid as dipping treatments for minimally processed potatoes (Mattila *et al.*, 1995a). 4-Hexylresorcinol is a good inhibitor of enzymatic browning for apples, potatoes and iceberg lettuce (Monsalve-Gonzalez *et al.*, 1993; Whitaker and Lee, 1995; Luo and Barbosa-Cánovas, 1995; Castañer *et al.*, 1996). It interacts with PPO and renders it incapable of catalysing the enzymatic reaction. 4-Hexylresorcinol has several advantages over the use of sulphites in foods, including the following (McEvily *et al.*, 1992):

- its specific mode of inhibitory action
- the lower levels required for effectiveness
- its inability to bleach preformed pigments
- chemical stability.

Ethylenediamine tetraacetic acid (EDTA), a complexing agent, has been used with potatoes (Cherry and Singh, 1990; Dennis, 1993) and iceberg lettuce (Castañer *et al.*, 1996) in combinations with other browning inhibitors. Sporix™, a chelating agent described by its supplier as an acidic polyphosphate, has been found to be an effective browning inhibitor in several fruits and vegetables (Gardner *et al.*, 1991; Sapers *et al.*, 1989). Sulfhydryl-containing amino acids like cysteine prevent brown pigment formation by reacting with quinone intermediates to form stable, colourless compounds (Dudley and Hotchkiss, 1989). Cysteine has been used as a browning inhibitor for potatoes, apples and iceberg lettuce (Molnar-Perl and Friedman, 1990; Castañer *et al.*, 1996) and it has also been used as an ingredient in a commercial browning inhibitor (Cherry and Singh, 1990).

Protease enzymes have been found to be effective browning inhibitors for apples and potatoes (Taoukis *et al.*, 1989; Labuza *et al.*, 1992; Luo, 1992). It is believed that an effective protease acts to hydrolyse and therefore inactivate the enzyme or enzymes responsible for enzymatic browning. Of the proteolytic enzymes tested so far, three plant proteases (ficin from figs, papain from papaya and bromelain from pineapple) in particular have proved to be effective. All the three proteases are sulfhydryl enzymes of broad specificity. According to Taoukis *et al.* (1989), ficin was as effective as sulphite for potatoes at 4°C, but slightly less effective than sulphite at 24°C. Papain was somewhat effective for potatoes at 4°C. Papain treatment can prevent enzymatic browning of apples about as well as sulphite treatment at both temperatures (4°C and 24°C).

Since there is no one substitute for sulphites in preventing browning, alternatives are usually ascorbic acid-based combinations. A typical combination may include:

- a chemical reductant (e.g. ascorbic acid)
- an acidulant (e.g. citric acid)
- a chelating agent (e.g. EDTA).

In using such combinations, or developing new ones, it is important to take an integrated approach by choosing proper raw materials, peeling method, processing and packaging conditions (Laurila *et al.*, 1998).

9.8 Biocontrol agents

As well as enzymatic browning, a key issue with minimally processed foods is microbiological safety. An emerging technology in controlling pathogen growth is the use of biocontrol technology such as lactic acid bacteria (LAB) which compete with, and thus inhibit, pathogen growth (Breidt and Fleming, 1997). LAB can produce both metabolites, such as lactic and acetic acids, which lower pH, and bacteriocins. Although they are not sufficient in isolation, bacteriocins such as nisin can contribute to dealing with certain cold-tolerant, Gram-positive bacteria (Bennik, 1997; Torriani *et al.*, 1997). Studies of the use of lactic acid bacteria have suggested using it in combination with other preservation techniques (Breidt and Fleming, 1997):

- reduction of the total microflora in the product by such procedures as washing using sanitisers, heat treatment or irradiation
- addition of a bacteriocin-producing biocontrol culture to achieve a target initial bacterial count (CFU/mL)
- storage of the product under refrigerated conditions.

Product shelf-life would then be determined by the growth of the biocontrol culture. If the product suffered temperature abuse during storage or distribution, for example, the biocontrol culture would grow more rapidly, thus preventing pathogen growth. Such cultures will be a fruitful source of further research.

9.9 Packaging

A key operation in producing minimally processed fruits and vegetables is packaging. The most studied packaging method for prepared raw fruits and vegetables is modified atmosphere packaging (MAP). The basic principle in MAP is that a modified atmosphere can be created passively by using properly permeable packaging materials, or actively by using a specified gas mixture together with permeable packaging materials. The aim of both is to create an optimal gas balance inside the package, where the respiration activity of a product is as low as possible while ensuring that oxygen concentration and carbon dioxide levels are not detrimental to the product. In general, the aim is to have a gas composition where there is 2–5% CO_2, 2–5% O_2 and the rest nitrogen (Kader *et al.*, 1989; Day, 1994).

High O_2 MAP treatment has been found to be particularly effective at inhibiting enzymatic browning, preventing anaerobic fermentation reactions, and inhibiting aerobic and anaerobic microbial growth (Day, 1997). High O_2 levels may cause substrate inhibition of PPO, or the high levels of colourless quinones subsequently formed may cause feedback production of PPO. Carbon monoxide (CO) gas atmosphere has also been found to inhibit mushroom PPO reversibly. Use of this compound in a MAP system would, however, require measures to ensure the safety of packing plant workers.

Achieving the right gas mixture is one of the most difficult tasks in manufacturing raw ready-to-use or ready-to-eat fruit and vegetable products. The main problem has been the lack of sufficiently permeable packaging materials (Day, 1994). Most films do not result in optimal O_2 and CO_2 atmospheres, especially when the produce has high respiration. However, one solution is to make microholes of defined sizes and a defined quantity in the material in order to avoid anaerobis (Exama *et al.*, 1993). This procedure significantly improves, for example, the shelf-life of grated carrots (Ahvenainen *et al.*, 1994). Other solutions are to combine ethylene vinyl acetate with orientated polypropylene and low density polyethylene or combine ceramic material with polyethylene. Both composite materials have significantly higher gas permeability than polyethylene or the oriented polypropylene much used in the packaging of salads, even though gas permeability should ideally be higher. These materials have good heat-sealing properties, and they are also commercially available (Ahvenainen and Hurme, 1994). The shelf-life of shredded cabbages and grated carrots packed in these materials is seven to eight days at 5°C and therefore two or three days longer than in the orientated polypropylene which is generally used in the vegetable industry (Hurme *et al.*, 1994; Ahvenainen *et al.*, 1994). A new breathable film has also been patented, which has a three-layer structure consisting of a two-ply blown coextrusion about 25 μm thick with an outer layer of K-Resin KR10 and an inner metallocene polyethylene layer. It is claimed that this film gives 16 days' shelf-life at 1–2°C for fresh salads washed in chlorine solution (Anon., 1996). Examples of suitable packaging materials for vegetables are shown in Table 9.4 (Ahvenainen *et al.*, 1994).

In dealing with fresh respiring products, it is advantageous to have film permeability alter to match product respiration rate to avoid the anaerobic conditions favoured by some pathogens. In practice, this can be achieved by linking permeability to temperature change. While the permeation rates of most packaging films are only modestly affected by changes in temperature, newer films have been developed with a temperature 'switch' point at which the film's permeation changes rapidly. This technology uses long-chain fatty alcohol-based polymeric chains. Under a given temperature these remain within a crystalline state. Once the temperature is exceeded, the side chains melt to a gas-permeable amorphous state (Anon., 1992; Anon., 1998). An alternative technology is to use a film with two differing layers, or two identical layers of differing thicknesses, both with minute cuts. As the temperature increases, the layers expand at differing temperatures causing the holes to enlarge, increasing

Table 9.4 Packaging materials for vegetables (Ahvenainen *et al.*, 1994).

Vegetable	Packaging material and thickness
Peeled potato, both whole and sliced	PE-LD, 50μm (also PA/PE, 70–100μm or comparable)
Grated carrot	PP-O, 40μm, microholed PP-O, PE/EVA/PP-O, 30–40μm
Sliced swede	PE-LD, 50μm
Grated swede	PE/EVA/PP-O, 40μm
Sliced beetroot	PE-LD, 50μm (also PA/PE, 70–100μm or comparable)
Grated beetroot	PP-O, 40μm, microholed PP-O, PE/EVA/PP-O, 30–40μm
Shredded Chinese cabbage	PP-O, 40μm, PE/EVA/PP-O, 30–40μm
Shredded white cabbage	PP-O, 40μm, PE/EVA/PP-O, 30–40μm
Shredded onion	PP-O, 40μm (also PA/PE, 70–100μm or comparable)
Shredded leek	PE-LD, 50μm, PP-O 40μm (also PA/PE, 70–100μm or comparable)

the film's permeability (Anon., 1994). Safety valve systems have also been proposed to prevent excessive oxygen depletion and carbon dioxide accumulation when a temporary temperature increase occurs (Exama *et al.*, 1993).

One interesting MAP method is moderate vacuum packaging (MVP) (Gorris *et al.*, 1994). In this system, respiring produce is packed in a rigid, airtight container under 40 kPa of atmospheric pressure and stored at refrigerated temperature (4–7°C). The initial gas composition is that of normal air (21% O_2, 0.04% CO_2 and 78% N_2) but at a reduced partial gas pressure. The lower O_2 content stabilises the produce quality by slowing down the metabolism of the produce and the growth of spoilage microorganisms. Gorris *et al.* (1994) have compared the storage of several whole and lightly processed fruits and vegetables under ambient conditions to MVP, and found that MVP improved the microbial quality with red bell pepper, chicory endive, sliced apple, sliced tomato, the sensory quality of apricot and cucumber and the microbial and sensory quality of mung bean sprouts and a mixture of cut vegetables. Gorris *et al.* (1994) also conducted pathogen challenge tests with *Listeria monocytogenes*, *Yersinia enterocolitica*, *Salmonella typhimurium* and *Bacillus cereus* on mung bean sprouts at 7°C. All of the pathogens lost viability quickly during the course of storage.

One of the greatest challenges is designing MAP for ready-to-eat products such as prepared, mixed salads, where respiration rates of each component differ. Experiments have been undertaken on combinations such as carrot, cucumber, garlic and green pepper, using a pouch form package made of low-density polyethylene film. These have demonstrated some improvement in product quality compared with other types of MAP package (Lee *et al.*, 1996).

9.9.1 Edible coatings

One possible 'packaging' method for extending the post-harvest storage of minimally processed fruit and vegetables is the use of edible coatings. These are thin layers of material that can be eaten by the consumer as part of the whole food product. Coatings have the potential to reduce moisture loss, restrict oxygen entrance, lower respiration, retard ethylene production, seal in flavour volatiles and carry additives (such as antioxidants) that retard discoloration and microbial growth (Baldwin *et al.*, 1995).

9.10 Storage conditions

Chilling is an important preservative hurdle, as is the control of humidity. Storage at 10°C or above allows most bacterial pathogens to grow rapidly on fresh-cut vegetables. Storage temperature is also important when MAP or vacuum packaging is used. Toxin production by *Clostridium botulinum*, or growth of other pathogens such as *Listeria monocytogenes*, is possible at temperatures above 3°C because of increased oxygen consumption in the package (Francis and O'Beirne, 1997). Processing, transport, display and intermediate storage should all be at the same low temperature (preferably 2–4°C) for produce not vulnerable to chilling injury. Changes in temperature should be avoided. Higher temperatures speed up spoilage and facilitate pathogen growth. Fluctuating temperatures cause in-pack condensation which also accelerates spoilage.

Temperature abuse is a widespread problem in the distribution chain, whether in storage, transportation, retail display or consumer handling. Where this is a significant problem, it may be necessary to restrict shelf-life, for example to five to seven days at a temperature of 5–7°C, when psychrotrophic pathogens have insufficient time to multiply and produce toxin. If the shelf-life of vacuum or MAP products is greater than ten days, and there is a risk that the storage temperature will be over 3°C, then products should meet one or more of the following controlling factors:

- a minimum heat treatment such as 90°C for ten minutes
- a pH of 5 or less throughout the food
- a salt level of 3.5% (aqueous) throughout the food
- a_w of 0.97 or less throughout the food
- any combination of heat and preservative factors which has been shown to prevent growth of toxin production by C. *botulinum*.

Practically, if the aim is to keep minimally processed produce in fresh-like state, the last mentioned factors, and mainly various preservative factors are the only possibilities to increase shelf-life and assure microbiological safety of MA – or vaccum-packed fresh produce (FAIR Concerted Action, 1999).

9.11 Processing guidelines for particular vegetables

Processing and packaging guidelines for pre-peeled and sliced potato, pre-peeled, sliced and grated carrot, shredded Chinese cabbage and white cabbage, cut onion and leek are given in Tables 9.5 to 9.10 at the end of this chapter.

9.12 Future trends

Much research is still to be done in order to develop minimally processed fruit and vegetable products with high sensory quality, microbiological safety and nutritional value. It is currently possible to reach seven to eight days' shelf-life at refrigerated temperatures (5°C), but, for some products, two or three weeks' shelf-life may be necessary. More information about the growth of pathogenic bacteria or nutritional changes in minimally processed fruits and vegetables with long shelf-life is needed.

A characteristic feature of minimal processing is the need for an integrated approach, where raw material, handling, processing, packaging and distribution must each be properly managed to make shelf-life extension possible. Hurdle technology using natural preservatives, e.g. inhibitors produced by lactic acid bacteria, and the matching of correct processing methods and ingredients to each other, needs to be developed further in the minimal processing of fresh produce. It is probable that in the future, fruits and vegetables intended for minimal processing will be cultivated under specified controlled conditions, and that plant geneticists will develop selected and created cultivars or hybrids adapted to the specific requirements of minimal processing (Varoquaux and Wiley, 1994; Martinez and Whitaker, 1995). Unit operations such as peeling and shredding need further development to make them more gentle. There is no sense in disturbing the quality of produce by rough treatment during processing, and then try to limit the damage by subsequent use of preservatives. Active packaging systems and edible films, as well as more permeable plastic films which better match with the respiration of fruits and vegetables, are particularly active areas for development.

9.13 References

AHVENAINEN, R. (2000) Ready-to-use fruit and vegetables. Flair-Flow Europe Technical Manual F-FE 376A/00. Teagasc (The National Food Centre), Dublin.

AHVENAINEN, R. and HURME, E. (1994) Minimal processing of vegetables. In:

Ahvenainen, R., Mattila-Sandholm, T. and Ohlsson, T. (eds) *Minimal Processing of Foods*. VTT Symposium 142, pp. 17–35, VTT, Espoo, Finland.

AHVENAINEN, R., HURME, E., HÄGG, M. and SKYTTÄ, E. (1998) Shelf-life of pre-peeled potato cultivated, stored and processed by various methods. *J Food Prot.* **61**: 591–600.

AHVENAINEN, R., HURME, E., KINNUNEN, A., LUOMA, T. and SKYTTÄ, E. (1994) Factors affecting the quality retention of minimally processed carrot. In: Proceedings of the Sixth International Symposium of the European Concerted Action Program COST 94 'Post-harvest treatment of fruit and vegetables'. *Current Status and Future Prospects*, Oosterbeek, 19–22 October, Commission of the European Communities, Brussels.

ANON. (1991) Sulphites banned. *Food Ingredients Process*, No. 11, 11.

ANON. (1992) Temperature compensating films for produce. *Prepared Foods.* **161**: 95.

ANON. (1994) Permeable plastics film for respiring food produce. *Food, Cosmetics Drug Pack.* **17**: 7.

ANON. (1996) Bags extend salad shelf time. *Packaging Digest.* January; 66–70.

ANON. (1998) 'Membrane' controls veggie tray's MAP permeation. *Packaging Digest.* October; 3.

BALDWIN, E.A., NISPEROS-CARRIEDO, M.O. and BAKER, R.A. (1995) Use of edible coatings to preserve quality of lightly (and slightly) processed products. *Crit. Rev. Food Sci. Nutr.* **35**: 509–24.

BENNIK, M.H.J. (1997) *Biopreservation in Modified Atmosphere Packaged Vegetables.* Thesis. Wageningen, The Netherlands: Agricultural University; 96.

BOLIN, H.R. and HUXSOLL, C.C. (1991) Effect of preparation procedures and storage parameters on quality retention of salad-cut lettuce. *J. Food Sci.* **56**: 60–7.

BRACKETT, R.E. (1994) Microbiological spoilage and pathogens in minimally processed refrigerated fruits and vegetables. In: Wiley, R.C. (ed.) *Minimally Processed Refrigerated Fruits & Vegetables*, pp. 269–312, Chapman & Hall, New York.

BREIDT, F. and FLEMING, H.P. (1997) Using lactic acid bacteria to improve the safety of minimally processed fruit and vegetables. *Food Technol.* **51**: 44–51.

CASTAÑER, M., GIL, M.I., ARTES, F. and TOMAS-BARBERAN, F.A. (1996) Inhibition of browning of harvested head lettuce. *J. Food Sci.* **61**: 314–16.

CHERRY, J.H. and SINGH, S.S. (1990) Discoloration preventing food preservative and method. Patent Number: 4,937,085.

DAY, B.P.F. (1994) Modified atmosphere packaging and active packaging of fruits and vegetables. In: Ahvenainen, R., Mattila-Sandholm, T. and Ohlsson, T. (eds) *Minimal Processing of Foods*. VTT Symposium 142, pp. 173–207, VTT, Espoo, Finland.

DAY, B.P.F. (1997) High oxygen modified atmosphere packaging: a novel approach for fresh prepared produce packaging. In: Blakistone, B. (ed.) *Packaging Yearbook 1996*. NFPA National Food Processors Association, pp. 55–65.

DENNIS, J.A.B. (1993) The effects of selected antibrowning agents, selected packaging methods, and storage times on some characteristics of sliced raw potatoes. Dissertation, Oklahoma State University, Stillwater.

DUDLEY, E.D. and HOTCHKISS, J.H. (1989) Cysteine as an inhibitor of polyphenol oxidase. *J. Food Biochem.* **13**: 65.

EXAMA, A., ARUL, J., LENCKI, R.W., LEE, L.Z. and TOUPIN, C. (1993) Suitability of plastic films for modified atmosphere packaging of fruits and vegetables. *J. Food Sci.* **58**: 1365–70.

FAIR CONCERTED ACTION (1999) FAIR CT96-1020. Harmonization of Safety Criteria for Minimally Processed Foods. Rational and Harmonization Report. November, 79 pp.

FRANCIS, G.A. and O'BEIRNE, D. (1997) Effects of gas atmosphere, antimicrobial dip and temperature on the fate of *Listeria innocua* and *Listeria monocytogenes* on minimally processed lettuce. *Int J. Food Sci Technol.* **32**: 141–51.

GARDNER, J., MANOHAR, S. and BORISENOK, W.S. (1991) Sulfite-free preservative for fresh peeled fruits and vegetables. Patent Number: 4,988,523.

GARG, N., CHUREY, J.J. and SPLITTSTOESSER, D.F. (1990) Effect of processing conditions on the microflora of fresh-cut vegetables. *J. Food Protect.* **53**: 701–3.

GORRIS, L. (1996) Safety and quality of ready-to-use fruit and vegetables (AIR1-CT92-0125). EU Research Results Ready for Application (RETUER) 21 May, Dublin, Ireland, 12 pp.

GORRIS, L.G.M., DE WITTE, Y. and BENNIK, M.J.H. (1994) Refrigerated storage under moderate vacuum. *ZFL Focus Int.* **45**, 6: 63–66.

HÄGG, M., HÄKKINEN, H., KUMPULAINEN, J., HURME, E. and AHVENAINEN, R. (1996) Effects of preparation procedures and packaging on nutrient retention in shredded Chinese cabbage. In: Fenwick, G.R., Richards, R.L. and Khokhar, S. (eds) *Agri-Food Quality: An Interdisciplinary Approach*, Cambridge, Royal Society of Chemistry, pp. 332–5.

HURME, E., AHVENAINEN, R., KINNUNEN, A. and SKYTTÄ, E. (1994) Factors affecting the quality retention of minimally processed Chinese cabbage. In: Proceedings of the Sixth International Symposium of the European Concerted Action Program COST 94 'Post-harvest treatment of fruit and vegetables'. *Current Status and Future Prospects*, Oosterbeek, 19–22 October, Commission of the European Communities, Brussels.

HURME, E., KINNUNEN, A., HEINIÖ, R.L., AHVENAINEN, R., *et al.* (1999) The sensory shelf-life of packed shredded iceberg lettuce dipped in glycerine betaine solutions. *J Food Prot.* **4**: 363–7.

HUXSOLL, C.C. and BOLIN, H.R. (1989) Processing and distribution alternatives for minimally processed fruits and vegetables. *Food Technol.* **43**: 124–8.

KABIR, H. (1994) Fresh-cut vegetables. In: Brody, A.L. (ed.) *Modified*

Atmosphere Food Packaging, pp. 155–160, Institute of Packaging Professionals, Herndon, VA.

KADER, A.A., ZAGORY, D. and KERBEL, E.L. (1989) Modified atmosphere packaging of fruits and vegetables. *Crit. Rev. Food Sci. Nutr.* **28**, 1: 1–30.

LABUZA, T.P., LILLEMO, J.H. and TAOUKIS, P.S. (1992) Inhibition of polyphenol oxidase by proteolytic enzymes. *Fruit Process.* **2**: 9–13.

LAURILA, E., HURME, E. and AHVENAINEN, R. (1998) Shelf-life of sliced raw potatoes of various cultivar varieties – substitution of bisulphites. *J Food Prot.* **61**: 1363–71.

LAURILA, E., KERVINEN, R. and AHVENAINEN, R. (1998), The inhibition of enzymatic browning in minimally processed vegetables and fruits. *Postharvest News Information* **4**: 53–66.

LEE, K.S., PARK, I.S. and LEE, D.S. (1996) Modified atmosphere packaging of a mixed prepared vegetable salad dish. *Int J Food Sci Technol.* **31**: 7–13.

LEISTNER, L. and GORRIS, L.G.M. (1995) Food preservation by hurdle technology. *Trends Food Sci.Technol.* **6**: 41–6.

LOZANO-DE-GONZÁLEZ, P.G., BARRETT, D.M., WROLSTAD, R.E. and DURST, R.W. (1993) Enzymatic browning inhibited in fresh and dried apple rings by pineapple juice. *J. Food Sci.* **58**: 399–404.

LUO, Y. (1992) Enhanced control of enzymatic browning of apple slices by papain. Dissertation, Washington State Univ., Pullman.

LUO, Y. and Barbosa-Cánovas, G.V. (1995) Inhibition of apple-slice browning by 4-hexylresorcinol. In: Lee, C.Y. and Whitaker, J.R. (eds) *Enzymatic Browning and Its Prevention*, Washington, DC, American Chemical Society, pp. 240–50.

MCEVILY, A.J., IYENGAR, R. and OTWELL, W.S. (1992) Inhibition of enzymatic browning in foods and beverages. *Crit. Rev. Food Sci. Nutr.* **32**: 253–73.

MANZANO, M., CITTERIO, B., MAIFRENI, M., PAGANESSI, M. and COMI, G. (1995) Microbial and sensory quality of vegetables for soup packaged in different atmospheres. *J. Sci. Food Agric.* **67**: 521–9.

MARCHETTI, R., CASADEI, M.A. and GUERZONI, M.E. (1992) Microbial population dynamics in ready-to-use vegetable salads. *Ital. J. Food Sci.* **2**: 97–108.

MARKHOLM, V. (1992) Intact carrots and minimally processed carrots: microflora and shelf life. *Reretning* nr. S 2190, Landbrugsministeriet Statens (Denmark) Planteavlsforsøg, 24 pp.

MARTINEZ, M.V. and WHITAKER, J.R. (1995) The biochemistry and control of enzymatic browning. *Trends Food Sci. Technol.* **6**: 195–200.

MATTILA, M., AHVENAINEN, R. and HURME, E. (1995a) Prevention of browning of pre-peeled potato. In: De Baerdemaeker, J., McKenna, B., Janssens, M., Thompson, A., Artes Calero, F., Höhn, E. and Somogyi, Z. (eds) *Proceedings of Workshop on Systems and Operations for Post-Harvest Quality*, pp. 225–34, COST 94 'Post-harvest treatment of fruit and vegetables'. Commission of the European Communities, Brussels.

MATTILA, M., AHVENAINEN, R., HURME, E. and HYVÖNEN, L. (1995b) Respiration rates of some minimally processed vegetables. In: De Baerdemaeker, J.,

McKenna, B., Janssens, M., Thompson, A., Artes Calero, F., Höhn, E. and Somogyi, Z. (eds) *Proceedings of Workshop on Systems and Operations for Post-harvest Quality*, pp. 135–45, COST 94 'Post-harvest treatment of fruit and vegetables', Commission of the European Communities, Brussels.

MEZA, J., LOZANO-DE-GONZÁLEZ, P., ANZALDÚA-MORALES, A., TORRES, J.V. and JIMÉNES, J. (1995) Addition of pineapple juice for the prevention of discoloration and textural changes of apple slices. IFT Annual Meeting 1995. Book of Abstracts, p. 68.

MOLNAR-PERL, I. and FRIEDMAN, M. (1990) Inhibition of browning by sulfur amino acids. Part 3. Apples and potatoes. *J.Agric. Food Chem.* **38**: 1652–6.

MONSALVE-GONZALEZ, A., BARBOSA-CÁNOVAS, G.V., CAVALIERI, R.P., MCEVILY, A.J. and IYENGAR, R. (1993) Control of browning during storage of apple slices preserved by combined methods. 4-Hexylresorcinol as anti-browning agent. *J. Food Sci.* **58**: 797–800, 826.

O'BEIRNE, D. (1995) Influence of raw material and processing on quality of minimally processed vegetables. *Progress Highlight* C/95 of EU Contract AIR1-CT92-0125 Improvement of the safety and quality of refrigerated ready-to-eat foods using novel mild preservation techniques. Commission of the European Communities, Brussels.

OHTA, H. and SUGAWARA, W. (1987) Influence of processing and storage conditions on quality stability of shredded lettuce. *Nippon Shokuhin Kogyo Gakkaishi* **34**: 432–8.

POWRIE, W.D. and SKURA, B.J. (1991) Modified atmosphere packaging of fruits and vegetables. In: Ooraikul, B. and Stiles, M.E. (eds) *Modified Atmosphere Packaging of Food*, pp. 169–245, Ellis Horwood Limited, Chichester.

PRETEL, M., FERNANDEZ, P., ROMOJARO, F. and MARTINEZ, A. (1998) The effect of modified atmosphere packaging on 'ready-to-eat' oranges. *Lebensm-Wiss u-Technol.* **31**: 322–8.

RIQUELME, F., PRETEL, M.T., MARTÍNEZ, G., SERRANO, M., AMORÓS, A. and ROMOJARO, F. (1994) Packaging of fruits and vegetables: recent results. In: Mathlouthi, M. (ed.) *Food Packaging and Preservation*, pp. 141–58, Blackie Academic & Professional, Glasgow.

SANTERRE, C.R., LEACH, T.F. and CASH, J.N. (1991) Bisulfite alternatives in processing abrasion-peeled Russet Burbank potatoes. *J.Food Sci.* **56**: 257–9.

SAPERS, G.M., HICKS, K.B., PHILLIPS, J.G., GARZARELLA, L., PONDISH, D.L., MATULAITIS, R.M., MCCORMACK, T.J., SONDEY, S.M., SEIB, P.A. and EL-ATAWY, Y.S. (1989) Control of enzymatic browning in apples with ascorbic acid derivatives, polyphenol oxidase inhibitors, and complexing agents. *J. Food Sci.* **54**: 997–1002, 1012.

SAPERS, G.M. and MILLER, R.L. (1992) Enzymatic browning control in potato with ascorbic acid-2-phosphates. *J. Food Sci.* **57**: 1132–5.

SAPERS, G.M. and MILLER, R.L. (1993) Control of enzymatic browning in pre-peeled potatoes by surface digestion. *J. Food Sci.* **58**: 1076–8.

SAPERS, G.M. and SIMMONS, G.F. (1998) Hydrogen peroxide disinfection of minimally processed fruit and vegetables. *Food Technol.* **52**(2): 48–52.

SKYTTÄ, E., KOSKENKORVA, A., AHVENAINEN, R., HEINIÖ, R.L. *et al.*, (1996) Growth risk of *Listeria monocytogenes* in minimally processed vegetables. *Proceedings of Food 2000 Conference on Integrating Processing, Packaging and Consumer Research.* Natick, MA, 19–21 October 1993. Hampton, VA: Science and Technology Corporation; 785–90.

TAOUKIS, P.S., LABUZA, T.P., LIN, S.W. and LILLEMO, J.H. (1989) Inhibition of enzymic browning. Patent WO 89/11227.

TORRIANI, S., ORSI, C. and VESCOVO, M. (1997) Potential of *Lactobacillus casei,* culture permeate, and lactic acid to control microorganisms in ready-to-use vegetables. *J Food Prot.* **60**: 1564–7.

VÁMOS-VIGYÁZÓ, L. (1981) Polyphenol oxidase and peroxidase in fruits and vegetables. *Crit. Rev. Food Sci. Nutr.* **15**: 49–127.

VAROQUAUX, P. and WILEY, R. (1994) Biological and biochemical changes in minimally processed refrigerated fruits and vegetables. In: Wiley, R.C. (ed.) *Minimally Processed Refrigerated Fruits & Vegetables*, pp. 226–68, Chapman & Hall, New York.

WHITAKER, J.R. and LEE, C.Y. (1995) Recent advances in chemistry of enzymatic browning. In: Lee, C.Y. and Whitaker, J.R. (eds) Enzymatic browning and its prevention. Washington, DC, ACS Symposium Series 600, pp. 2–7.

WILEY, R.C. (1994) *Minimally Processed Refrigerated Fruits & Vegetables*, Chapman & Hall, New York.

WILLOCX, F., HENDRICKX, M. and TOBBACK, P. (1994) The influence of temperature and gas composition on the evolution of microbial and visual quality of minimally processed endive. In: Singh, R.P. and Oliveira, F.A.R. (eds) *Minimal Processing of Foods and Process Optimization: An Interface*, pp. 475–92, CRC Press, Boca Raton, CA.

ZOMORODI, B. (1990) The technology of processed/prepacked produce: preparing the product for modified atmosphere packaging (MAP). *Proceedings of the 5th International Conference on Controlled/Modified Atmosphere/ Vacuum Packaging, CAP' 90*, San Jose, CA. 17–19 January 1990, pp. 301–30, Schotland Business Research, Princeton.

Appendix: Tables 9.5–9.10

Table 9.5 Processing guidelines for pre-peeled and sliced potato (Ahvenainen *et al.*, 1994)

Processing temperature	4–5°C
Raw material	Suitable variety or raw material lot should be selected using a rapid storage test of prepared produce at room temperature. Attention must be focused on browning susceptibility.
Pre-treatment	Careful washing with good quality water before peeling. Damaged and contaminated parts, as well as spoiled potatoes must be removed.
Peeling	1) One-stage peeling: knife machine. 2) Two-stage peeling: slight carborundum first, and then knife peeling.
Washing	Washing immediately after peeling. The temperature and amount of washing water should be 4–5°C and 3 l/kg potato. Washing time 1 min. Obs. The microbiological quality of washing water must be excellent. In washing water, in particular for sliced potato, it is preferable to use citric acid with ascorbic acid (max. concentration of both 0.5%) possibly combined with calcium chloride, sodium bentsoate or 4-hexyl resorcinol to prevent browning.
Slicing	Slicing should be done immediately after washing with sharp knives.
Straining off	Loose water should be strained off in a colander.
Packaging	Packaging immediately after washing in vacuum or in a gas mixture of 20% CO_2 + 80% N_2. The headspace volume of a package 2 l/1 kg potato. Suitable oxygen permeability of packaging materials is 70 cm^3/m^2 24 h 101.3 kPa, 23°C, RH 0% (80 μm nylon-polyethylene).
Storage	4–5°C, preferably in dark.
Other remarks	Good manufacturing practices must be followed (hygiene, low temperatures and disinfection).
Shelf-life	The shelf-life of pre-peeled whole potato is 7–8 days at 5°C. Due to browning, sliced potato has very poor stability, the shelf-life is only 3–4 days at 5°C.

Table 9.6 Processing guidelines for pre-peeled and sliced carrot

Processing temperature	4–5°C
Raw material	Suitable variety or raw material lot should be selected using a rapid storage test on prepared produce at room temperature. Attention must be focused on respiration activity and whitening of surfaces.
Pre-treatment	Careful washing with good quality water before peeling. Damaged and contaminated parts, as well as spoiled carrots must be removed.
Peeling and slicing	1) One-stage peeling: knife machine. 2) Two-stage peeling: slight carborundum first, and then knife peeling. Slicing should be done immediately after washing with sharp knives. Optimal size for slices is 5 mm.
Washing	Washing immediately after slicing. The temperature and amount of washing water should be 0–5°C and 3 l/kg carrot. Washing time 1 min. Obs. The microbiological quality of washing water must be excellent. In washing water, no additives are needed.
Straining off	Loose water should be strained off in a colander.
Packaging	Packaging immediately after washing in air. Suitable oxygen permeability of packaging materials is $2900\,cm^3/m^2$ 24 h 101.3 kPa, 23°C, RH 0% (e.g. $50\mu m$ LD polythene or corresponding material), but also material with oxygen permeability about $70\,cm^3/m^2$ 24 h 101.3 kPa, 23°C, RH 0% (e.g. $80\mu m$ nylon-polyethylene).
Storage	4–5°C, preferably in the dark.
Other remarks	Good manufacturing practices must be followed (hygiene, low temperatures and disinfection).
Shelf-life	Sliced carrot is quite preservable. The shelf-life is at least 7–8 days at 5°C.

Table 9.7 Processing guidelines for grated carrot (Ahvenainen *et al.*, 1994)

Processing temperature	0–5°C
Raw material	Suitable variety or raw material lot should be selected using a rapid storage test of prepared produce at room temperature.
Pre-treatment	Carrots must be washed carefully before peeling. Stems, damaged and contaminated parts, as well as spoiled carrots must be removed.
Peeling	Peeling with knife or carborundum machine.
Washing	Immediately after peeling. The temperature and amount of washing water: 0–5°C and 3 l/1 kg carrot, respectively. The washing time 1 min. Obs. The microbiological quality of washing water must be excellent. It is preferable to use active chloride 0.01% or 0.5% citric acid in washing water.
Grating	The shelf-life of grated carrot is the shorter the finer the shredding grade. The optimum grate degree is 3–5 mm.
Centrifugation	Immediately after grating. Grate may be lightly sprayed with water before centrifugation. The centrifugation rate and time must be selected, so that centrifugation only removes loose water, but does not break vegetable cells.
Packaging	Immediately after centrifugation. Proper packaging gas is normal air, and the headspace volume of a package 2 l/1 kg grated carrot. Suitable oxygen permeability of packaging materials is between 1,200 (e.g. oriented polypropylene) and 5,800, preferably 5,200–5,800 (e.g. polyethylene-ethylene vinyl acetate-oriented polypropylene) cm^3/m^2 24 h 101.3 kPa, 23°C, RH 0%. Perforation (one microhole/150 cm^3) of packaging material is advantageous. Diameter of microhole 0.4 mm.
Storage	0–5°C, preferably in dark.
Other remarks	Good manufacturing practices must be followed (hygiene, low temperatures and disinfection).
Shelf-life	7–8 days at 5°C.

Table 9.8 Processing guidelines for shredded Chinese cabbage and white cabbage

Processing temperature	0–5°C
Raw material	Suitable variety or raw material lot should be selected using a rapid storage test on prepared produce at room temperature.
Pre-treatment	Outer contaminated leaves and damaged parts, as well as stem and spoiled cabbage must be removed.
Shredding	The shelf-life of shredded cabbage is the shorter the finer the shredding grade. The optimum shredding degree is about 5 mm.
Washing of shredded cabbage	Immediately after shredding. The temperature and amount of washing water: 0–5°C and 3 l/1 kg cabbage, respectively. The washing time 1 min. Obs. The microbiological quality of the washing water must be excellent. Washing should be done in two stages: 1) Washing with water containing active chlorine 0.01% or 0.5% citric acid. 2) Washing with plain water (rinsing).
Centrifugation	Immediately after washing. The centrifugation rate and time must be selected so that centrifugation only removes loose water, but does not break vegetable cells.
Packaging	Immediately after centrifugation. Proper packaging gas is normal air, and the headspace volume of a package 2 l/1 kg cabbage. Suitable oxygen permeability of packaging material is between 1,200 (e.g. oriented polypropylene) and 5,800, preferably 5,200–5,800 (e.g. polyethylene-ethylene vinyl acetate-oriented polypropylene) cm^3/m^2 24 h 101.3 kPa, 23°C, RH 0%. For white cabbage, perforations (one microhole/150 cm^3) can be used. The diameter of the microhole is 0.4 mm.
Storage	0–5°C, preferably in the dark.
Other remarks	Good manufacturing practices must be followed (hygiene, low temperatures and disinfection).
Shelf-life	7 days for Chinese cabbage and 3–4 days for white cabbage at 5°C.

Table 9.9 Processing guidelines for cut onion

Processing temperature	0–5°C
Raw material	Suitable variety or raw material lot should be selected using a rapid storage test on prepared produce at room temperature.
Pre-treatment	Stems, damaged and contaminated parts, as well as spoiled onions must be removed.
Peeling	Peeling with knife or with pressurised air (dry onions).
Washing	Mild washing immediately after peeling. The temperature of washing water should be 0–5°C. Obs. The microbiological quality of washing water must be excellent. It is preferable to use active chlorine 0.01% in washing water.
Cutting	The cutting should be done immediately after washing with sharp knives. The shelf-life of cut onion is shorter the smaller the pieces.
Washing and centrifugation	No washing or centrifugation for cut onion.
Packaging	Immediately after cutting. Proper packaging gas is normal air or gas mixture 5% O_2 + 5–20% CO_2 + 75–90% N_2, and the headspace volume of a package 2 l/1 kg onion. Suitable oxygen permeability of packaging materials is between 1,200 (e.g. oriented polypropylene) and 2,900 (50μm LD-polyethylene) cm^3/m^2 24 h 101.3 kPa, 23°C, RH 0%.
Storage	0–5°C, preferably in the dark.
Other remarks	Good manufacturing practices must be followed (hygiene, low temperatures and disinfection).
Shelf-life	Cut onion has very poor stability, the shelf-life is only 3 days at 5°C.

Table 9.10 Processing guidelines for cut leek

Processing temperature	0–5°C
Raw material	Suitable variety of raw material lot should be selected using a rapid storage test on prepared produce at room temperature.
Pre-treatment	Stems, damaged and contaminated parts, as well as spoiled leeks must be removed. Careful washing with water.
Cutting	The cutting should be done immediately after washing with sharp knives. The shelf-life of cut leek is shorter the smaller the pieces.
Washing	Careful washing immediately after cutting. The temperature of washing water should be 0–5°C and washing time 1 min. Obs. The microbiological quality of washing water must be excellent. It is preferable to use active chlorine 0.01% in washing water.
Centrifugation	Careful centrifugation after washing is needed.
Packaging	Immediately after centrifugation. Proper packaging gas is normal air. The headspace volume of a package 2 l/1 kg leek. Suitable oxygen permeability of packaging materials is between 1,200 (e.g. oriented polypropylene) and 2,900 (50μm LD-polyethylene) cm^3/m^2 24 h 101.3 kPa, 23°C, RH 0%. If cutting grade is small (i.e. big cuts), quite impermeable materials can also be used, e.g. 80μm nylon-polyethylene, the permeability of which is 70 cm^3/m^2 24 h 101.3 kPa, 23°C, RH 0%. If packaging material is too permeable, the odour of leek can migrate from the package to other products.
Storage	0–5°C, preferably in the dark.
Other remarks	Good manufacturing practices must be followed (hygiene, low temperatures and disinfection).
Shelf-life	Cut leek has very poor stability, the shelf-life is only 3–4 days at 5°C.

10

Minimal processing in practice: seafood

M. Gudmundsson and H. Hafsteinsson, Technological Institute of Iceland (MATRA), Reykjavik

10.1 Introduction

The idea of minimal processing is to preserve food materials with minimum damage but still retain as much freshness in taste and other sensoric properties as possible. Many different food processing methods have the potential to be used in minimal processing either alone or in combination with other methods. Food irradiation is one such method that has been known for a long time but has not gained general acceptance. However, in the last decades some novel non-thermal technologies have been emerging which look promising for minimal processing without the detrimental effect of heat. These are methods like high pressure processing, high electric field pulses treatment and other less investigated methods like oscillation magnetic field treatment and use of light pulses for sterilisation. Modified atmosphere packaging can also be considered a minimal processing method but will not be dealt with in this chapter. The main emphasis in this chapter will be on the effect of high pressure and high electric field pulses on seafood.

10.2 High pressure processing of seafood: introduction

Food processing with high pressure is one of the latest methods for food preservation although it is still in its development phase. Foods preserved with high pressure look promising as they keep their natural appearance, taste and flavour. Even though high pressure processing has been commercialised for products like juices and jams (Farr 1990) it has still not found commercial application for marine products. Effects of high pressure on seafood have only been tried in a limited number of researches. They have focused on different kinds of fishes, fish mince, surimi and effects on fish proteins (Shoji and Saeki 1989, Shoji et al. 1990, Okamoto et al. 1990, Ohshima et al. 1992, Yoshioka et al. 1992, Murakami et al. 1992, Goto et al. 1993, Iso et al. 1993, Ohshima et al. 1993, Yukizaki et al. 1993, Yukizaki et al. 1994, Ledward 1998).

10.3 Impact on microbial growth

There have been many studies on the effect of high pressure on reduction of microbial growth and sterilisation of food and beverages (Johnson and ZoBell 1949, Jaenicke 1981, Hoover et al. 1989, Cheftel 1995). They show that inactivation of microorganisms by high pressure depends on the pressure level and the duration of treatment in order to reduce bacterial growth in many kinds of food like milk, meats, fruits and juices. The inactivation involves denaturation of proteins, i.e. protein unfolding, aggregation and gelation which can lead to inhibition of enzymatic activities (like ATPases) and destruction of vital intracellular organelles for the microorganism (Johnston et al. 1992, Suzuki et al. 1992, Ogawa et al. 1992, Cheftel 1995). High pressure can also affect the membrane structure, chemical reaction and release of intracellular constituents which can contribute to microbial inactivation (MacDonald 1992, Shimada et al. 1993, Mozhaev et al. 1994, Cheftel 1995). The extent of inactivation depends on the type of microorganism, the state of the microbes (i.e. growth phase, stationary phase or spores), the pressure level, the process time and temperature, and the composition of the dispersion medium (Carlez et al. 1994, Cheftel 1995).

The pH of the medium apparently has little influence on protection of microbes but salt, sugar and low water content seem to have strong baroprotective effect (Cheftel 1995). In most cases, at ambient temperatures it is necessary to apply pressures above 200 MPa in order to induce inactivation of vegetative microorganisms. Some pathogenic microorganisms need 275 MPa for 15–30 minutes at 20°C to be inactivated like *Yersinia enterocolitica* and others need 700 MPa pressurisation like *Salmonella enteritis*, *Escherichia coli* and *Staphylococcus aureus* (Patterson et al. 1995).

A few studies have been done specifically on the effect of high pressure on inactivation of microbes in seafood products. The effect of high pressure on total count of bacteria in tuna and squid samples treated with 450 MPa for 15 minutes at 25°C reduced the plate count between one and two log cycles. This treatment was not sufficient alone to reduce effectively or sterilise the sample (Shoji and Saeki 1989).

Perhaps it is necessary to use a combination of moderate pressure and heat treatment in order to inhibit or inactivate vegetative microbes. The combination of heat treatment from zero to 60°C and high pressure up to 400 MPa tested on *Lactobacillus casei* and *E. coli* showed that low temperature treatment (0°C) and high pressure were more effective than other treatments in inactivating microbes (Sonoike et al. 1992).

High pressure treatment (500 MPa/10 min.) has been shown to effectively kill bacteria like *Vibrio parahaemolyticus*, *Vibrio cholerae* and *Vibrio mimicus* in sea urchin eggs but the eggs still retained their original flavour and taste (Yukizaki et al. 1993). However, only 200 MPa pressure for five minutes at 0°C is needed to inactivate *Vibrio parahaemalyticus* in a buffer solution (Yukizaki et al. 1994), indicating that some compounds in sea urchin eggs have baroprotective effect. A study on oyster preservation by high pressure (López-Caballero et al. 2000a) showed that pressurisation of 400 MPa for five minutes reduced the total quantity of microorganisms up to five log cycles. The pressurised oysters were stable for 41 days at 2°C but 13 days for the control sample. Another study on preservation of chilled and vacuum packaged prawns by high pressure at 200 and 400 MPa showed that the shelf-life was extended by one and two weeks respectively compared to prawns that were only chilled and vacuum packaged (López-Caballero et al. 2000b).

10.4 Impact on quality

Flavour, taste and texture are important quality parameters regarding consumer acceptance for seafood products. These quality parameters are affected during storage by many factors that can reduce the freshness or spoil the product. These factors may involve protein denaturation, enzyme activities that produce off-flavour compounds and lipid oxidation. The consequences can be dripping of the fish muscle leading to dry and tough texture, rancidity and off-flavours. High pressure will affect these factors in such a way that some of the quality parameters will be improved compared to other preservation methods and others will decrease compared to fresh products.

10.4.1 Effects on microstructure, fish proteins and enzymes

High pressure is said to induce breakdown of ionic bonds due to electrostriction and at least partly on hydrophobic interactions. In contrast hydrogen bonds seem to strengthen somewhat under pressure and covalent bonds have low sensitivity towards pressure. It depends on the pressure level, whether denaturation of proteins and inactivation of enzymes occurs. Denaturation can involve dissociation of oligomeric structures, unfolding of monomeric structure, protein aggregation and protein gelation (Balny and Masson 1993, Gross and Jaenicke 1994, Funtenberger et al. 1995, Cheftel 1995). Denaturation of proteins is the main underlying factor in changes in the microstructure of muscle products and in the inactivation of enzymes. Seafood treated with high pressure is therefore prone to all these changes, which depend on the severity of the pressurisation. The effects of high pressure can be either reversible or irreversible depending on the pressure and temperature. In general reversible effects of high pressures are observed below 100–200 MPa (e.g. dissociation of proteins into subunits).

Above 200 MPa, non-reversible effects occur and they may include complete inactivation of enzymes and denaturation of proteins (Balny and Masson 1993). Denaturation depends also on external parameters, e.g. temperature, pH and solvent composition (sugar, salts and other additives). Hydrophobic interactions are first affected by pressure below 150 MPa and therefore the quaternary structure of proteins are first to change. Tertiary structure changes occur above 200 MPa and changes of secondary structure take place above 700 MPa (Balny and Masson 1993, Cheftel 1995). Thus protein denaturation caused by high pressure involves rearrangement and/or destruction of non-covalent bonds such as hydrogen, hydrophobic interaction and ionic bonds of quaternary and tertiary structure of proteins while covalent bonds are not affected (Okamoto et al. 1990, Balny and Masson 1993, Cheftel 1995).

The main proteins of fish muscle are myofibrillar and sarcoplasmic proteins. Myofibrillar proteins are the proteins that determine the structure of the muscle while sarcoplasmic proteins are water-soluble non-structural proteins. The myofibrillar proteins constitute 65–80% of the total proteins in the fish muscle. They are mainly composed of the contractile proteins actin and myosin, regulatory proteins, elastic proteins and some other minor proteins. Myosin denatures at 100–200 MPa and actin at 300 MPa. Only a few soluble proteins survive pressure of 800 MPa (Balny and Masson 1993).

High pressure treatment of carp myofibrils at 150 MPa for 30 minutes destroyed the arrangement of myofibrils and the striation pattern was lost (Ohshima et al. 1993). By contrast, myofibrils treated at 38°C for two hours still exhibited a striped appearance, although some unique structural changes had occurred. The mobilities of myosin heavy chain and actin are not changed by high pressure of 150 MPa or heating at 38°C (Shoji and Saeki 1989). However, when normal muscle from cod and mackerel is treated with high pressure, certain sarcoplasmic proteins become covalently linked together and are thus resistant to extraction with SDS (Ohshima et al. 1992).

10.5 Effects on enzymatic activity

Rigor mortis starts when the ATP level decreases post mortem in fish muscle. ATP is degraded into several compounds by dephosphorylases inherent in the fish muscle. Some of these compounds are intermediate compounds but others can accumulate in the fish during storage. The amount of these compounds is used to evaluate the freshness of the fish by the ratio of two of these compounds to ATP, which is called the k value (Saito et al. 1959, Sakaguchi and Koike 1992). When a carp muscle was treated with high pressures of 200, 350 and 500 MPa and subsequently stored at 5°C, further suppression of the decrease in inosine 5'monophosphate level (intermediate breakdown compound) was observed at 350 and 500 MPa (Shoji and Saeki 1989). These results strongly suggest that the enzymes involved in degradation of ATP undergo protein denaturation and are deactivated during high pressure treatment. On the other hand, heat promotes breakdown of ATP in fish, even in a very fresh one.

There is a difference between the inactivation of ATPase activity by heat and high pressure. The inactivation of Ca^{2+}ATPase activity by heat follows first order kinetics (Arai 1977). On the other hand, Ca^{2+}ATPase activity in carp myofibrils that were pressurised at 125 and 150 MPa showed a shift in linear relationship with time. After a certain time has lapsed there is an apparent breakpoint in the pressure–time relation of enzyme activity and the actvity decreases at a slower rate. This suggests that the mechanism of denaturation by heat is somewhat different from denaturation of proteins by high pressure treatment (Ohshima et al. 1993).

This is supported to some extent by research on seven different fish species (Iso et al. 1993). They measured the effect of heat and pressurisation (200 MPa for 13 h) on denaturation of myofibrillar proteins by differential scanning calorimetry (DSC). The thermograms of proteins heated in DSC showed three peaks, indicating the denaturation of the two different myosin chains at 44°C and 51°C and actin at 71°C. However, the pressurised fish proteins when heated showed only two peaks, one small peak for the myosin and a broad indistinct peak for actin. The pressurised fish proteins seem to have been partially denatured, as total enthalpy change was noticeably smaller for pressurised fish proteins than untreated proteins. The myosin and actin were both partially denatured by the pressure at the same time.

Studies have been done on the effect of high pressure on enzymes that contribute to the deterioration of seafood (Ashie and Simpson 1995). These enzymes were trypsin, chymotrypsin, cathepsin and collagenase. They found that all the enzymes studied were susceptible to pressure at 100–400 MPa and proportional to duration of pressure application. Trypsin was more susceptible to inactivation than chymotrypsin.

Lipases are still active in the fish muscle during storage at low temperatures and they will eventually release free fatty acids from glycolipids that accumulate with time in the fish (de Koning and Mol 1990). High pressure treatment of fish above 405 MPa before storage is needed to stop the increase of fatty acids and the decrease of phospholipids (Ohshima et al. 1993).

10.6 Effects on texture and microstructure

Researches on red meat show that pressurisation does not markedly affect ageing or conditioning of post-rigor meat at pressures below 200 MPa and at ambient temperature (Cheftel and Culioli 1997). Also collagen, which is mostly stabilised with hydrogen bonds, is little affected by pressure at ambient temperatures. There are, however, significant effects of pressure on the organisation and subsequent gelation of myofibrillar proteins in both meat and fish (Ledward 1998). In Fig. 10.1, one can see a comparison of untreated and pressure treated salmon muscle samples. The effect of 400 MPa pressure for 30 minutes on the

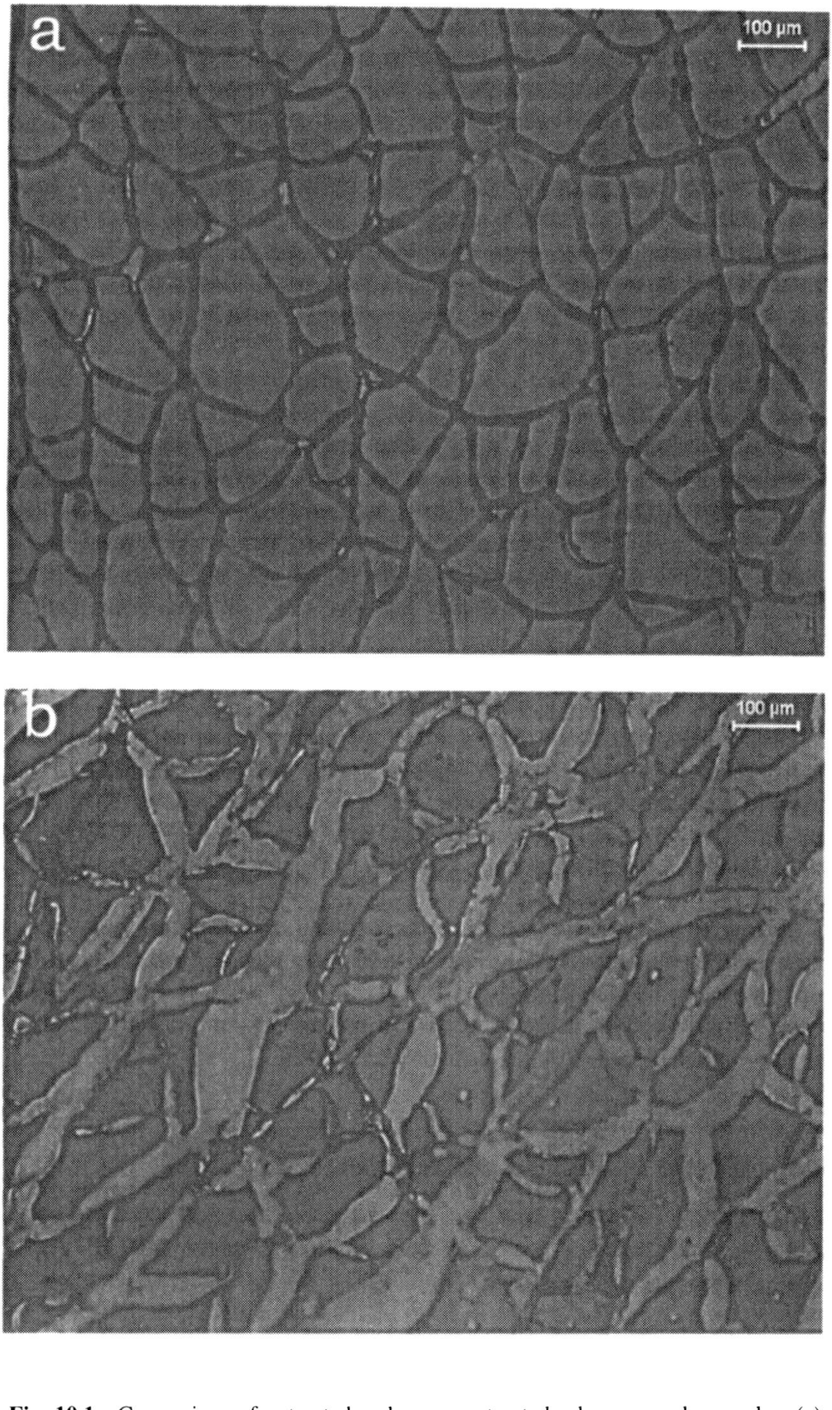

Fig. 10.1 Comparison of untreated and pressure-treated salmon muscle samples: (a) untreated salmon, (b) salmon treated with 400 MPa for 30 minutes. Data from the authors.

microstructure is clearly seen as the muscle has disintegrated and the cells have decreased in size compared to the untreated sample (data from the authors).

The stability of different myosins to pressure depends, as with their thermal stabilities, on the environment of the species. Thus, myosins from both turkey and pork are significantly more stable to pressure than myosin from cold-water fish like cod (Cheah and Ledward 1996, Angsupanich and Ledward 1998, Angsupanich et al. 1999). At relatively low pressures 100–200 MPa, myosin initially aggregates by two heads fusing together to form one headed structure, and they further aggregate and form a clump of heads with the tails extending radially outwards (Cheftel and Culioli 1997). A similar mechanism has been proposed for the initial stages in the thermal gelation of myosin (Yamamoto et al. 1990).

Although the initial stages of the aggregation of myosin, either through heat or pressure, may be similar, the subsequent gelation mechanisms are very different which is not surprising considering the relative stabilities of hydrogen bonds and hydrophobic interactions to pressure and temperature. Thus, on heat treatment the myosin tails readily unfold and form a gel network or aggregate depending on the conditions and are primarily stabilised by disulphide linkages and hydrophobic interactions (Yamamoto et al. 1990). On pressure treatment there is a formation of new or modified hydrogen bonded structures which, when subjected to DSC, melt a few degrees below the temperature at which native myosin denatures (Angsupanich and Ledward 1998). The hydrogen bonds melt at relatively low temperatures but are relatively insensitive towards pressure in the range of 200–800 MPa (Angsupanich and Ledward 1998). In addition, it has been established that at pressures above 400 MPa, the myosin heavy chain can form disulphide linkages (Angsupanich and Ledward 1998). Disulphide bonds are also formed on heat gelation in these systems (Lee and Lanier 1995).

In short, the thermally produced myosin gels are stabilised primarily by disulphide linkages and hydrophobic interactions. However, the pressure-induced myosin gel results in a gel network, which is stabilised by both disulphide linkages and a significant number of hydrogen bonds, which can subsequently be broken on heat treatment.

The texture of pressure-treated fish is therefore markedly different from heat-treated fish (Angsupanich and Ledward 1998, Angsupanich et al. 1999). Pressure-treated fresh cod muscle at 400 MPa showed much greater hardness determined by texture profile analysis than heated cod muscle at 50°C. Cod treated with pressures below or above 400 MPa showed less hardness than at 400 MPa. If, however, the pressure-treated cod muscle is heated it will be similar in hardness to a heat-treated one. The texture of pressure-treated fish is relatively heat sensitive and will soften up on heat treatment at low temperatures. Largest changes were seen in adhesiveness, chewiness and gumminess of cod muscle at pressure treatment below 400 MPa compared to untreated samples. However, the results of high pressure treatment on bluefish showed that pressure of 101 MPa increased the firmness of the fish muscle during storage at 4–7°C but pressurisation of 202 and 303 MPa had the opposite effect (Ashie et al. 1997). The use of evaluation method matters as Ashie and Simpson (1996) got different results with sensoric analysis and Instron compression probe on pressurised bluefish at 300 MPa. The pressurised bluefish was judged harder in sensoric analysis than the control sample but the reverse with the compression method.

Oysters did also show increased shear strength after pressurisation compared to untreated oysters (López-Caballero et al. 2000a). Shrimps treated with high pressure of 200 or 400 MPa were somewhat harder than control samples (López-Caballero et al. 2000b). It has also been found that high pressure treatment improved gel-forming ability of poor performing fish minces (Pérez-Mateos and Montero 2000). In surimi and other fish meat products, high pressure treatment readily induces gelation at low temperatures (Ohshima et al. 1993, Shoji et al. 1994).

Hydrostatic pressure of 200 MPa for 30 minutes at 25°C was needed to induce gel from carp crude actomyosin that could support its own weight and maintain its shape. The pressure-induced gel kept its original colour and flavour and was glossy and soft in comparison with heat-induced gels (Ohshima et al. 1993). The gels tended to increase in hardness and to decrease in adhesiveness as the applied pressure was increased. However, they were still soft and had large extendibility and were not fractured by high stress. There are significant differences in appearance and textural properties between pressure- and heat-induced gels. Heat-induced carp gels swell a little and are relatively hard but lack adhesiveness. These results are further indications that the gelation mechanism is different between pressure- and heat-induced gels.

Study on the difference between heat- and pressure-treated fish gels has showed that a blue whiting gel formed with high pressure of 200 MPa (10°C and ten minutes) had greater breakforce and more cohesiveness than heat-induced gels (Borderias et al. 1997). Pressurisation at 4°C prior to incubation at 25°C or 40°C increased the gel strength two- to threefold in uncooked surimi gels that contained transglutaminase (TGase). High pressure rendered protein substrates more accessible to TGase, thereby enhancing intermolecular cross-link formation and gel strength. The TGase enzyme was not affected by high pressure up to 300 MPa at 4°C (Ashie and Lanier 1999).

The effects of high pressure on sarcoplasmic proteins of sardine, walleye pollack, marble sole and horse mackerel has also been investigated (Okamoto et al. 1990). The sarcoplasmic proteins become insoluble and precipitate at pressure above 140 MPa and when the concentration of sarcoplasmic was above 50 mg/ml, the proteins formed gels. The properties of the gels were affected by many factors such as fish species, pH, protein concentration, pressure and treatment time. The hardness of pressure-induced gels was highest for the sole but the breaking strength was highest for horse mackerel. The strength of the gels increased with increased applied pressure, where the sole gel showed the greatest increase. Water-holding capacity of the gels generally decreased with increased pressure for the sole and pollack but was relatively constant for the sardine and horse mackerel at applied pressure up to 370 MPa. Breaking strength showed maximum at pH between 5 and 6 but water-holding capacity was then at minimum (Okamoto et al. 1990). An observation by electronic microscope showed that the gels were porous and rheological measurements showed that they were elastic and quite different from heat-induced gels. The breaking strength of pressure-induced gels made at 470 MPa was much greater than for heat-induced gels of the same proteins (Okazaki and Nakamura 1992). The results for sarcoplasmic fish proteins show similar trends when pressurised as for the myofibrillar proteins.

High pressure has also been applied to surimi of both Pacific whiting and Alaska pollack (Shoji et al. 1990, Chung et al. 1994). Surimi analogues are traditionally made from heat-induced gels at temperatures near 90°C (Lee 1984, Lanier and Lee 1992). A number of fish species may undergo a weakening of gel structure during normal heating regimes because of endogenous proteases in the muscle tissue (Niwa 1992). As the temperature increases during the cooking, it will cause gelation of the surimi product. However, the product is heated through an interval 50–60°C, where these proteases are most active. At present this potential weakening of the gel is bypassed with the use of protein inhibitors like beef plasma proteins or egg whites (Matsumoto and Noguchi 1992). High pressure-treated Pacific whiting and Alaska pollack surimi gels with added protease inhibitors showed greatly increased elasticity at all pressure/temperature combinations (100–280 MPa and temperatures between 28°C and 50°C) when compared to heat-induced surimi gels. However, the gel strength varied. The gel strength of Alaska pollack surimi gel with added inhibitor was higher than control except gels formed at 50°C and highest pressure used. A pressure-treated Pacific whiting surimi gel without inhibitor had threefold increase in strain and stress values compared to heat-induced gels except at 50°C where the pressurised surimi did not form a gel. This indicates that protease activity is increased under pressure at that temperature (Chung et al. 1994).

Another study (Shoji et al. 1990) on surimi showed that surimi with 2.5% salt formed strong gels when treated with pressure between 200 and 400 MPa at 0°C for ten minutes. The gel formed at 300 MPa formed the strongest gel of the high pressure-induced surimi gels. The pressure-induced surimi gels as in the study above formed gels with greater gel strength than heat-induced gels and were more transparent.

Fractionation and electrophoresis of the surimi gel proteins suggested that the gel formation by high pressure depends largely on the cross-linking of myosin heavy chains (Shoji et al. 1990).

10.7 Effects on lipid oxidation

The marine lipids are characterised by high levels of polyunsaturated fatty acids (PUFA) (Ackman 1990). The PUFAs are generally susceptible to autoxidation and oxidative degradation of lipids in foods and foodstuffs during processing and subsequent storage directly affects the quality of products, including flavour, colour, texture and nutritional value (Eriksson 1982). Highly purified fats and oils are believed to be relatively stable to oxidation when subjected to high pressure. However, with commercial fats the relationship between high pressure and sensitivity of the fat to lipid oxidation is a complex function of water activity (Cheah and Ledward 1995).

There are few studies on the effect of high pressure on fish oils. When extracted sardine oil was treated with hydrostatic pressure of 506 MPa for 60 minutes, the oxidation indicators, peroxide value (POV) and thiobarbituric acid (TBA) did not change (Tanaka et al. 1991). On the other hand, when cod muscles were exposed to high hydrostatic pressure of 202, 404 and 608 MPa for 15 and 30 minutes, the POV of the extracted oils increased with increased hydrostatic pressure and processing time. Even more pronounced effects were observed for mackerel muscle lipids (Ohshima et al. 1992). These results indicate that pure oils are stable after high pressure treatment but not lipids in the fish muscle probably due to release of metal ions that act as catalysators. This is supported by the work of other researchers (Cheah and Ledward 1995, Angsupanich and Ledward 1998). They found also that applying high pressure above 400 MPa decreased the oxidative stability of the lipids in cod.

Lipid oxidation appears to be catalysed in the range of water activities that are most common in meat and fish products, when subjected to pressures over 400 MPa (Cheah and Ledward 1995). The pressure induces changes in fat and tissues that probably release metal ions from specific complexes, which are then able to catalyse the oxidation. This has been supported by a study on incorporation of appropriate antioxidants and specific metal chelators that effectively inhibit oxidation (Cheah and Ledward 1997). It is not clear from what compounds the ions are released. Haem compounds are considered unlikely as a catalytic effect is also seen in cod muscle but a complex like haemosiderin and other insoluble complexes are likely candidates as oxidation still happened despite removal of soluble metal complexes (Cheah and Ledward 1996, Ledward 1998).

10.8 Effects on appearance and colour

High pressure can be used in food processing even though it causes denaturation of proteins as it inactivates microorganisms without changing the flavour, colour, vitamins and tastes of foods (Hayashi 1993). The only noticeable colour change in white fish such as cod and also in mackerel is the loss of translucency and that the fish becomes opaque due to denaturation of proteins and the fish looks similar to cooked fish (Ohshima et al. 1992, Shoji et al. 1990, Cheah and Ledward 1996, Angsupanich et al. 1999). These changes take place at pressures between 100 and 200 MPa for cod (Angsupanich and Ledward 1998).

10.8.1 Other uses of high pressure

It has been mentioned that high pressure can be used to produce surimi gels and other fish products of better quality (Okamoto et al. 1990).

High pressure below sub-zero temperature can also be used to rapidly produce small ice crystals (microcrystallisation) in a product, which would be less detrimental to microstructure and the texture of the product than traditional freezing (Karino et al. 1994, Cheftel 1995). This could be of great advantage in frozen fish products.

Another potential application is thawing of product under pressure between zero and 20°C because water is not frozen in this range at pressure of 210 MPa. Thus it is possible to thaw product at sub-zero temperatures with the help of pressurisation (Kalichevsky et al. 1995, Cheftel 1995).

10.9 Future trends of high pressure treatment

As it is difficult to destroy both microorganisms and microbial spores even at as high pressure as 450 MPa (Miyao et al. 1993), it seems necessary to use high pressure treatment in combination with temperatures either below −20°C or at moderately high heating temperature to obtain acceptable preservation and quality of the fish product. This effect could also be obtained in combination with another minimal processing method. Further studies on the use of high pressure in combination with other minimal processing methods are therefore necessary. High pressure has also been shown to successfully inhibit some inherent enzymatic activities above 405 MPa which are undesirable for seafood quality, which supports the use of high pressure to increase fish product quality during storage.

High pressure treatment has been shown to produce kamaboko with very fine surface and induce fish gels with very interesting properties. The pressure-induced fish gels give an indication that a range of novel products could be produced from fish or other marine products that have an appearance and texture that is different from traditional products.

10.10 The use of high electric field pulses

Basically high intensity pulsed electric fields (PEF) is a non-thermal preservation method like high pressure treatment and it has the potential to be used in minimal processing (Knorr 1995, Knorr et al. 1998, Barbosa-Cánovas et al. 1998). The possible uses of PEF as a food preservation method have been investigated for a number of years. The main emphasis has been on inactivation of different types of microorganisms in different phases, i.e. growth and stationary phases or as spores (Sale and Hamilton 1967, Castro et al. 1993, Hülsheger and Nieman 1980, Wouters et al. 1999). A few studies have been done on the use of PEF to improve the yield of juices (Flaumenbaum 1968, Knorr et al. 1994). For the consumer of fish products the sensoric experience is important besides safety issues and the texture is a large part of that experience. Traditional processes for fish products such as frozen storage, drying, salting and canning have from moderate to severe effect on the microstructure of the product compared to fresh product (Duerr and Dyer 1952, Connell 1964, Chu and Sterling 1970, Dunajski 1979, Bello et al. 1982, Fennema 1990, Mackie 1993, Sikorski and Kotakowska 1994, Greaser and Pearson 1999). There are only limited researches on the effect of PEF treatment on microstructure of food (Barsotti et al. 1999, Fernandez-Diaz et al. 2000) and only one is available on fish products as far as we are aware (Gudmundsson and Hafsteinsson 2001).

10.11 Impact on microbial growth

The lethal action of electric fields on living cells has been explained by dielectric breakdown of the cell membrane (Zimmermann et al. 1976, Zimmermann 1986, Sale and Hamilton 1967). The applied external electric field induces transmembrane potential, which above certain critical value of 1V causes pore formation that can be lethal to microorganisms. The irreversible changes occur to a cell when external electric field between 1 and 10kV/cm is used for more than 10–15 minutes (Zimmermann et al. 1976).

It has been shown that relative rate of killing bacteria is related to the field strength, time and also number of pulses and pulse width. It is also clear that PEF inactivation is a function of type of microorganism and the microbial growth stage, initial amount of microbes, ionic concentration and conductivity of the suspension (Hülsheger et al. 1981, Wouters and Smelt 1997). Bacteria in the growth phase are more sensitive to electric field than stationary bacteria and spores are the most resistant (Sale and Hamilton 1967, Wouters and Smelt 1997). Some factors in the suspension media or in the sample seem to have protective effect, for example cations, proteins and lipids (Hülsheger et al. 1981, Zhang et al. 1994, Grahl and Märkl 1996, Martín et al. 1997). The bactericidal effect of PEF decreases with increased ionic strength (Hülsheger et al. 1981). Therefore it is more difficult to inactivate microorganisms in semi-solid or solid food materials than in dilute buffer solutions as they are rich in ions and other protective substances (Hülsheger et al. 1981, Zhang et al. 1994). It has also been pointed out that many foods are heterogeneous with areas of different electrical resistivity, which can alter the effects of PEF treatment as some areas will be untreated and others over-treated in such material (Barsotti et al. 1999).

To our knowledge no specific study has been published on the effect of PEF treatment on different types of bacteria in seafood. The effect of PEF treatment on total bacteria count has, however, been reported on lumpfish roes, where treatment of 11 kV/cm and seven pulses (2μs in width) reduced the total bacteria count by one log cycle (Gudmundsson and Hafsteinsson 2001).

In seafood the most potential bacterial pathogens are those of the *Vibrionaceae* family and the most important of these are *Vibrio cholerae*, *Vibrio parhaemolyticus* and *Vibrio vulnificus* and one can also mention *Aeromonadas hydrophila* (Wekell et al. 1994). Other pathogenic bacteria that can be present in seafood for various reasons are *Salmonella* species, *E. coli*,

Shigella, *Campylobacter*, *Yersinia enterocolitica*, *Clostridium botulinum*, *Listeria monocytogenes*, *Staphylococcus aureus* and *Bacillus cereus* (Liston 1990).

Listeria monocytogenes in a stationary phase was reduced by two log cycles and between two and three log cycles in the growth phase at 20 kV/cm and 30 pulses respectively (Hülsheger et al. 1983). *Staphylococcus aureus* and *E. coli* in the stationary phase treated in the same way were reduced between three and four log cycles. Other researches show reduction from two to nine log cycles for *E. coli* using fields from 20 kV/cm up to 70 kV/cm (Dunn and Pearlman 1987, Zhang et al. 1994). The reduction for *Staphylococcus aureus* was two log cycles at 27.5 kV/cm (Hamilton and Sale 1967) and reduction of four log cycles for *Salmonella dublin* at 18 kV/cm (Dunn and Pearlman 1987). No studies are available for *Vibrio* species, *Shigella* or *Campylobacter*. However, it is clear that external electric fields above 20 kV/cm are needed to inactivate most types of microorganisms for a minimum of two to three log cycles.

10.12 Effects on protein and enzymatic activity

It has been shown that ovalbumin and other egg white proteins do not denature when treated with PEF with electric field of 27–33 kV/cm and using 50–400 pulses (Jeantet et al. 1999, Fernandes-Diaz et al. 2000). There are no studies on PEF dealing with denaturation of proteins in seafood, except that it has been reported that treatment of cod with PEF treatment of up to 18.6 kV/cm and seven pulses (2μs width) did not affect the proteins from cod (Gudmundsson and Hafsteinsson 2001). According to their results from SDS electrophoresis no changes were seen in molecular bands in PEF-treated cod proteins compared to untreated samples.

Studies on the effect of PEF treatment on enzyme activity show that many enzymes are unaffected even at electric fields above 30 kV/cm. These are enzymes like amylases, lipase, NADH dehydrogenase, succinic hydrogenase and hexogenase (Hamilton and Sale 1967). However, PEF treatment inactivates some enzymes like proteases from *Pseudomonas fluorescens* at 15 kV/cm and 98 pulses and plasmin at 30 kV/cm and 50 pulses (Vega-Mercado et al. 1995a, 1995b). The enzymes α-amylase, lipase and glucose oxidase were markedly inactivated at very high electric fields of 64–87 kV/cm, whereas peroxidase and polyphenoloxidase were more resistant (Ho et al. 1997). No studies are available that deal specifically with enzymes from marine sources.

10.13 Effects on texture and microstructure

There are only a few studies on the effect of PEF treatment on the quality of foods, mainly on juices and other pumpable foods, which show that sensoric properties are not affected (Knorr et al. 1994, Qin et al. 1995, Barbosa-Cánovas et al. 1996). Research on meat and seafood is very limited. Only one publication on the effect of texture and microstructure of muscle foods has been published to our knowledge (Gudmundsson and Hafsteinsson 2001).

Changes in microstructure and texture can be expected as a consequence of the permiabilisation caused by PEF treatment which can induce changes in water-holding properties of the muscle.

Salmon treated with PEF of 1.36 kV/cm and 40 pulses caused gaping in the fish muscle and collagen leakage into the extra-cellular gap between the muscle cells as seen in microscope (Gudmundsson and Hafsteinsson 2001). Treatment with high pressure at 300 MPa also caused gaping. Teleostic fishes including salmon contain a low amount of connective tissue in the muscle (0.66%) (Dunajski 1979, Eckhoff et al. 1998). On the other hand, for example, chicken meat contains about 2% connective tissue (Baily and Light 1989). It is also known that increased size of cells makes them more vulnerable to PEF treatment (Sale and Hamilton 1967, Hülsheger et al. 1983). The muscle cells of salmon are considerably larger than any bacteria as muscle cells are usually 50–100μm in diameter but bacteria are 0.3–2.0μm (Nester et al. 1983, Wong 1989). These two facts could explain why salmon and fish in general do not tolerate even a mild PEF treatment without damage to the microstructure.

The impact of PEF treatment on microstructure of fish muscle cannot be the result of protein denaturation as far too low intensity of electric field was used. The probable explanation is punctuation of the cell membranes which causes leakage of cell fluids into extra-cellular space.

Fresh lumpfish roes treated with 12 kV/cm and 12 pulses (2μs) were intact after the treatment except for a very low percentage of the roes, as can be seen in Fig. 10.2. Firmness of PEF-treated roes measured with a compression test showed also that the PEF treatment only marginally affected the firmness of the roes (Gudmundsson and Hafsteinsson 2001). Another study (Craig and Powrie 1988) on frozen and then thawed salmon roes showed that 46% less energy is needed to rupture such roes than fresh roes. The three-layer membrane of the roes probably gives them the strength to tolerate the PEF treatment. The roes can probably tolerate even stronger PEF treatment, which could then make it more plausible to use PEF treatment on roes for preservation.

10.14 Future trends of PEF treatment

Preservation of fish products with PEF treatment does not seem plausible as relatively low intensity of electric field pulses has a detrimental effect on the fish microstructure and at the same time the low field voltage does not effectively reduce the growth of bacteria. Roes, on the other hand, seem to tolerate PEF treatment without a visible effect on the microstructure or texture. A PEF treatment could therefore be valuable as a pre-treatment for roes but that needs to be further investigated. Other possible uses of PEF treatment in the fish industry have not been explored but it could be possible to use PEF treatment in

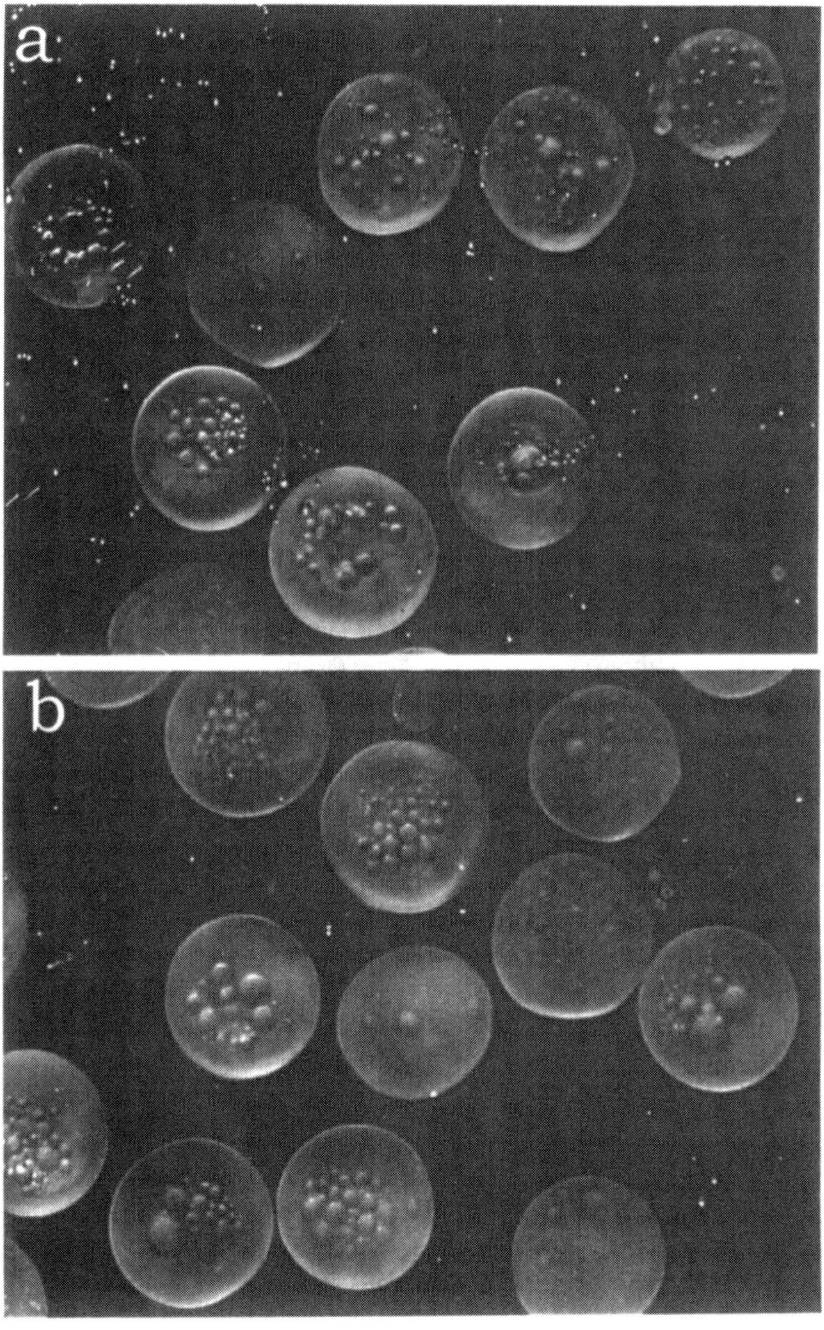

Fig. 10.2 Comparison of untreated and PEF-treated lumpfish roes: (a) untreated roes, (b) roes treated with 12 kV/cm and 12 pulses. Data from the authors.

a similar way as in juice production where the PEF treatment ruptures the tissue cells, which makes it easier to extract the valuable substances. This could be done on waste material and byproducts from the fish industry and possible products could be, for example, enzymes or fish oil.

10.15 References

ACKMAN R G (1990), 'Seafood lipids and fatty acids', *Food Review International*, **6**(4), 617–46.

ANGSUPANICH K, EDDE M and LEDWARD D A (1999) 'The effects of high pressure on the myofibrillar proteins of cod and turkey', *Journal of Agricultural and Food Chemistry*, **47**(1), 92–9.

ANGSUPANICH K and LEDWARD D A (1998), 'High pressure treatment effects on cod (Gadus morhua) muscle', *Food Chemistry*, **63**(1), 39–50.

ARAI K (1977), 'Fish muscle proteins', Tokyo, *Japanese Society of Scientific Fisheries.* Kouseisha-kouseikaku, 75–90.

ASHIE I N A and LANIER T C (1999), 'High pressure effects on gelation of surimi and turkey breast muscle enhanced by microbial transglutaminase', *Journal of Food Science*, **64**(4), 704–8.

ASHIE I N A and SIMPSON B K (1995), 'High pressure effects on some seafood enzymes', IFT Annual Meeting, Session 71D-9.

ASHIE I N A and SIMPSON B K (1996), 'Application of hydrostatic pressure control enzyme-related seafood texture deterioration', *Food Res Int*, **29**, 564–75.

ASHIE I N A, SIMPSON B K and RAMASWAMY H S (1997), 'Changes in texture and microstructure of pressure-treated fish muscle tissue during chilled storage', *Journal of Muscle Foods*, **8**, 13–32.

BAILY A J and LIGHT N D (1989), *Connective Tissue in Meat and Meat Products*, London, Elsevier Applied Science.

BALNY C and MASSON P (1993), 'Effects of high pressure on proteins', *Foods Review International*, **9**(4), 611–28.

BARBOSA-CÁNOVAS G V, POTHAKAMURY U R, PALOU E and SWANSON B G (1998), *Non-Thermal Preservation of Foods*, New York, Marcel Dekker, pp. 1–276.

BARBOSA-CÁNOVAS G V, QIN B L and SWANSON B G (1996), 'Preservation of foods by pulsed electric fields: system design and key components', in Rodrigo M, Martinez A, Fiszman S M, Rodrigo C and Mateu A, *Proceedings of the International Symposium on Advanced Technologies in Sterilization and Safety of Foods and Non Food Products*, Valencia, Instituto de Agroquímica y Technologica de Alimentos, 273–87.

BARSOTTI L, MERLE P and CHEFTEL J C (1999), 'Food processing by pulsed electric fields. I. Physical aspects', *Food Review International*, **15**(2), 163–80.

BELLO R A, LUFT J H and PIGOTT G M (1982) 'Ultrastructural study of skeletal fish muscle after freezing at different rates', *Journal of Food Science*, **47**, 1389–94.

BORDERIAS A J, PÉREZ-MATEOS M, SOLAS M and MONTERO P (1997), 'Frozen storage of high-pressure and heat induced gels of blue whiting (Micromesistius poutassou) muscle: rheological, chemical and ultrastructure studies', *Z Lebensm Unters Forch A*, **205**, 335–42.

CARLEZ A, ROSEC J P, RICHARD N and CHEFTEL J C (1994), 'Bacterial growth during chilled storage of high pressure-treated minced meat', *Lebensm Wiss Technol*, **27**, 48–54.

CASTRO A J, BARBOSA-CÁNOVAS G V and SWANSON B G (1993), 'Microbial inactivation of foods by pulsed electric fields', *J Food Proc Pres*, **17**, 47–73.

CHEAH P B and LEDWARD D A (1995), 'High pressure effects on lipid oxidations', *Journal of American Oil Chemist Society*, **72**, 1059–63.

CHEAH P B and LEDWARD D A (1996), 'High pressure effects on lipid oxidation in minced pork', *Meat Science*, **43**, 123–34.

CHEAH P B and LEDWARD D A (1997), 'Catalytic mechanism of lipid oxidation following high pressure treatment of pork fat and meat', *Journal of Food Science*, **62**, 1135–8, 1141.

CHEFTEL J C (1995), 'Review: high pressure, microbial inactivation and food preservation', *Food Science and Technology International*, **1**, 75–90.

CHEFTEL J C and CULIOLI J (1997), 'Effects of high pressure on meat: a review', *Meat Science*, **46**, 211–36.

CHU G H and STERLING C (1970), 'Parameters of texture change in processed fish: myosin denaturation', *Journal of Texture Studies*, **2**, 214–22.

CHUNG Y C, GEBREHIWOT A, FARKAS D F and MORRISSEY M T (1994), 'Gelation of surimi by high hydrostatic pressure', *Journal of Food Science*, **59**(3), 523–4.

CONNELL J J (1964), 'Fish muscle proteins and some effects on them of processing', in Schultz H W and Anglemier A F (eds) *Proteins and their Reactions*, Connecticut, Avi Westport, pp. 255–94.

CRAIG C L and POWRIE W D (1988), 'Rheological properties of fresh and frozen chum salmon eggs with and without treatment by cryoprotectants', *Journal of Food Science*, **53**(3), 684–7.

DE KONING A J and MOL T H (1990), 'Rates of free fatty acid formation from phospholipids and neutral lipids in frozen cape hake (Merluccius spp) mince at various temperatures', *Journal of Science of Food and Agriculture*, **50**, 391–8.

DUERR J D and DYER W J (1952), 'Protein in fish muscle. IV. Denaturation by salt', *J. Fish Res Bd Can*, **8**, 325–31.

DUNAJSKI E (1979), 'Texture of fish muscle', *Journal of Texture Studies*, **10**, 301–9.

DUNN J E and PEARLMAN J S (1987), 'Methods and apparatus for extending the shelf life of fluid food products', US Patent 4,695,472.

ECKHOFF K M, AIDOS I, HEMRE G I and LIE Ø (1998), 'Collagen content in farmed Atlantic salmon (Salmo salar, L.) and subsequent changes in solubility during storage on ice', *Food Chemistry*, **62**, 197–200.

ERIKSSON C E (1982), 'Lipid oxidation catalysis and inhibitions in raw materials and processed foods', *Food Chemistry*, **9**, 3–19.

FARR D (1990), 'High pressure technology in the food industry', *Trends in Food Science and Technology*, **1**(1), 14–16.

FENNEMA O R (1990), 'Comparative water holding properties of various muscle foods', *Journal of Muscle Foods*, **1**, 363–81.

FERNANDES-DIAZ M D, BARSOTTI L, DUMAY E and CHEFTEL J C (2000), 'Effects of pulsed electric fields on ovalbumin solutions and dialyzed egg white', J *Agric Food Chem*, **48**, 2332–9.

FLAUMENBAUM B L (1968), 'Anwendung der Elektroplasmolyse bei der Herstellung von Fruchtscäften', *Flüssiges Obst*, **35**, 19–22.

FUNTENBERGER S, DUMAY E and CHEFTEL J C (1995), 'Pressure aggregation of β-lactoglobulin isolate in different pH 7 buffers', *Lebensm Wiss Technol*, **28**, 410–18.

GOTO H, KAJIYAMA N and NOGUCHI A (1993), 'Changes of soy and fish proteins under high pressure', in Hayashi R (ed.) *High pressure Bioscience and Food Science*, Kyoto, San-Ei Publications, 315–21.

GRAHL T and MÄRKL H (1996), 'Killing of microorganisms by pulsed electric fields', *Appl Microbiol Biotechnol*, **45**, 148–57.

GREASER M L and PEARSON A M (1999), 'Flesh foods and their analogues', in Rosenthal A J (ed.) *Food Texture*, Gaithersburg, MD, Aspen Publishers Inc, 228–58.

GROSS M and JAENICKE R (1994), 'Protein under pressure. The influence of high hydrostatic pressure on structure, function and assembly of proteins and protein complexes', *Eur J Biochem*, **221**, 617–30.

GUDMUNDSSON M and HAFSTEINSSON H (2001), 'Effect of electric field pulses on microstructure of muscle foods and roes', *Trends in Food Science and Technology*, **12**, 122–8.

HAMILTON W A and SALE A J H (1967), 'Effects of high electric fields on microorganisms. II. Mechanism of action of the lethal effect', *Biochim Biophys Acta*, **148**, 789–800.

HAYASHI R (1993), *High Pressure Bioscience and Food Science*, Kyoto, San-Ei Suppan.

HO S Y, MITTAL G S and CROSS J D (1997), 'Effects of high field electric pulses on activity of selected enzymes', *Journal of Food Engineering*, **31**, 69–84.

HOOVER D G, METRICK C, PAPINEAU A M, FARKAS D F and KNORR D (1989), 'Biological effects of high hydrostatic pressure on food microorganisms', *Food Technology*, **43**, 99–107.

HÜLSHEGER H and NIEMANN E G (1980), 'Lethal effects of high voltage pulses on E. coli K12', *Radiat Environ Biophys*, **18**, 281–8.

HÜLSHEGER H, POTEL J and NIEMANN E G (1981), 'Killing of bacteria with electric pulses of high field strength', *Raditiat Environ Biophys*, **20**, 53–65.

HÜLSHEGER H, POTEL J and NIEMANN E G (1983), 'Electric field effects on bacteria and yeast cells', *Raditiat Environ Biophys*, **22**, 149–62.

ISO S, MIZUNO H, OGAWA H, MOCHIZUKI Y and ISO N (1993), 'Differential scanning calorimetry of pressurized fish meat', *Fisheries Science*, **60**(1), 127–8.

JAENICKE R (1981), 'Enzymes under extreme conditions', *Ann Rev Biophys Bioeng*, **10**, 1–67.

JEANTET R, BARON F, NAU F, ROIGNANT M and BRUTÉ G (1999), 'High intensity pulsed electric fields applied to egg white: effect on Salmonella enteritis inactivation and protein denaturation', *J Food Protect*, **62**, 1381–6.

JOHNSON F H and ZOBELL C E (1949), 'The retardation of thermal disinfection of Bacillus subtilis spores by hydrostatic pressure', *J Bacteriol*, **57**, 353–8.

JOHNSTON D E, AUSTIN B A and MURPHY R J (1992), 'The effects of high pressure treatment on skim milk', in Balny C, Hayashi R, Heremans K and Masson P (eds) *High Pressure and Biotechnology*, Colloque INSERM/ John Libbey Ltd, 243–7.

KALICHEVSKY M T, KNORR D and LILLFORD P J (1995), 'The effects of high pressure on water and potential food applications', *Trends in Food Science and Technology*, **6**, 253–9.

KARINO S, HANE H and MAKITA T (1994), 'Behavior of water and ice at low temperature and high pressure', in Hayashi R (ed.) *High Pressure Bioscience*, Kyoto, San-Ei Suppan, 54–65.

KNORR D (1995), 'Advances and limitations of non-thermal food preservation methods', in Ahvenainen R, Mattila-Sandholm T and Ohlsson T (eds) *New Shelf-life Technologies and Safety Assessments*, Helsinki, VTT Symposium 148, 7–18.

KNORR D, GEULEN M, GRAHL T and SITZMANN W (1994), 'Food application of high electric field pulses', *Trends in Food Science and Technology*, **5**, 71–5.

KNORR D, HEINZ V, UN-LEE D, SCHLÜTER O and ZENKER M (1998), 'High pressure processing of foods: introduction', in Autio K (ed.) *Fresh Novel Foods by High Pressure*, Espoo, VTT Symposium 186, Finland.

LANIER T C and LEE C M (1992), *Surimi Technology*, New York, Marcel Dekker.

LEDWARD D A (1998), 'High pressure processing of meat and fish', in Autio K (ed.) *Fresh Novel Foods by High Pressure*, Espoo, VTT Symposium 186, 165–76.

LEE C M (1984), 'Surimi process technology', Food Technology, **38**(11), 69–80.

LEE H and LANIER T C (1995), 'The role of crosslinking in the texturizing of muscle protein sols', *Journal of Muscle Foods*, **6**, 125–38.

LISTON J (1990), 'Microbial hazards of seafoods consumption', *Food Technol*, **44**(2), 56–62.

LÓPEZ-CABALLERO M E, PÉREZ-MATEOS M, MONTERO P and BORDERÍAS J A (2000a), 'Oyster preservation by high-pressure treatment', *Journal of Food Protection*, **63**(2), 196–201.

LÓPEZ-CABALLERO ME, PÉREZ-MATEOS M, BORDERÍAS J A and MONTERO P (2000b), 'Extension of the shelf life of prawns (Penaeus japonicus) by vacuum packaging and high-pressure treatment', *Journal of Food Protection*, **63**(10), 1381–8.

MACDONALD A G (1992), 'Effects of high hydrostatic pressure on natural and artificial membranes', in Balny C, Hayashi R, Heremans K and Masson P, *High Pressure and Biotechnology*, Colloque INSERM/ John Libbey, 67–74.

MACKIE I M (1993), 'The effects of freezing on flesh proteins', *Food Reviews International*, **9**, 575–610.

MARTÍN O, QIN B L, CHANG F J, BARBOSA-CÁNOVAS G V and SWANSON B G (1997), 'Inactivation of Echerichia coli in skim milk by high intensity pulsed electric fields', *J Food Process Eng*, **20**, 317–36.

MATSUMOTO J J and NOGUCHI S F (1992), 'Cryostabilization of protein in surimi', in Lanier T C and Lee C M, *Surimi Technology*, New York, Marcel Dekker, 357–88.

MIYAO S, SHINDOH T, MIYAMORI K and ARITA T (1993), 'Effects of high pressurization on the growth of bacteria derived from surimi (fish paste)', *Nippon Shokuhin Kogyo Gakkaishi*, **40**(7), 478–84.

MOZHAEV V V, HEREMANS K, FRANK J, MASSON P and BALNY C (1994), 'Exploiting the effects of high hydrostatic pressure in biotechnological applications', *Trends in Biotechnol*, **12**, 493–501.

MURAKAMI T, KIMURA I, YAMAGISHI T, YAMASHITA M and SATAKE M (1992), 'Thawing of frozen fish by hydrostatic pressure', in Balny C, Hayashi R, Heremans K and Masson P (eds) *High Pressure and Biotechnology*, Colloque INSERM/ John Libbey, 329–31.

NESTER E, ROBERTS C E, LIDSTROM M E, PEARSALL N N and NESTER M T (1983), *Microbiology*, Philadelphia, Holt-Saunders International Edition.

NIWA E (1992), 'Chemistry of surimi gelation', in Lanier T C and Lee C M, *Surimi Technology*, New York, Marcel Dekker, 389–428.

PATTERSON M F, QUINN M, SIMPSON R and GILMOUR M (1995), 'The sensitivity of vegetative pathogens to high hydrostatic pressure treatment in phosphate-buffer saline and foods', *J Food Protect*, **58**, 524–9.

PÉREZ-MATEOS M and MONTERO P (2000), 'Response surface methodology multivariate analysis of properties of high-pressure induced fish mince gel', *Eur Food Res Technol*, **211**, 79–85.

OGAWA H, FUKUHISA K and FUKUMOTO H (1992), 'Effect of high hydrostatic pressure on sterilization and preservation of citrus juice', in Balny C, Hayashi R, Heremans K and Masson P (eds) *High Pressure and Biotechnology*, Colloque INSERM/ John Libbey, 269–78.

OHSHIMA T, NAKAGAWA T and KOIZUMI C (1992), 'Effect of high hydrostatic pressure on the enzymatic degradation of phospholipid in fish muscle during storage', in Blight, E G (ed.) *Seafood Science and Technology*, Oxford, Fishing News Books, 64–75.

OHSHIMA T, USHIO H and KOIZUMI C (1993), 'High-pressure processing of fish and fish products', *Trends in Food Science and Technology*, **4**, 370–5.

OKAMOTO M, KAWAMURA Y and HAYASHI R (1990), 'Application of high pressure to food processing: textural comparison of pressure- and heat-induced gels of food proteins', *Agric Biol Chem*, **54**(1), 183–9.

OKAZAKI E and NAKAMURA K (1992), 'Factors influencing texturization of sarcoplasmic protein of fish by high pressure treatment', *Nippon Suisan Gakkaishi*, **58**(11), 2197–206.

QIN B L, POTHAKAMURY U R, VEGA H, MARTÍN O, BARBOSA-CÁNOVAS G V and

SWANSON B G (1995), 'Food pasteurization using high-intensity pulsed electric fields', *Food Technology*, **49**(12), 55–60.

SAITO T, ARAI K and MATSUYOSHI M (1959), 'A new method for estimating the freshness of fish', *Bull Jpn Soc Sci Fish*, **24**, 749–50.

SAKAGUCHI M and KOIKE A (1992), 'Freshness assessment of fish fillets using the Torrymeter and k-value', in Huss H H, Jakobsen M and Liston J (eds) *Quality Assurance in the Fish Industry*, Amsterdam, Elsevier, 333-8.

SALE A J H and HAMILTON W A (1967), 'Effects of high electric fields on micro-organisms I. Killing of bacteria and yeasts', *Biochim Biophys Acta*, **148**, 781–8.

SHIMADA S, ANDOU M, NAITO N, YAMADA N, OSUMI M and HAYASHI R (1993), 'Effects of hydrostatic pressure on the ultrastructure and leakage of internal substances in the yeast Saccharomyces cerevisiae', *Appl Microbiol Biotechnol*, **40**, 123–31.

SHOJI T and SAEKI H (1989), 'Processing and preservation of fish meat by pressurization', in Hayashi R (ed.) *Use of High Pressure in Food*, Kyoto, San-Ei Publications, 75–87.

SHOJI T, SAEKI H, WAKAMEDA A, NAKAMURA M and NONAKA M (1990), 'Gelation of salted paste of Alaska pollack by high hydrostatic pressure and change in myofibrillar protein in it', *Nippon Suisan Gakkaishi*, **56**(12), 2069–76.

SHOJI T, SAKEI H, WAKAMEDA A and NONAKA M (1994), 'Influence of ammonium sulfate on the formation of pressure-induced gel formation walleye pollack surimi', *Nippon Suisan Gakkaaishi*, **60**, 101–9.

SIKORSKI A E and KOTAKOWSKA A (1994), 'Changes in protein in frozen stored fish', in Sikorski Z E, Pan B S and Shahidi F (eds) *Seafood Proteins*, New York, Chapman & Hall, 99–112.

SONOIKE K, SETAYMA T, KUMA Y and KOBAYASHI S (1992), 'Effects of pressure and temperature on the death rate of Lactobacillus casei and E. coli', in Balny C, Hayashi R, Heremans K and Masson P (eds) *High Pressure and Biotechnology*, Colloque INSERM/ John Libbey, 297–301.

SUZUKI A, KIM K, HONMA N, IKEUCHI Y and SAITO M (1992), 'Acceleration of meat conditioning by high pressure treatment', in Balny C, Hayashi R, Heremans K and Masson P (eds) *High Pressure and Biotechnology*, Colloque INSERM/John Libbey, 217–27.

TANAKA M, XUEYI Z, NAGASHIMA Y and TAGUCHI T (1991), 'Effect of high pressure on lipid oxidation in sardine meat', *Nippon Suisan Gakkaishi*, **57**(5), 957–63.

VEGA-MERCADO H, POWERS J, BARBOSA-CÁNOVAS G V and SWANSON B G (1995a),'Plasmin inactivation with pulsed electric fields', *Journal of Food Science*, **60**, 1143–6.

VEGA-MERCADO H, POWERS J, BARBOSA-CÁNOVAS G V, SWANSON B G and LÜDECKE L (1995b), 'Inactivation of protease from Pseudomonas fluorescens M3/6 using high voltage pulsed electric fields', Anaheim CA, Annual IFT 1995 Meeting, 3–7 June, Book of abstracts, paper no. 89–3, p. 267.

WEKELL M M, MANGER R, COLBURN K, ADAMS A and HILL W (1994), 'Microbiological quality of seafoods: bacteria and parasites', in Shahidi F and Botta J D (eds) *Seafoods: Chemistry, Processing Technology and Quality*, London, Blackie Academic & Professional, 196–219.

WONG D W S (1989), *Mechanism and Theory in Food Chemistry*, New York, AVI Books, 48–109.

WOUTERS P C, DUTREUX W, SMELT J P P M and LELIEVELD H L M (1999), 'Effects of pulsed electric fields on inactivation kinetics of Listeria innocua', *Applied and Environmental Microbiology*, **65**(12), 5364–71.

WOUTERS P C and SMELT J P P M (1997), 'Inactivation of microorganisms with pulsed electric fields: potential for food preservation', *Food Biotechnology*, **11**(3), 193–229.

YAMAMOTO K, MIURA T and YASUI T (1990), 'Gelation of myosin filament under high hydrostatic pressure', *Food Structure*, **9**, 269–77.

YOSHIOKA K, KAGE Y and OMURA H (1992), 'Effect of high pressure on texture and ultrastructure of fish and chicken muscles and their gels', Balny C, Hayashi R, Heremans K and Masson P (eds) *High Pressure and Biotechnology*, Colloque INSERM/ John Libbey, 325–7.

YUKIZAKI C, KANO M and TSUMAGARI H (1993), 'The sterilization of sea urchin eggs by hydrostatic pressure', in Hayashi R (ed.) *High Pressure Bioscience and Food Science*, Kyoto, San-Ei Publications, 225–8.

YUKIZAKI C, KAWANO M, JO N and KAWANO K (1994), 'High pressure sterilization and processing of sea urchin eggs', in Hayashi R, Kunagi S, Shimada S and Suzuki A, *High Pressure Bioscience*, Kyoto, San-Ei Suppan, 248–58.

ZHANG Q, CHANG F J, BARBOSA-CÁNOVAS G V and SWANSON B G (1994), 'Inactivation of microorganisms in a semisolid model food using high electric voltage pulsed electric fields', *Lebensm Wiss u Technol*, **27**, 538–43.

ZIMMERMANN U (1986), 'Electrical breakdown, electropermeabilization and electrofusion', *Rev Physiol Biochem Pharmacol*, **105**, 176–256.

ZIMMERMANN U, PILWAT G, BECKERS F and RIEMANN F (1976), 'Effects of external electrical fields on cell membranes', *Bioelectrochem Bioenergetics*, **3**, 58–83.

10.16 Acknowledgement

Experimental studies were carried out at Matra (Icetec) in Iceland and were supported by the Icelandic Research Council, the Nordic Industry Fund and the European Commission under the FAIR programme.

11

Minimal processing in the future: integration across the supply chain

R. Ahvenainen, VTT Biotechnology, Espoo

11.1 Introduction

The concept of minimal processing has developed much during recent years. Nowadays minimal processing is regarded as the mildest possible preservation technique tailored to a particular raw material and food. The aim is to maintain natural properties of a foodstuff by non-obtrusive processing using a minimum of preservatives. According to that, preservation methods in the minimal processing of foods can basically be classified into three different categories:

1. Optimised traditional preservation methods (e.g. canning, blanching, freezing, drying and fermentation) in order to improve sensory quality and to save energy, not forgetting microbiological safety.
2. Novel mild preservation techniques, such as high pressure technology, electric field pulses (so-called non-thermal processing), various mild heat treatments, sous-vide cooking, post-harvest technologies, protective microbiological treatment, etc.
3. Combinations of various methods and techniques (hurdle effect), whereby a synergistic effect is obtained.

To reach the aims of minimal processing, for very short shelf-life products, e.g. fresh-cut vegetables, preservation methods that will prolong the shelf-life are used, whereas long shelf-life products, e.g. canned food, need methods that give improved sensory and nutritional quality even with shorter shelf-life.

The concept of minimal processing is still progressing. The results of a research programme at VTT in Finland concluded that the future minimal processing concept covers the food production from the field to the table (Ahvenainen *et al.* 2000). The characteristic feature in future minimal

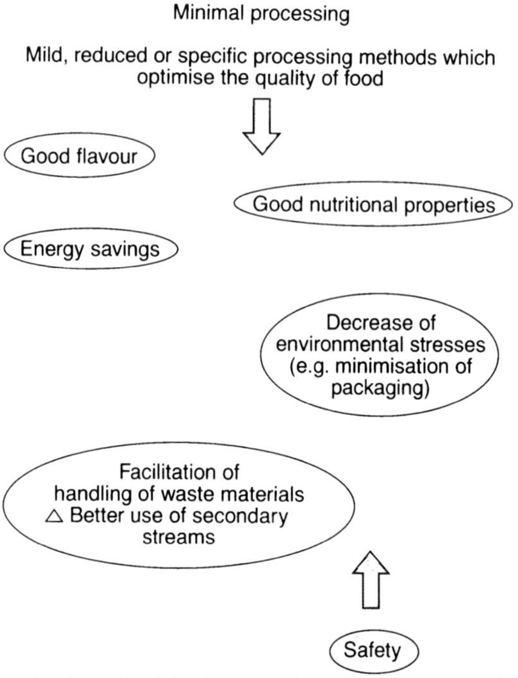

Fig. 11.1 The main aims of minimal processing without compromising the safety of foodstuffs (Ahvenainen *et al.*, 2000).

processing is an integrated approach, in which raw materials, handling, processing, packaging and distribution must all be properly considered to make shelf-life extension possible. At its best, minimal processing uses raw material suitable for minimal processing, mild processing methods without chemical preservatives and packaging techniques increasing shelf-life (see Figs 11.1 and 11.2). On the other hand, during product development much focus is put on the safety of the final product without forgetting the other quality attributes and proper shelf-life of the product (Table 11.1). The concept will support the idea of sustainable development.

In particular, the integrated approach from the field to the table has been found to be the most promising way to maintain high quality and ensure the safety of fresh foodstuffs, such as strawberries, vegetables and fish (Ahvenainen *et al.* 2000). In the integrated approach considerable emphasis is placed on the treatment of food during and immediately after harvesting. The correct degree of ripeness, gentle treatment in harvesting, careful and rapid chilling, low temperature in storage and during distribution and as little mechanical stress as possible after harvesting are key points to ensure the quality and shelf-life.

Minimal processing will be increased in the future. Already nowadays the food production is being divided into global and local food. The significance of local food is increasing all the time due to various global food catastrophes (BSE, foot-and-mouth disease, dioxin, etc). This gives totally new possibilities

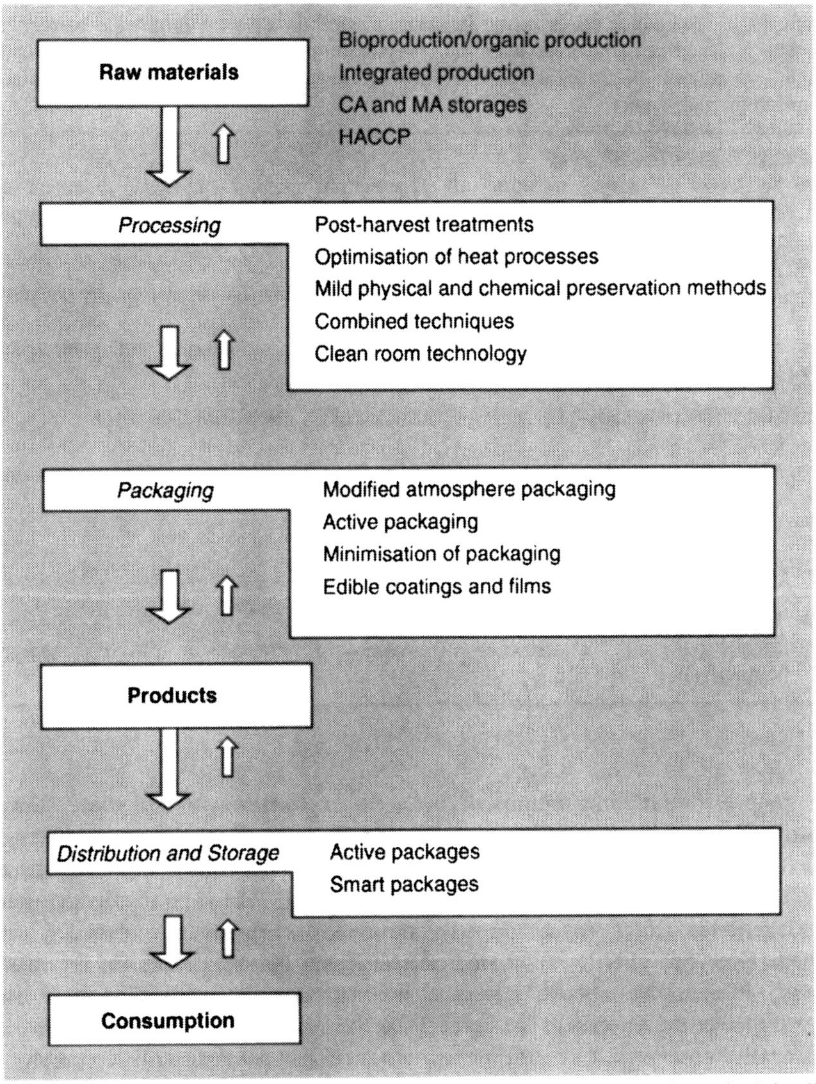

Fig. 11.2 An integrated approach is essential for the quality and safety of minimally processed food products (Ahvenainen *et al.*, 2000).

for minimal processing of food raw material and food components, because long shelf-life is not necessary.

11.2 Key issues in an integrated approach

Key issues in an integrated approach are organic or integrated primary production, certificated environment and quality system, traceable raw materials and conditions, minimised/optimised processes and reduced amount of additives

Table 11.1 Procedure for ensuring the microbiological safety of minimally processed foodstuffs, i.e. some facts to be taken into consideration in the management of safety and quality of microbiologically spoiling foodstuffs in product development, production, distribution and storage

PRODUCT DEVELOPMENT
- Tailoring of recipe, selecting of preservation method (physical, chemical or combination) and packaging method. Optimisation of methods and recipe (statistical design is often needed to decrease the number of experiments)
- Tailoring of sensory and nutritional quality
- Shelf-life studies including sensory, nutritional and microbiological quality (several repeats should be carried out)
- Predictive microbiology and inoculation studies for evaluation of pathogenic risk
- Consumer tests

PRODUCTION (HANDLING OF RAW MATERIALS, PROCESSING AND PACKAGING, ETC.)
- Good manufacturing practices and high hygiene (HACCP, factory layout, cleaning, correct temperatures, etc.)
- On-line non-destructive leak testing of packages

DISTRIBUTION AND STORAGE OF PRODUCTS
- Protection of packages for transportation
- Correct temperature
- Smart indicators for detection of breakage of packages or too high storage temperatures

in product formulation, optimised packages, exploitation of secondary flow, distribution conditions and the use of the products (cooking, warming, eating). According to that, the criteria for minimally processed foods were created at VTT Research Programme (Ahvenainen *et al.* 2000). These can be divided into basic criteria, which are not directly connected to any specific product, and product-specific criteria. All basic criteria should be valid and, on the other hand, of the product-specific criteria at least criteria 1, 2, 3, 4 and 7 should also be valid (see the criteria in Table 11.2), before the product can be regarded as minimally processed. In the following some criteria are dealt with separately.

11.3 Raw materials

Future minimal processing is based on raw materials produced by organic or integrated production with high hygiene level (e.g. irrigation based on pure water). Furthermore, it is probable that, in the future, fruits, vegetables and berries intended for minimal processing will be cultivated under specified controlled conditions, and furthermore, plant geneticists will develop selected and created cultivars or hybrids adapted to the specific requirements of minimal processing (Varoquaux and Wiley, 1994; Martinez and Whitaker, 1995).

Table 11.2 Criteria for minimally processed foods (Ahvenainen *et al.*, 2000)

Basic criteria:

1. Valid certificated quality system in the company producing minimally processed foods.
2. Valid certificated environment system in the company producing minimally processed foods.
3. Traceable main raw materials and process conditions.

Product-specific criteria:

1. Main raw materials should be produced according to good agricultural practices (integrated production or organic production, high hygiene level).
2. As good as possible preservation of nutritional and sensory quality without compromising microbiological safety in the process.
3. Reduction or complete abandonment of the use of preservatives, or their replacement by natural preservatives/compounds. Minimal processing must not increase the use of preservatives.
4. When using mild methods, the use of energy should not increase, but rather an energy saving should be attained in the process as well as in the whole food chain.
5. Reduction in the number of processing stages can also be regarded as minimisation of a process (minimal treatment).
6. Saving of costs in the process as well as in the value chain (optimised value chain).
7. Optimised packaging according to the precision packaging concept (Lyijynen *et al.*, 1998; Hurme and Pullinen, 1998).
8. Exploitation of secondary flow (e.g. peeling waste) and method of utilisation.

11.4 Mild and optimised processes

In order to fulfil all criteria set to processes, optimisation is a solution that satisfies all needs: consumers' preference for good taste, good nutritional quality and safe food products and, on the other hand, profitability of production (Fig. 11.3).

Optimisation of processes has an influence on the sensorial, nutritional and microbiological quality of food, yield and energy consumption. However, optimisation requires systematic knowledge of the complex correlation of raw material, recipe and process variables and product parameters, e.g. food quality and manufacturing costs. In addition, optimisation of raw material consumption improves profitability of production, because the costs of raw materials make up a remarkable part of production costs.

When using minimal processing in heating and cooking processes, it has an effect on the weight losses and energy consumption, too. However, microbiological safety of the product is a premise that cannot be compromised. Therefore recognition of the factors that have an effect on the microbiological quality is a necessity when reducing cooking processes.

When industrial-scale processes are optimised, experiments for studying the correlation among different variables have to be run on an industrial scale as well. Because implementation of experiments is expensive and hinders primary

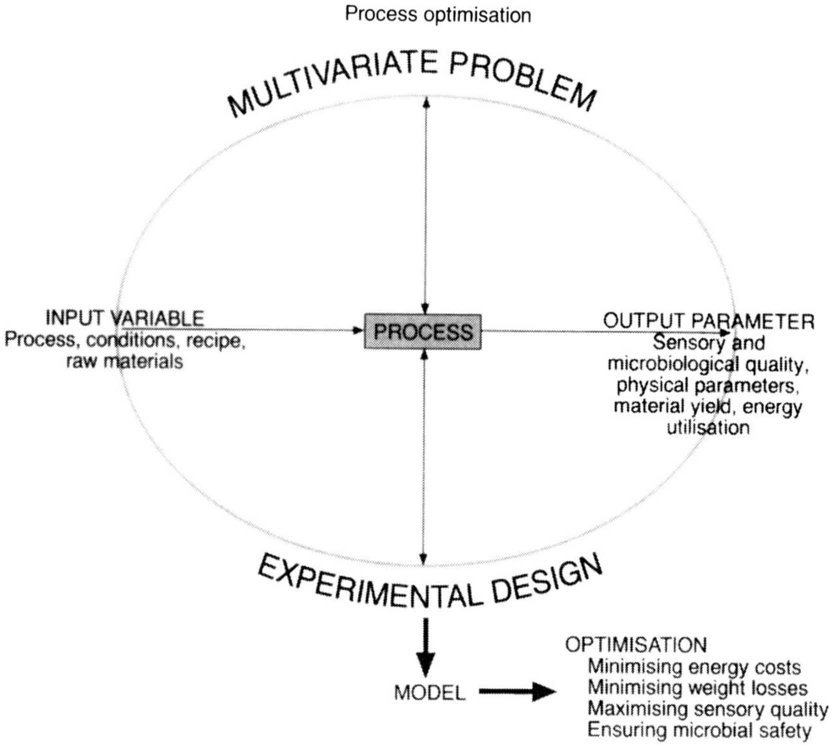

Fig. 11.3 The scheme for process optimisation (Ahvenainen *et. al.*, 2000).

production, the number of experiments must be minimised. In that case statistical methods are a solution for experimental design. When applying statistical experimental design it is possible to use a limited number of experiments and still study effects of several variables and handle both recipe and process variables at the same time. With statistical models production can be optimised, accurate processing parameters can be defined and the whole manufacturing process can be analysed. In addition, the knowledge based on the models can be utilised in product and processing development work.

Application of statistical methods is based on the numerical processing of results. That provides the use of reliable measurement and analysing methods. Besides traditional analysing methods there are several predictive microbiological programs available on the market, which are useful for the evaluation of the microbiological safety of food products. Overall, process optimisation is a complex task where you have to pay attention to a large number of variables.

11.5 Reduction of the number of processing stages

In order to guarantee good product quality it is also important to optimise (most often to minimise) the number of processing stages and also to avoid stoppages and delays between different processing and treatment stages. One example is the handling of fresh strawberries (see section 11.8 and Fig. 11.4). Another example is the Cryomix system (Cryomix is a trademark of Armor Inox, used by Air Liquide under licence) (Mermelstein, 2000). The system uses sequential injections of sauce and liquid nitrogen to freeze layers of sauce onto IQF (individually quick frozen) products, coating each piece of product completely and uniformly, at coating ratios as high as 50%. The mixing of sauce and IQF food product (e.g. vegetables with or without meat, fish, poultry; pasta; meat balls, etc.) is easy and thawing of the product before coating is not necessary in order to have a certain amount of sauce to the product. On the other hand, the use of the coated product is easy and convenient. The Cryomix system coats individual pieces of food with the correct amount of sauce, so that no matter how many pieces of food the user removes from the package, the correct amount of sauce, as determined by the product manufacturer, will be present.

11.6 Package optimisation

For a long time packaging has also had an active role in processing, preservation and retaining quality of foods. Many cooking and preservation processes have been and are still largely based on proper packaging. For example, packaging is an integral part of canning, aseptic, sous-vide and baking processes. On the other hand, the benefits of several preservation methods (e.g. drying, freezing) would be lost without protective packaging after processing. Most of the dried and frozen food products need more or less oxygen, light and water vapour impermeable packaging materials. Packaging development has also changed the preservation methods used for food products. Ten to fifteen years ago, all poultry products and industrially prepared raw minced meat were sold as frozen. Nowadays, thanks to modified atmosphere packaging based on protective gas and novel gas impermeable packaging materials, they are mainly sold as chilled. The industrial preparation of fresh-cut fruit and vegetables for retail sale is also possible today, due to respirable packaging films.

Now packaging plays an increasingly important role in the whole food chain: from the field to the consumer's table. For example, many fresh agricultural products such as berries and mushrooms are picked in the field or in the greenhouse directly to consumer packages and plastic- or fibre-based trays. Therefore the product is touched only once before it reaches the consumer. Another example is ready-to-eat foods and snack products which are packed in microwaveable trays that serve as an eating dish.

Packaging has developed during recent years, mainly due to increased demands on product safety, shelf-life extension, cost efficiency, environmental issues and consumer convenience. In order to improve the performance of the packaging, innovative active and smart packaging concepts are being developed and tested in laboratories and companies around the world. Active and smart packaging have great commercial potential to assure the quality and safety of food without (or at least with less) additives and preservatives, thus reducing food wastage, food poisoning and allergies. Smart packaging will also give technological help to the food industry and business to carry out in-house control required by food law and find critical control points in the food chain from the factory to the consumer's table.

In today's competitive market optimal packages are a major advantage when persuading consumers to buy a certain brand. Packaging has to satisfy various requirements effectively and economically. The objective is to design an optimised package that satisfies all marketing and functional requirements sufficiently and fulfils environmental and cost demands as well as possible (Hurme and Pullinen, 1998). An answer to this complex equation is, for example, a packaging optimisation method called VTT Precision Packaging Concept. Industry can use it for designing efficient, functional and cost-effective packages. Required shelf-life, logistics and market-driven thinking are the base elements of the concept. An optimised package designed according to the Concept gives the required shelf-life to the packed product (Lyijynen *et al.* 1998, Hurme and Pullinen, 1998).

11.7 Sustainable production

In the future minimal processing will support sustainable development even more than nowadays. An essential part of this is proper utilisation of secondary flow, e.g. peeling waste from the vegetable and fruit industry and seeds from the berry, vegetable and fruit industry. Many healthy compounds are located in the peel, just under the peel and in the seeds. By various extracting methods, e.g. supercritical carbon dioxide extraction or enzymatic methods, it is possible to extract flavonoids and other healthy compounds from peels and seeds. The remaining waste can be composted and used, for example, as soil conditioner.

11.8 Examples of food products manufactured using an integrated approach

As an example, both the shelf-life and 'sell by date' of strawberries has been prolonged significantly with minimal processing. The shelf-life of strawberries, cultivated in Finland and picked when ripe, varies when chilled and cold stored from one to seven days. The shelf-life varies depending on the crop season, the time of the picking, the effectiveness of chilling and the variety of strawberry picked. When the quality and shelf-life of strawberries are to be optimised, attention needs to be paid to every phase of the strawberry's life, from picking to

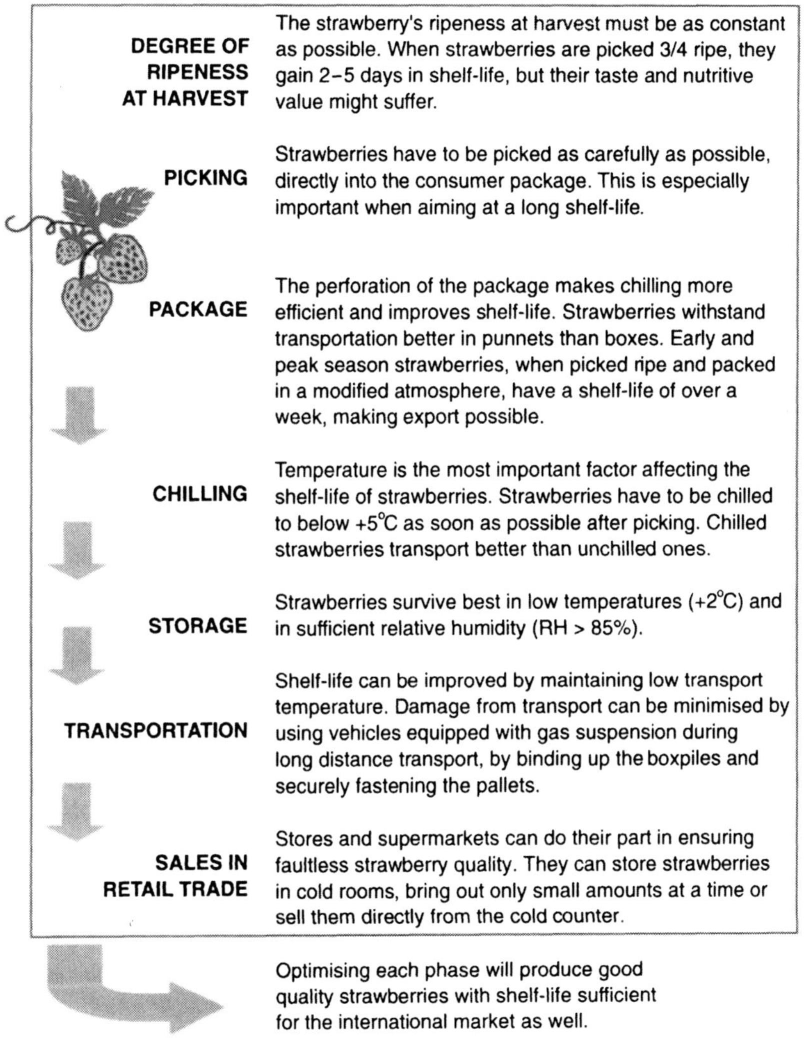

Fig. 11.4 Integration of every phase from the field to the consumer's table is essential to guarantee the good quality of strawberries.

consumption. Reduction of mechanical stress is important and it means that strawberries should be picked directly into the consumer package in order to avoid extra treatment of strawberries (Fig. 11.4).

The Finnish food company Lännen Tehtaat has optimised its production process comprehensively. Presently they use the minimal processing concept. Consequently, the company has been successful in reducing its energy, water and material consumption as well as the volume of waste. Lännen Tehtaat has been granted the ISO 9001 Quality Certificate and the ISO 14001 Environmental Certificate (FINPRO, 1999).

11.9 Future trends

Healthy, convenient and environmentally friendly foodstuffs are becoming more important for consumers. Safety is a self-evident criterion. As competition and product assortment increase, high sensory quality will become an increasingly important competitive advantage. The minimal processing concept supports all these demands set by consumers.

Much research and development work must still be carried out in order to develop minimally processed food products that satisfy consumers and which have high sensory quality, microbiological safety and nutritional value.

11.9.1 Quality aspects

The unit operations, such as peelers and shredders for vegetables and potatoes, need further development to make them more gentle. There is no sense in disturbing the quality of produce by rough treatment during processing, only to patch it up later by preservatives. One possible solution is a combination of enzymes (cellulases, hemicellulases and pectinases) with mechanical peeling (carborundum or knife peeling).

Ultrasound, electric pulses and light pulses are also promising technologies, from the point of view of both shelf-life and product modification. However, the problem is that light pulses in particular are still very expensive and in their infancy. One reason for their high cost may be that there is only one manufacturer for this technology. In any case, profitable, inexpensive physical methods for lowering microbial counts as well as for modification of food structure and formulation of new flavour properties are needed in the future for the production of fresh foods (also raw foods) and novel food products of high quality.

Consumer attitudes towards food products manufactured by novel methods should be studied. For example, high pressure technology is a promising way to increase the shelf-life of raw meat, fish and vegetables, but it changes the sensory quality so much that consumers may not accept it. However, for sterile foods, high pressure technology combined with chilling technology is a way to produce high quality products with low microbiological risks compared to traditional foods. In particular, vegetables and hard roe are potential raw food products to be treated by high pressure (Ahvenainen *et al.*, 2000).

11.9.2 The safety of novel processing technologies

There are several important problems to be solved before the novel processing technologies such as high pressure, ultrasound, electric pulses and light pulses may receive wide use in food productions (FAIR, 1999; Ahvenainen *et al.*, 2000):

1. As these technologies are new, however, there is not yet a large volume of scientific information concerning (among others) their modes of action, effects of foods and food constituents, microbiological effects, monitoring and post-process storage issues. In keeping with the novel foods regulation, this scientific information is necessary to assess and ascertain properly the safety of food products prepared using such technologies.
2. The effect(s) of novel processes on the occurrence and fate of pathogenic and spoilage microorganisms in a new food product is only one part of the question. Information is also required in areas such as enzymes, vitamins, micronutrients, allergens and toxins as well as on the potential formation of residues in the food (all this during and after processing). On the other hand, there are indications that some of these methods are promising for removal/ denaturation of allergenic compounds, making the food suitable for allergenic people.
3. Consideration should be given to adequate monitoring of process parameters, particularly at critical control points of the process. Examples of such process parameters are the thermometers or enzyme markers such as phosphatase in the case of heat treatment, or effective dosimeters in the case of irradiation. This is particularly important in the context of HACCP.
4. Many technologies will not be used in isolation, but rather in combination with other processing systems such as chilled storage, heating and preservatives. In the case of novel technologies, information is required on the effect of these combination treatments. Furthermore, in future shelf-life and safety improvement as well as reduction of preservatives and other harmful compounds will be of interest, e.g. in the substitution of nitrites, phosphates or glutamates in meat products. Thus, the search for potential natural compounds and the study of their combination with physical methods should still continue. Such information should be quantitative and not only qualitative, as the availability of quantitative data will help to apply the concept of substantial equivalence.
5. In general, future research should be primarily focused on the microbial ecology of minimally processed foods and the growth chraracteristics and interactions of pathogenic and spoilage organisms during and after combined treatments. The microbe–food interactions are always product related, as are the microbial metabolites produced during spoilage and prolonged shelf-life. Future innovations for food product safety are therefore dependent on the basic research on microbial ecology and metabolism in foods.
6. Because the requirements for approval of novel foods under the novel foods regulation are quite complex, it would be helpful if clear guidelines were devised so that manufacturers can comprehend and use them to evaluate the consequences of using novel technologies and in preparing their novel foods

application. Guidelines that are much appreciated by manufacturers have been issued before on topics such as HACCP, modified atmosphere packaging and sous-vide cooking and seem to contribute to the success in practice of these technologies.

7. Guidance for manufacturers would need to help them to assess whether they have enough detailed information concerning the new technology and its various effects or (equally importantly) to identify the knowledge gap in considering the microbiological or chemical safety and quality of the product. It should also help ensure that manufacturers implement proper process monitoring and post-processing storage.

Nowadays, minimal processing methods are mainly applied for reduction of the harmful effects of severe cooking processes and preservatives (such as sulphites, sodium chloride), or to increase the shelf-life of short-shelf food products (e.g. marinated meat, raw and smoked fish, vegetables), without compromising microbiological safety or sensory quality. Naturally, in the future shelf-life and safety improvement as well as reduction of preservatives and other harmful compounds will be of interest, e.g. in the substitution of nitrites, phosphates or glutamates in meat products. Thus, the search for potential natural compounds and the study of their combination with physical methods should still continue. Ozone, chitosan and mustard oil are among the promising compounds, but there are also many others, e.g. natural plant volatiles, bioactive peptides and antimicrobial enzymes. The combined methods should be studied with real foodstuffs, not only in laboratory media.

The antimicrobial packaging materials are also one potential way to decrease the amount of preservatives and focus the function of preservatives more precisely where microbial growth and spoilage mainly occur, on the surface of the food. However, these materials are still rather rare on the market and need much research and development work.

Edible coating technology also has promising possibilities as a ‘precision weapon’ in many applications, e.g. on the surface of smoked fish products, but needs still further development. Coating technology combined with high local sodium chloride content is a promising method to hinder the growth of *Clostridium botulinum*. A similar approach should be studied against *Listeria* which is, together with *Clostridium botulinum*, one of the most typical pathogenic bacteria in fish products (Lindström *et al.*, 1999).

11.9.3 Health aspects

The minimal processing concept supports the megatrend to produce health-promoting foods because, by the selection of proper processing methods and conditions, it is possible to preserve nutritional compounds in foods. There are also indications that certain minimal processing methods can be used for reduction of allergenic compounds in food products (Kunugi, 1999). Therefore, when developing health-promoting foods emphasis should also be placed on the preservation of beneficial compounds already existing in the raw food material as well as on the removal of harmful compounds. Often, from the legislative point of view, this might even be an easier way to develop health-promoting foodstuffs than the approach in which health-promoting compounds are added to the food product.

11.9.4 Logistics

The microbiological safety of minimally processed foods is one of the most important demands. Most minimally processed foods need cold chain during the distribution. Every measure should be used for the assurance of safety during transport and retailing (Table 11.1). Smart packaging (package indicators), which give information on the microbial quality and safety of packaged food products, is one of the possible measures. (Smolander, 2000). The key idea of smart package indicators is their ability to give information about the product quality directly, the package and its headspace gases and the storage conditions of the package. A freshness indicator indicates directly the microbial quality of the product by reacting to the metabolites produced in the growth of microorganisms. If perishable food products are stored above the suggested storage temperature, a rapid microbial growth takes place. The product is spoiled before the estimated 'use by' date and, in the worst case, growth or toxin production of pathogenic bacteria takes place. Time–temperature indicators attached to the package surface integrate the time–temperature history of the package throughout the whole distribution chain and hence give indirect information on the product quality. For many perishable products exclusion of oxygen and high concentration of CO_2 improve the stability of the product as the growth of aerobic microorganisms is prevented. As a result of potential package leak, the protective atmosphere is lost. Package leak can also increase the microbial spoilage by enabling product contamination with harmful microorganisms. It is possible to check the package integrity immediately after the packaging procedure, but a leak indicator attached to the package gives information on the package integrity throughout the whole distribution chain (Mattila-Sandholm *et al.*, 1998).

Some intelligent packaging concepts are already commercially available and their use increases constantly. New concepts of leak indicators and freshness indicators are patented and it can be expected that new commercially available products will be available in the near future. In some future visions it has even been predicted that the use of quality and safety indicating tags in certain easily perishable packaged products would become statutory within a few decades (Smolander, 2000).

Today the commercially available smart packaging concepts are labels reacting with a visible change in response to time and temperature (TTIs) or the presence of certain chemical compounds (leak indicators, freshness indicators). These kinds of visible indicators are ideal in many cases; however, in the future it can be expected that a smart package may contain more complex invisible messages, which can be read at a distance. A label could be introduced as a chip but advances in ink technology might enable the use of clever printed circuits as well. The security tags and radio frequency identity tags are the first examples of electronic labelling. In addition to information on product identification, date of manufacture, price, etc., electronic tags could also function as a time–temperature, leak and/or freshness indicator, and it could be expected that the advances in electronics and biotechnology (e.g. biosensors, immunodiagnostics) would be followed by the emergence of totally new concepts of smart packaging.

11.9.5 Legislative aspects

One problem restricting the implementation of new minimal processing methods is the current food legislation. For example, malic acid combined with ascorbic acid and modified atmosphere packaging was found to be a promising method to substitute sulphites in the browning inhibition of pre-peeled potatoes. Another example is the use of carbon monoxide gas in packaging, which already at very low concentrations (0.1–0.3%) allows the packaging of retail red meat without high oxygen levels, retaining the natural colour of meat as valued by consumers (Ahvenainen *et al.*, 2000). Current legislation does not allow the use of malic acid for pre-peeled potatoes or carbon monoxide for packaging. However, it is worth noticing that potato itself contains malic acid, which is also a typical compound in apples. Carbon monoxide has been used for meat packing in Norway for several years (Sørheim *et al.*, 1997). Change of legislation or the granting of a special permit for use from the authorities is very slow in the EU. The attendance of representatives of the authorities in research and development projects aiming at novel methodologies for food preservation will be essential to accelerate the approval of novel methods in the future.

11.9.6 Trade marks/brands

Trade marks are one of the most important marketing and promoting tools nowadays. Because the minimal processing concept supports the image of health-promoting properties of foodstuffs, the combination of a message indicating both minimal processing and health-promoting properties of the food product in the same trade mark might be relevant in the future. However, the legislation concerning the marketing of health-promoting foods should first be developed. Furthermore, no trade mark available in the market nowadays contains a message from all the aspects that consumers appreciate in food from field to table: safety, health promotion, good sensory quality, environment, national. If new trade marks are to be prepared, these aspects should also be included in the mark.

11.10 References

AHVENAINEN, R, AUTIO, K, HELNADER, I HONKAPÄÄ, K, KERVINEN, R, KINNUNEN, A, LUOMA, T, LYIJYNEN, T, LÄHTEENMÄKI, L, MATTILA-SANDHOLM, T, MOKKILA, M and SKYTTÄ, E (2000), VTT Research programme on minimal processing. Final report. VTT Research Notes 2052, VTT Technical Research Centre of Finland, Espoo.

FAIR CONCERTED ACTION FAIR CT96-1020 (1999), *Harmonization of Safety Criteria for Minimally Processed Foods: Rational and Harmonization Report,* European Commission.

FINPRO (1999), *Pure Pleasure Throughout the Finnish Food Chain: From the Field to the Table.* FINPRO, Helsinki.

HURME E and PULLINEN T (1998), 'Precision packaging concept: advantages and practical applications', *The European Food & Drink Review.* Winter, 61–4.

KUNUGI S (1999), 'Emerging technologies in Japan', Eur. Conf. Emerging Food Science and Technology. Tampere, 22–24 November. Abstracts of papers and posters. European Federation of Food Science and Technology, 40.

LINDSTRÖM M, LYIJYNEN T, MYLLÄRINEN P, AHVENAINEN R and KORKEALA H (1999) 'Edible coating for inhibition of *Clostridium botulinum* Type E in vacuum-packaged hot-smoked rainbow trout', in Tuijtelaars A C J, Samson R A, Rombouts F M and Notermans S (eds) *Proceedings of the Seventeenth International Conference of the International Committee on Food Microbiology and Hygiene 'Food Microbiology and Food Safety into the Next Millennium'*, 63–4.

LYIJYNEN T, HURME E, HEISKA K and AHVENAINEN R (1998), *Towards Precision Food Packaging by Optimization.* Espoo VTT, VTT Tiedotteita – Meddelanden – Research Notes, 1915.

MARTINEZ M V and WHITAKER J R (1995), 'The biochemistry and control of enzymatic browning', *Trends Food Sci. Technol.* **6**, 195–200.

MATTILA-SANDHOLM T, AHVENAINEN R, HURME E and JÄRVI-KÄÄRIÄINEN T (1998), 'Oxygen-sensitive colour indicator for detecting leaks in gas-protected food packages', EP 0 666 977 B1.

MERMELSTEIN N H (2000), 'Coating IQF foods with sauce', *Food Technol.* **54**(12), 68–70.

SMOLANDER M (2000), 'Principles of smart packaging', *Packaging Technology,* March/April, 9–12.

SØRHEIM O, AUNE T and NESBAKKEN T (1997), 'Technological, hygienic and toxicological aspects of carbon monoxide used in modified-atmosphere packaging of meat', *Trends in Food Science & Technology*, September, **8**, 307–12.

VAROQUAUX P and WILEY R (1994), 'Biological and biochemical changes in minimally processed refrigerated fruits and vegetables', in Wiley R C (ed.) *Minimally Processed Refrigerated Fruits & Vegetables*, New York, USA, Chapman & Hall, 226–8.

Index

Lightning Source UK Ltd.
Milton Keynes UK
11 February 2011

167370UK00001B/8/P

9 781855 735477